TATA INSTITUTE OF FUNDAMENTAL RESEARCH
STUDIES IN MATHEMATICS

General Editor : S. RAMANAN

1. M. Herve : SEVERAL COMPLEX VARIABLES
2. M.F. Atiyah and others: DIFFERENTIAL ANALYSIS
3. B. Malgrange: IDELAS OF DIFFERENTIABLE FUNCTIONS
4. S.S. Abhyankar and others: ALGEBRAIC GEOMETRY
5. D. Mumford: ABELIAN VARIETIES
6. L. Schwartz: RANDON MEASURES ON ARBITRARY TOPLOGICAL SPACES AND CYLINDIRICAL MEASURES
7. W.L. Baily Jr. and others: DISCRETE SUBGROUPS OF LIE GROUPS AND APPLICATIONS TO MODULI
8. C.P. RAMANUJAM - A TRIBUTE
9. C.L. Siegel: ADVANCED ANALYTIC NUMBER THEORY
10. S. Gelbert and others: AUTOMORPHIC FORMS, REPRESENTATION THEORY AND ARITHMETIC
11. M.F. Atiyah and others: VECTOR BUNDLES ON ALGEBRAIC VARIETIES
12. R. Askey and others: NUMBER THEORY AND RELATED TOPICS
13. S.S. Abhyankar and others: GEOMETRY AND ANALYSIS

GEOMETRY AND ANALYSIS

Papers presented at the Bombay Colloquium 1992, by

ABHYANKAR ADIMURTHI BEAUVILLE
BIRKENHAKE DE CONCINI DEMAILLY
HABOUSH HITCHIN JOSEPH KAC
LAKSHMIBAI LANGE MARUYAMA MEHTA
NITSURE OKAMOTO PROCESI SIU SUBRAMANIAN

Published for the
TATA INSTITUTE OF FUNDAMENTAL RESEARCH, BOMBAY
OXFORD UNIVERSITY PRESS
1995

Oxford University Press, Walton Street, Oxford 0X2 6DP

OXFORD NEW YORK

ATHENS AUCKLAND BANGKOK BOMBAY
CALCUTTA CAPE TOWN DAR ES SALAAM DELHI
FLORENCE HONG KONG ISTANBUL KARACHI
KUALA LUMPUR MADRAS MADRID MELBOURNE
MEXICO CITY NAIROBI PARIS SINGAPORE
TAIPEI TOKYO TORONTO

and associates in
BERLIN IBADAN

ISBN 0 19 563740 2

Printed at Konam Printers, Bombay
Published by Neil O'Brien, Oxford University Press, Oxford House,
Apollo Bunder, Bombay 400 001.

INTERNATIONAL COLLOQUIUM ON GEOMETRY AND ANALYSIS

BOMBAY, 6-14 JANUARY 1992

REPORT

AN INTERNATIONAL COLLOQUIUM on 'Geometry and Analysis' was held at the Tata Institute of Fundamental Research, Bombay from January 6 to January 14, 1992. Professors M. S. Narasimhan and C. S. Seshadri turned sixty at about this time. In view of the crucial role both of them played in the evolution of the School of Mathematics as a centre of excellence, it was considered appropriate to devote the Colloquium to recent developments in areas of geometry and analysis close to their research interests, and to felicitate them on this occasion. The range of topics dealt with included vector bundles, moduli theory, complex geometry, algebraic and quantum groups, and differential equations. The Colloquium was co-sponsored by the International Mathematical Union and the Tata Institute of Fundamental Research, and was financially supported by them and the Sir Dorabji Tata Trust.

The Organizing Committee for the Colloquium consisted of Professors Lakshmibai, David Mumford, Gopal Prasad, R. Parthasarathy (Chairman), M.S. Raghunathan, T.R. Ramadas, S. Ramanan and R.R. Simha. The International Mathematical Union was represented by Professor Mumford.

The following mathematicians gave one-hour addresses at the Colloquium: S.S. Abhyankar, A. Adimurthy, A. Beauville, F.A. Bogomolov, C. de Concini, J.P. Demailly, W.J. Haboush, G. Harder, A. Hirschowitz, N.J. Hitchin, S.P. Inamdar, K.T. Joseph, G.R. Kempf, V. Lakshmibai, H. Lange, M. Maruyama, D. Mumford, M.P. Murthy, N. Nitsure, M.V. Nori, K. Okamoto, C. Procesi, N. Raghavendra, S. Ramanan, Y.T. Siu, V. Srinivas, S. Subramanian, G. Trautmann.

Besides the members of the School of Mathematics of the Tata Institute, mathematicians from universities and educational institutions in India and abroad were also invited to attend the Colloquium.

The social programme for the Colloquium included a Tea Party on January 9, a documentary film on 6 January, a Carnatic Flute Recital on January 8, a Hindustani Vocal Recital on January 11, a Kathakali Dance on January 13, a Dinner at Gallops Restaurant, Mahalaxmi on January 12, an Excursion to Elephanta Caves on January 14 , and a Farewell Dinner Party on January 14, 1992.

CONTENTS

Fundamental Group of the Affine Line in Positive Characteristic

Shreeram S. Abhyankar *

1 Introduction

I have known both Narasimhan and Seshadri since 1958 when I had a nice meal with them at the Student Cafeteria in Cité Universitaire in Paris. So I am very pleased to be here to wish them a Happy Sixtieth Birthday. My association with the Tata Institute goes back even further to 1949-1951 when, as a college student, I used to attend the lectures of M. H. Stone and K. Chandrasekharan, first in Peddar Road and then at the Yacht Club. Then in the last many years I have visited the Tata Institute numerous times. So this Conference is a nostalgic homecoming to me.

To enter into the subject of Fundamental Groups, let me, as usual, make a

2 High-School Beginning

So consider a polynomial

$$f = f(Y) = Y^n + a_1Y^{n-1} + \ldots + a_n$$

with coefficients $a_1, \ldots, a_n$ in some field K; for example, K could be the field of rational numbers. We want to solve the equation $f = 0$, i.e., we want to find the roots of f. Assume that f is irreducible and has no multiple roots. Suppose somehow we found a root y_1 of f. Then to make the problem of finding the other roots easier, we achieve a decrease in degree by "throwing away" the root y_1 to get

$$f_1 = f_1(Y) = \frac{f(Y)}{(Y - y_1)} = Y^{n-1} + b_1Y^{n-2} + \ldots + b_{n-1}.$$

*Invited Lecture delivered on 8 January 1992 at the International Colloquium on Geometry and Analysis in TIFR in Bombay. This work was partly supported by NSF Grant DMS-91-01424

If $f_1(Y)$ is also irreducible and if somehow we found a root y_2 of f_1, then "throwing away" y_2 we get

$$f_2 = f_2(Y) = \frac{f_1(Y)}{(Y - y_2)} = Y^{n-2} + c_1 Y^{n-3} + \ldots + c_{n-2}.$$

Note that the coefficients $b_1, \ldots, b_{n-1}$ of f_1 do involve y_1 and hence they are not in K, but they are in $K(y_1)$. So, although we assumed f to be irreducible in $K[Y]$, when we said "if f_1 is irreducible," we clearly meant "if f_1 is irreducible in $K(y_1)[Y]$." Likewise, irreducibility of f_2 refers to its irreducibility in $K(y_1, y_2)[Y]$. And so on. In this way we get a sequence of polynomials $f_1, f_2, \ldots, f_m$ of degrees $n-1, n-2, \ldots, n-m$ in Y with coefficients in $K(y_1), K(y_1, y_2), \ldots, K(y_1, \ldots, y_m)$, where f_i is irreducible in $K(y_1, \ldots, y_i)[Y]$ for $i = 1, 2, \ldots, m-1$. If f_m is reducible in $K(y_1, \ldots, y_m)[Y]$ then we stop, otherwise we proceed to get f_{m+1}, and so on. Now we may ask the following

Question: Given any positive integers $m < n$, does there exist an irreducible polynomial f of degree n in Y with coefficients in some field K such that the above sequence terminates exactly after m steps, i.e., such that $f_1, f_2, \ldots, f_{m-1}$ are irreducible but f_m is reducible?

I presume that most of us, when asked to respond quickly, might say: "Yes, but for large m and n it would be time consuming to write down concrete examples." However, the **SURPRISE OF THE CENTURY** is that the **ANSWER** is **NO**. More precisely, it turns out that

(2.1) f_1, f_2, f_3, f_4, f_5 irreducible $\Rightarrow f_6, f_7, \ldots, f_{n-3}$ irreducible.

In other words, if $f_1, \ldots, f_5$ are irreducible then $f_1, \ldots, f_{n-1}$ are all irreducible except that f_{n-2}, which is a quadratic, may or may not be irreducible. This answers the case of $m \geq 6$. Going down the line to $m \leq 5$ and assuming $m < n-2$, for the case of $m = 5$ we have that

(2.2) f_1, f_2, f_3, f_4 irreducible but f_5 reducible $\Rightarrow n = 24$ or 12

and for the case of $m = 4$ we have that

(2.3) f_1, f_2, f_3 irreducible but f_4 reducible $\Rightarrow n = 23$ or 11.

Going further down the line, for the case of $m = 3$ we have that

(2.4) f_1, f_2 irreducible but f_3 reducible $\Rightarrow$ Refined FT of Proj Geom

i.e., if f_1, f_2 are irreducible but f_3 is reducible, then there are only a few possibilities and they are suggested by the Fundamental Theorem of Projective Geometry, which briefly says that "the underlying division ring of a synthetically defined desarguesian projective plane is a field if and only if any three points of a projective line can be mapped to any other three points of that projective line by a unique projectivity." Going still further down the line, for the case of $m = 2$ we have that

(2.5) f_1 irreducible but f_2 reducible $\Rightarrow$ known but too long

i.e., if f_1 is irreducible but f_2 is reducible, the answer is known but the list of possibilities is too long to write down here. Finally, for the case of $m = 1$ we have that

(2.6) f_1 has exactly two irreducible factors $\Rightarrow$ Pathol Proj Geom + Stat

i.e., if f_1 has exactly two irreducible factors, then again a complete answer is known, which depends on Pathological Projective Geometry and Block Designs from Statistics! Here I am reminded of the beautiful course on Projective Geometry which I took from Zariski (in 1951 at Harvard), and in which I learnt the Fundamental Theorem mentioned in (2.4). At the end of that course, Zariski said to me that "projective geometry is a beautiful dead subject, so don't try to do research in it" by which he implied that the ongoing research in that subject at that time was rather pathological and dealt with non-desarguesian planes and such. But in the intervening thirty or forty years, this "pathological" aspect, under the guise of Block Designs of Statistics, has made great strides in the hands of pioneers from R. C. Bose [21] and S. S. Shrikhande [58] to P. Dembowski [29] and D. G. Higman [34], and has led to a complete classification of Rank 3 groups, which from our view-point of the theory of equations is synonymous to case (2.6). So realizing how even a great man like Zariski could be wrong occasionally, I have learnt to drop one of my numerous prejudices, namely my prejudice against Statistics.

Note that a permutation group is said to be *transitive* if any point (of the permuted set) can be sent to any other, via a permutation in that group. Likewise, a permutation group is *m–fold transitive* (briefly: m–transitive) if any m points can be sent to any other m points, via a permutation in that group. *Doubly Transitive* = 2–transitive, *Triply Transitive* = 3–transitive, and so on. By the *one point stabilizer* of a transitive permutation group we mean the subgroup consisting of those permutations which keep a certain point fixed; the *orbits* of that subgroup are the minimal subsets of the permuted set which are mapped to themselves by every permutation in that subgroup; the number of these orbits is called the *rank* of the group, and the lengths of the nontrivial orbits are called the *subdegrees* of the group, so that the number of subdegrees is one less than the rank. Thus a *Rank 3* group is a transitive permutation group whose one point stabilizer has three orbits; the lengths of the two nontrivial orbits are the subdegrees. Needless to say that a Rank 2 group is nothing but a Doubly Transitive permutation group. At any rate, in case (2.6), the degrees of the two irreducible factors of f_1 correspond to the subdegrees of the relevant Rank 3 group. Now CR3 (= the Classification Theorem of Rank 3 groups) implies that very few pairs of integers can be the subdegrees of Rank 3 groups, and very few nonisomorphic Rank 3 groups can have the same subdegrees; see Kantor–Liebler [42] and Liebeck [44]. Hence (2.6) says that if f_1 has exactly two irreducible factors then their degrees (and hence also n) can have only certain very selective values.

Here, by the relevant group we mean the *Galois group* of f over K, which we denote by $\text{Gal}(f, K)$ and which, following Galois, we define as the group of those permutations of the roots $y_1, \ldots, y_n$ which retain all the polynomial relations between them with coefficients in K. This definition makes sense without f being irreducible but still assuming f to have no multiple roots. Now our assumption of f being irreducible is equivalent to assuming that $\text{Gal}(f, K)$ is transitive. Likewise, $f_1, \ldots, f_{m-1}$ are irreducible iff $\text{Gal}(f, K)$ is m–transitive. Moreover, as already indicated, f_1 has exactly two irreducible factors iff $\text{Gal}(f, K)$ has rank 3. To match this definition of Galois with the modern definition, let L be the *splitting field* of f over K, i.e., $L = K(y_1, \ldots, y_n)$. Then according to the modern definition, the *Galois group* of L over K, denoted by $\text{Gal}(L, K)$, is defined to be the group of all automorphisms of L which keep K pointwise fixed. Considering $\text{Gal}(f, K)$ as a permutation group of the subscripts $1, \ldots, n$ of $y_1, \ldots, y_n$, for every $\tau \in \text{Gal}(L, K)$ we have a unique $\sigma \in \text{Gal}(f, K)$ such that $\tau(y_i) = y_{\sigma(i)}$ for $1 \leq i \leq n$. Now we get an isomorphism of $\text{Gal}(L, K)$ onto $\text{Gal}(f, K)$ by sending each τ to the corresponding σ.

Having sufficiently discussed case (2.6), let us note that (2.5) is equivalent to CDT (=Classification Theorem of Doubly Transitive permutation groups) for which we may refer to Cameron [22] and Kantor [41]. At any rate, CDT implies that if f_1 is irreducible but f_2 is reducible then we must have: either $n = q$ for some prime power q, or $n = (q^l - 1)/(q - 1)$ for some integer $l > 1$ and some prime power q, or $n = 2^{2l-1} + 2^{l-1}$ for some integer $l > 2$, or $n = 2^{2l-1} - 2^{l-1}$ for some integer $l > 2$, or $n = 15$, or $n = 176$, or $n = 276$. Likewise (2.4) is equivalent to CTT (= Classification of Triply Transitive permutation groups) which is subsumed in CDT, and as a consequence of it we can say that if f_1, f_2 are irreducible but f_3 is reducible then we must have: either $n = 2^l$ for some positive integer l, or $n = q + 1$ for some prime power q, or $n = 22$.

Similarly, (2.3) is equivalent to CQT (= Classification of Quadruply Transitive permutation groups) which is subsumed in CTT, and as a consequence of it we can say that if f_1, f_2, f_3 are irreducible but f_4 is reducible then we must have: either $n = 23$ and $\text{Gal}(f, K) = M_{23}$ or $n = 11$ and $\text{Gal}(f, K) = M_{11}$, where M stands for Mathieu. Likewise, (2.2) is equivalent to CFT (= Classification of Fivefold Transitive permutation groups) which is subsumed in CQT, and as a consequence of it we can say that if f_1, f_2, f_3, f_4 are irreducible but f_5 is reducible then we must have: either $n = 24$ and $\text{Gal}(f, K) = M_{24}$ or $n = 12$ and $\text{Gal}(f, K) = M_{12}$.

Note that, M_{24} and M_{12} are the only 5–fold but not 6–fold transitive permutation groups other than the *symmetric group* S_5 (i.e., the group of all permutations on 5 letters) and the *alternating group* A_7 (i.e., the subgroup of S_7 consisting of all *even* permutations). Moreover, M_{23} and M_{11} are the respective one point stabilizers of M_{24} and M_{12} and they are the only 4–fold but not 5–fold transitive permutation groups other than S_4 and A_6. Here the subscript denotes the degree, i.e., the number of letters being permuted. The four groups $M_{24}, M_{23}, M_{12}, M_{11}$, were constructed by Mathieu [46] in 1861 as examples of highly transitive permutation groups. But the fact that they are the only 4–transitive permutation groups other than the symmetric groups and the alternating groups, was proved only in 1981 when CDT, and hence also CTT, CQT, CFT and CST, were deduced from CT (= Classification Theorem of finite simple groups); see Cameron [23] and Cameron–Cannon [24]. Recall that a group is *simple* if it has no nonidentity normal subgroup other than itself; it turns out that the five *Mathieu groups* $M_{24}, M_{23}, M_{22}, M_{12}, M_{11}$, where M_{22} is the one point stabilizer of M_{23}, are all simple. Now CST refers to the Classification Theorem of Sixfold Transitive permutation groups, according to which the symmetric groups and the alternating groups are the only 6–transitive permutation groups; note that S_m is m–transitive but not $(m + 1)$–transitive, whereas A_m is $(m - 2)$–transitive but not $(m - 1)$–transitive for $m \geq 3$. In (2.2) to (2.6) we had assumed $m < n - 2$ to avoid including the symmetric and alternating groups; dropping this assumption, (2.1) is equivalent to CST with the clarification that, under the assumption of (2.1), the quadratic f_{n-2} is irreducible or reducible according as $\text{Gal}(f, K) = S_n$ or A_n.

We have already hinted that CR3 was also deduced as a consequence of CT; see Liebeck [44]. The proof of CT itself was completed in 1980 (see Gorenstein [32]) with the *staggering statistics*: 30 years; 100 authors; 500

papers; 15,000 pages! Add some more pages for CDT and CR3 and so on.

All we have done above is to translate this group theory into the language of theory of equations where K is ANY field. So are we still talking High–School? Not really, unless we admit CT into High–School!

3 Galois

Summarizing, to compute the Galois group $\mathrm{Gal}(f, K)$, say when K is the field $k(X)$ of univariate rational functions over an algebraically closed ground field k, by throwing away roots and using some algebraic geometry we find some multi–transitivity and other properties of the Galois group and feed these into the group theory machine. Out comes a list of possible groups. Reverting to algebraic geometry, sometimes augmented by High–School manipulations, we successively eliminate various members from that list until, hopefully, one is left. That then is the answer. I say hopefully because we would have a contradiction on our hand if none was left. This reminds me of Hindu Philosophy in which the ultimate reality (**Brahman**) is described by *Neti Neti*, not this, not that. If you practice pure *Advaita*, then nothing is left, which is too austere. So we fall back on the kinder *Dvaita* according to which the unique **God** remains.

4 Riemann and Dedekind

In case $K = \mathbf{C}(X)$ and $a_i = a_i(X) \in \mathbf{C}[X]$ for $1 \le i \le n$, where $\mathbf{C}$ is the field of complex numbers, following Riemann [53] we can consider the *monodromy group* of f thus.

Fix a nondiscriminant point μ, i.e., value $\mu \in \mathbf{C}$ of X for which the equation $f = 0$ has n distinct roots. Then, say by the Implicit Function Theorem, we can solve the equation $f = 0$ near μ, getting n analytic solutions $\eta_1(X), \ldots, \eta_n(X)$ near μ. To find out how these solutions are intertwined, mark a finite number of values $\alpha_1, \ldots, \alpha_w$ of X which are different from μ but include all the discriminant points, and let $\mathbf{C}_w$ be the complex X–plane minus these w points. Now by making analytic continuations along any closed path Γ in $\mathbf{C}_w$ starting and ending at μ we get a permutation of $\eta_1, \ldots, \eta_n$, i.e., a permutation Γ' of $1, \ldots, n$ so that η_i continues into η_j with $\Gamma'(i) = j$ for $1 \le i \le n$. As Γ varies over all closed paths in $\mathbf{C}_w$ starting and ending at μ, the permutations Γ' span a subgroup of S_n called the *monodromy group* of f which we denote by $M(f)$.

By identifying the analytic solutions $\eta_1, \ldots, \eta_n$ with the algebraic roots $y_1, \ldots, y_n$, the monodromy group $M(f)$ gets identified with the Galois group

$\text{Gal}(f, \mathbf{C}(X))$, and so these two groups are certainly isomorphic as permutation groups.

To get generators for $M(f)$, given any $\alpha \in \mathbf{C}_w$, let Γ_α be the path in $\mathbf{C}_w$ consisting of a line segment from μ to a point very near α followed by a small circle around α and then back to μ along the said line segment. Let us write the corresponding permutation Γ'_α as a product of disjoint cycles, and let $e_1, \ldots, e_h$ be the lengths of these cycles. To get a tie–up between these Riemannian considerations and the thought of Dedekind [28], let v be the valuation of $\mathbf{C}(X)$ corresponding to α, i.e., $v(g)$ is the order of zero at α for every $g \in \mathbf{C}[X]$. Then, as remarked in my 1957 paper [3], the cycle lengths $e_1, \ldots, e_n$ coincide with the ramification exponents of the various extensions of v to the *root field* $\mathbf{C}(X)(y_1)$, and their LCM equals the ramification exponent of any extension of v to the splitting field $\mathbf{C}(X)(y_1, \ldots, y_n)$. In particular, Γ'_α is the identity permutation iff α is not a *branch point*, i.e., if and only if the ramification exponents of the various extensions of v to the root field $\mathbf{C}(X)(y_1)$ (or equivalently to the splitting field $\mathbf{C}(X)(y_1, \ldots, y_n)$) are all 1. At any rate, a branch point is always a discriminant point but not conversely. Indeed, the difference between the two is succinctly expressed by Dedekind's Theorem according to which the ideal generated by the Y–derivative of f equals the product of the *different* and the *conductor*. In this connection you may refer to pages 423 and 438 of my Monthly Article [5] which constitutes some of my *Ramblings* in the woods of algebraic geometry. You may also refer to pages 65 and 169 of my recent book [6] for Scientists and Engineers into which these Ramblings have now been expanded.

Having given a tie–up between the ideas of Riemann and Dedekind (both of whom were pupils of Gauss) concerning branch points, ramification exponents, and so on, it is time to say that these things actually go back to Newton [47]. For an excellent discussion of the seventeenth century work of Newton on this matter, see pages 373-397 of Part II of the 1886 Textbook of Algebra by Chrystal [27]. For years having recommended Chrystal as the best book to learn algebra from, from time to time I decide to take my own advice and look into the vast material which I had skipped before. Indeed, what a wealth of information it contains! At any rate, à la Newton, we can use fractional power series in X to factor f into linear factors in Y, and then combine conjugacy classes to get a factorization $f = \prod_{i=1}^{h} \phi_i$ where ϕ_i is an irreducible polynomial of degree e_i in Y whose coefficients are power series in $X - \alpha$. If the field $\mathbf{C}$ were not algebraically closed then the degree of ϕ_i would be $e_i f_i$ with f_i being certain "residue degrees" and we would get the famous formula $\sum_{i=1}^{h} e_i f_i = n$ of Dedekind–Domain Theory. See Lectures 12 and 21 of Scientists [6].

Geometrically speaking, i.e., following the ideas of Max Noether [48], if we consider the curve $f = 0$ in the (X, Y)–plane, then the discriminant points correspond to vertical lines which meet the curve in less than n points, the branch points correspond to vertical tangents, and the "conductor points" are the singularities. See figure 9 on page 429 of Ramblings [5].

Getting back to finding generators for $M(f)$, with the refinement of discriminant points into branch points and conductor points in hand, it suffices to stipulate that $\alpha_1, \ldots, \alpha_w$ include all the branch points rather than all the discriminant points. Now by choosing the *base point* μ suitably, we may assume that the lines from μ to $\alpha_1, \ldots, \alpha_w$ do not meet each otherexcept at μ. Now it will turn out that the permutations $\Gamma'_{\alpha_1}, \ldots, \Gamma'_{\alpha_w}$ generate $M(f)$. This follows from the *Monodromy Theorem* together with the fact that the *(topological) fundamental group* $\pi_1(\mathbf{C}_w)$ of $\mathbf{C}_w$ (also called the Poincaré group of $\mathbf{C}_w$) is the *free group* $\mathbf{F}_w$ on w generators. Briefly speaking, the monodromy Theorem says that two paths, which can be continuously deformed into each other, give rise to the same analytic continuations. The fundamental group itself may heuristically be described as that incarnation of the monodromy group which works for all functions whose branch points are amongst $\alpha_1, \ldots, \alpha_w$. More precisely, $\pi_1(\mathbf{C}_w)$ consists of the equivalence classes of closed paths beginning and ending at μ where equivalent means they can be continuously deformed into each other. Now the (equivalence classes of the) paths $\Gamma_{\alpha_1}, \ldots, \Gamma_{\alpha_w}$ are free generators of $\pi_1(\mathbf{C}_w)$, and we have an obvious epimorphism of $\pi_1(\mathbf{C}_w) = \mathbf{F}_w$ onto $M(f)$ and hence the permutations $\Gamma'_{\alpha_1}, \ldots, \Gamma_{\alpha_w}$ generate $M(f)$; for relevant pictures etc., you may see pages 442-443 of Ramblings [5] or pages 171-172 of Scientists [6]. So the curve $f = 0$, or equivalently the *Galois extension* $L = \mathbf{C}(X)(y_1, \ldots, y_n)$, is an *unramified covering* of $\mathbf{C}_w$, and the Galois group $\mathrm{Gal}(L, \mathbf{C}(X)) = \mathrm{Gal}(f, \mathbf{C}(X))$ is generated by w generators. Surprisingly, to this day *there is no algebraic proof of this algebraic fact.*

The Riemann Existence Theorem says that conversely, every finite homomorphic image of $\pi_1(\mathbf{C}_w)$ can be realized as $M(f)$ for some f. Thus by defining the *algebraic fundamental group* $\pi_A(\mathbf{C}_w)$ as the set of all finite groups which are the Galois groups of finite unramified coverings of $\mathbf{C}_w$, we can say that $\pi_A(\mathbf{C}_w)$ coincides with the set of all finite groups generated by w generators. *Needless to say that, a fortiori, there is no algebraic proof of the converse part of this algebraic fact either.*

Now, in the complex (X, Y)–plane, $f = 0$ is a curve C_g of some *genus* g, i.e., if from C_g we delete a finite number of points including all its singularities, then what we get is homeomorphic to a sphere with g handles minus a finite number of points. For any nonnegative integer w, let $C_{g,w}$ be obtained by adding to C_g its *points at infinity*, then *desingularizing* it, and finally removing $w + 1$ points from the desingularized version. Then $C_{g,w}$ is homeomorphic to a sphere with g handles minus $w+1$ points, and hence it can be seen that $\pi_1(C_{g,w}) = \mathbf{F}_{2g+w}$; for instance see the excellent topology book of Seifert and Threlfall [54]. The above monodromy and existence considerations generalize from the genus zero case to the case of general g, and we get the *result* that the *algebraic fundamental group* $\pi_A(C_{g,w})$ coincides with the set of all finite groups generated by $2g+w$ generators, where $\pi_A(C_{g,w})$ is *defined* to be the set of all finite groups which are the Galois groups of finite unramified coverings of $\mathbf{C}_{g,w}$.

5 Chrystal and Forsyth

Just as Chrystal excels in explaining Newtonian (and Eulerian) ideas, Forsyth's 1918 book on Function Theory [31] is highly recommended for getting a good insight into Riemannian ideas. Thus it was by absorbing parts of Forsyth that, in my recent papers [8] and [10], I could algebracize some of the monodromy considerations to formulate certain "Cycle Lemmas" which say that under such and such conditions the Galois group contains permutations having such and such cycle structure.

Now the Riemann Existence Theorem was only surmised by Riemann [52] by appealing to the *Principle* of his teacher Dirichlet which, after Weierstrass' criticism was put on firmer ground by Hilbert in 1904 [35]. In the meantime another classical treatment of the Riemann Existence Theorem was carried out culminating in the Klein–Poincaré–Koebe theory of automorphic functions, for which again Forsyth's book is a good source. A modern treatment of the Riemann Existence Theorem using coherent analytic sheaves was finally given by Serre in his famous GAGA paper [55] of 1956.

6 Serre

Given any algebraically closed ground field k of any nonzero characteristic p, in my 1957 paper [3], all this led me to *define* the *algebraic fundamental group* $\pi_A(C_{g,w})$ of $C_{g,w} = C_g$ minus $w+1$ points, where w is a nonnegative integer and C_g is a nonsingular projective curve of genus g over k, to be the set of all finite groups which can be realized as Galois groups of finite unramified coverings of $C_{g,w}$. In that paper, I went on to *conjecture* that $\pi_A(C_{g,w})$ coincides with the set of all finite groups G for which $G/p(G)$ is generated by $2g+w$ generators, where $p(G)$ is the subgroup of G generated by all its p–Sylow subgroups. The $g = w = 0$ case of this conjecture, which may be called the *quasi p–group conjecture*, says that for the affine line L_k over k we have $\pi_A(L_k) = Q(p)$ where $Q(p)$ denotes the set of all quasi p–

groups, i.e., finite groups which are generated by their p–Sylow subgroups. It may be noted every finite simple group whose order is divisible by p is obviously a quasi p–group. Hence in particular the alternating group A_n is a quasi p–group whenever either $n \geq p > 2$ or $n - 3 \geq p = 2$. Likewise the symmetric group S_n is a quasi p–group provided $n \geq p = 2$.

In support of the quasi p–group conjecture, in the 1957 paper I wrote down several equations giving unramified coverings of the affine line L_k and suggested that their Galois groups be computed. This included the equation $\overline{F}_{n,q,s,a} = 0$ with

$$\overline{F}_{n,q,s,a} = Y^n - aX^sY^t + 1 \text{ and } n = q + t$$

where $0 \neq a \in k$ and q is a positive power of p and s and t are positive integers with $t \not\equiv 0(p)$, and we want to compute its Galois group $\overline{G}_{n,q,s,a} = \mathrm{Gal}(\overline{F}_{n,q,s,a}, k(X))$.

By using a tiny amount of the information contained in the above equation, I showed that $\pi_A(L_k)$ contains many unsolvable groups, and indeed by taking homomorphic images of subgroups of members of $\pi_A(L_k)$ we get all finite groups; see Result 4 and Remark 6 on pages 841-842 of [3]. This was somewhat of a surprise because the complex affine line is simply connected, and although $\pi_A(L_k)$ was known to contain p–cyclic groups (so called Artin–Schreier equations), it was felt that perhaps it does not contain much more. This feeling, which turned out to be wrong, might have been based on the facts that L_k is a "commutative group variety" and the fundamental group of a topological group is always abelian; see Proposition 7 on page 54 of Chevalley [26].

To algebracize the fact that the complex affine line is simply connected, by the genus formula we deduce that the affine line over an algebraically closed ground field of characteristic zero has no nontrivial unramified coverings. In our case of characteristic p, the same formula shows that every member of $\pi_A(L_k)$ is a quasi p–group; see Result 4 on page 841 of [3].

Originally I found the above equation $\overline{F}_{n,q,s,a} = 0$ by taking a section of a surface which I had constructed in my 1955 Ph.D. Thesis [1] to show that Jung's classical method [40] of surface desingularization does not work for nonzero characteristic because the local fundamental group above a normal crossing of the branch locus need not be solvable, while in the complex case it is always abelian. This failure of Jung's method led me to devise more algorithmic techniques for desingularizing surfaces in nonzero characteristic, and this formed the positive part of my Ph.D. Thesis [2].

Soon after the 1957 paper, I wrote a series of articles [4] on "tame coverings" of higher dimensional algebraic varieties, and took note of Grothendieck [33] proving the "tame part" of the above conjecture which says that the members of $\pi_A(C_{g,w})$ whose order is prime to p are exactly all the finite groups of order prime to p generated by $2g + w$ generators.

But after these two things, for a long time I forgot all about coverings and fundamental groups.

Then suddenly, after a lapse of nearly thirty years, Serre pulled me back into the game in October 1988 by writing to me a series of letters in which he briefly said: "I can now show that if $t = 1$ then $\overline{G}_{n,q,s,a} = PSL(2,q)$. Can you compute $\overline{G}_{n,q,s,a}$ for $t \geq 2$? Also, can you find unramified A_n coverings of L_k?"

Strangely, the answers to both these questions turned out to be almost the same. Namely, with much prodding and prompting by Serre (hundred e-mails and a dozen s-mails = snail-mails) augmented by group theory lessons first from Kantor and Feit and then from Cameron and O'Nan, and by using the method of throwing away roots, CT in the guise of CDT, the Cycle Lemmas, the Jordan–Marggraff Theorems on limits of transitivity (see Jordan [39] and Marggraff [45] or Wielandt [60]), and finally some High–School type factorizations, in the papers [8] to [10] I proved that:

(6.1) $t = 1 \Rightarrow \overline{G}_{n,q,s,a} = PSL(2,q)$.

(6.2) $q = p > 2 \leq t$ and $(p,t) \neq (7,2) \Rightarrow \overline{G}_{n,q,s,a} = A_n$.

(6.3) $q = p > 2 \leq t$ and $(p,t) = (7,2) \Rightarrow \overline{G}_{n,q,s,a} = PSL(2,8)$.

(6.4) $q = p = 2 \Rightarrow \overline{G}_{n,q,s,a} = S_n$.

(6.5) $q < t$ and $p > 2 \Rightarrow \overline{G}_{n,q,s,a} = A_n$.

(6.6) $p = 2 < q < t \Rightarrow \overline{G}_{n,q,s,a} = A_n$.

(6.7) $p = 2 < q = 4$ and $t = 3$ (and $n = 7$) $\Rightarrow \overline{G}_{n,q,s,a} = PSL(3,2)$.

(6.8) $p = 3 < q = 9$ and $t = 2$ (and $n = 11$) $\Rightarrow \overline{G}_{n,q,s,a} = M_{11}$.

Note that $PSL(m,q) = SL(m,q)/(\text{scalar matrices})$ where $SL(m,q) =$ the group of all m by m matrices whose determinant is 1 and whose entries are in the field GF(q) of q elements. Now my proof of (6.1) uses the Zassenhaus–Feit–Suzuki Theorem which characterizes doubly transitive permutation groups for which no 3 points are fixed by a nonidentity permutation; see Zassenhaus [62], Feit [30] and Suzuki [59]. As Serre has remarked, his proof of (6.1) may be called a "descending" proof as opposed to my "ascending" proof. Serre's proof may be found in his November 1990 letter to me which appears as an Appendix to my paper [8]. Actually, when [8] was already in press, Serre found that a proof somewhat similar to his was already given by Carlitz [25] in 1956.

Throwing away one root of $\overline{F}_{n,q,s,a}$ and then applying Abhyankar's Lemma (see pages 181-186 of Part III of [4]) and deforming things conveniently, we get the monic polynomial $\overline{F}'_{n,q,s,a,b,u}$ of degree $n-1$ in Y with coefficients in

$k(X)$ given by

$$\overline{F}'_{n,q,s,a,b,u} = t^{-2}[(Y+t)^t - Y^t](Y+b)^q - aX^{-s}Y^u$$

with $0 \neq b \in k$ and positive integer $u < n-1$. Now upon letting

$$r = (q+t)\mathrm{LCM}\left(t, \frac{q-1}{\mathrm{GCD}(q-1, q+t)}\right)$$

and $\overline{G}'_{n,q,s,a,b,u} = \mathrm{Gal}(\overline{F}'_{n,q,s,a,b,u}, k(X))$, in the papers [8] and [10] I also proved that, in the following cases, $\overline{F}'_{n,q,s,a,b,u} = 0$ gives an unramified covering of L_k with the indicated Galois group:

(6.1′) $b = u = t > 2 \neq q = p$ and $s \equiv 0(p-1)$ and $s \equiv 0(t) \Rightarrow \overline{G}'_{n,q,s,a,b,u} = A_{n-1}$.

(6.2′) $b = u = t = 2$ and $q = p \neq 7$ and $s \equiv 0(p-1) \Rightarrow \overline{G}'_{n,q,s,a,b,u} = A_{n-1}$.

(6.3′) If $t = 2$ and $q = p > 5$ then u can be chosen so that $1 < u < (p+1)/2$ and $\mathrm{GCD}(p+1, u) = 1$, and for any such u, upon assuming $b = u/(u-1)$ and $s \equiv 0(u(p+1-u))$, we have $\overline{G}'_{n,q,s,a,b,u} = A_{n-1}$.

(6.4′) $b = u = t$ and $q = p = 2$ and $s \equiv 0(t) \Rightarrow \overline{G}'_{n,q,s,a,b,u} = S_{n-1}$.

(6.5′) $b = u = t > q$ and $p > 2$ and $s \equiv 0(r) \Rightarrow \overline{G}'_{n,q,s,a,b,u} = A_{n-1}$.

(6.6′) $b = u = t > q = p = 2$ and $s \equiv 0(r) \Rightarrow \overline{G}'_{n,q,s,a,b,u} = S_{n-1}$.

(6.7′) $b = u = t > q > p = 2$ and $s \equiv 0(r) \Rightarrow \overline{G}'_{n,q,s,a,b,u} = A_{n-1}$.

Another equation written down in the 1957 paper giving an unramified covering of L_k is $\widetilde{F}_{n,t,s,a} = 0$ where n, t, s are positive integers with

$$t < n \equiv 0(p) \text{ and } \mathrm{GCD}(n,t) = 1 \text{ and } s \equiv 0(t)$$

and $\widetilde{F}_{n,t,s,a}$ is the polynomial given by

$$\widetilde{F}_{n,t,s,a} = Y^n - aY^t + X^s \text{ with } 0 \neq a \in k.$$

Again upon letting $\widetilde{G}_{n,t,s,a} = \mathrm{Gal}(\widetilde{F}_{n,t,s,a}, k(X))$, in the papers [8] and [10] I proved that:

(6.1*) $1 < t < 4$ and $p \neq 2 \Rightarrow \widetilde{G}_{n,t,s,a} = A_n$.

(6.2*) $1 < t < n-3$ and $p \neq 2 \Rightarrow \widetilde{G}_{n,t,s,a} = A_n$.

(6.3*) $1 < t = n-3$ and $p \neq 2$ and $11 \neq p \neq 23 \Rightarrow \widetilde{G}_{n,t,s,a} = A_n$.

(6.4*) $1 < t < 4 < n$ and $p = 2 \Rightarrow \widetilde{G}_{n,t,s,a} = A_n$.

(6.5*) $1 < t < n-3$ and $p = 2 \Rightarrow \widetilde{G}_{n,t,s,a} = A_n$.

In Proposition 1 of the 1957 paper I discussed the polynomial

$$Y^{hp+t} + aXY^{hp} + 1 + \sum_{i=1}^{h-1} a_i Y^{(h-i)p} \text{ with } t \not\equiv 0(p) \text{ and } 0 \neq a \in k \text{ and } a_i \in k$$

giving an unramified covering of L_k. The polynomial $\overline{F}$ studied in (6.1) to (6.6) is the $hp = q$ and $a_1 = \ldots = a_{h-1} = 0$ case of this after "reciprocating" the roots and changing X to X^s. Considering the $p = 2 = h = t - 1$ and $a = a_1 = 1$ case of this we get the polynomial

$$F^{\circ} = Y^7 + XY^4 + Y^2 + 1$$

and it can be shown that:

(6.1°) For $p = 2$ the equation $F^{\circ} = 0$ gives an unramified covering of L_k with $\mathrm{Gal}(F^{\circ}, k(X)) = A_7$.

By throwing away a root of F° and then invoking Abhyankar's Lemma we obtain the polynomial

$$F'^{\circ} = Y^6 + X^{27}Y^5 + X^{54}Y^4 + (X^{18} + X^{36})Y^3 + X^{108}Y^2 + (X^{90} + X^{135})Y + X^{162}$$

and therefore by (6.1°) we see that:

(6.2°) For $p = 2$ the equation $F'^{\circ} = 0$ gives an unramified covering of L_k with $\mathrm{Gal}(F'^{\circ}, k(X)) = A_6$.

Now it was Serre who first prompted me to use CT in calculating the various Galois groups discussed above. But after I had done this, again it was Serre who prodded me to try to get around CT. So, as described in the papers [8] and [10], by traversing a suitable path in items (6.1) to (6.2°) we get a *complete equational proof* of the following Facts *without CT*:

Facts. (6.i) For all $n \geq p > 2$ we have $A_n \in \pi_A(L_k)$. (6.ii) For all $n \geq p = 2$ we have $S_n \in \pi_A(L_k)$. (6.iii) For all $n \geq p = 2$ with $3 \neq n \neq 4$ we have $A_n \in \pi_A(L_k)$; (note that A_3 and A_4 are not quasi 2-groups).

While attempting to circumvent CT, once I got a very amusing e-mail from Serre saying "About the *essential* removal of CT from your A_n-determinations: what does essential mean? (Old story: a noble man had a statue of himself made by a well-known sculptor. The sculptor asked: do you want an equestrian statue or not? The noble man did not understand the word. He said: oh, yes, equestrian if you want, but not too much...). This is what I feel about *non essential use of CT*."

In any case, learning and adopting (or adapting) all this group theory has certainly been very rejuvenating to me. To state CT very briefly: Z_p (= the cyclic group of prime order p), A_n (excluding $n \leq 4$), $PSL(n+1, q)$ (excluding $n = 1$ and $q \leq 3$) together with 15 other related and reincarnated infinite families, and the 26 *sporadics* including the 5 Mathieus is a complete list of finite simple groups; for details see Abhyankar [8] and Gorenstein [32].

7 Jacobson and Berlekamp

Concerning items (6.6), (6.7′), (6.4*) and (6.5*), when I said that I proved them in [8] and [10], what I actually meant was that I proved their weaker version asserting that the Galois group is the alternating group or the symmetric group, and then thanks to Jacobson's Criterion, the symmetric group possibility was eliminated in my joint paper with Ou and Sathaye [14]. What I am saying is that the classical criterion, according to which the Galois group of an equation is contained in the alternating group iff its modified discriminant is a square, is not valid for characteristic two. A version of such a criterion which is valid for all characteristics including two was given by Jacobson in his Algebra books published in 1964 [37] and 1974 [38]; an essentially equivalent version may also be found in the 1976 paper [20] of Berlekamp with some preliminary work in his 1968 book [19]; both these criteria have a bearing on the Arf invariant of a quadratic form [18] which itself was inspired by some work of Witt [61]. This takes care of A_n coverings for characteristic two provided $6 \neq n \neq 7$.

This leaves us with the A_6 and A_7 coverings for characteristic two described in items (6.2°) and (6.1°). Again using the Jacobson's Criterion, these are dealt with in my joint paper with Yie [17].

Thus, although Facts (6.i) and (6.ii) are indeed completely proved in my papers [8] and [10], but for Fact (6.iii) I was lucky to enjoy the active collaboration of my former (Sathaye) and present (Ou and Yie) students.

Likewise, item (6.7) is not proved in my papers [8] to [10], but was communicated to me by Serre (e-mail of October 1991) and is included in my joint paper [17] with Yie. Similarly, item (6.8) is not in my papers [8] to [10], but is proved in my joint paper [15] with Popp and Seiler.

In [8] I used CT for proving some parts of (6.1*) to (6.5*). Now I shall "merge" the polynomial $\widetilde{F}$ into the polynomial $\overline{F}$ and thereby get a proof of a stronger version of (6.1*) to (6.5*) without CT. Let us start by modifying the Third Irreducibility Lemma in Section 19 of [8] thus [in the proof of that lemma, once $\xi_\lambda(1,Z)$ has been mistakenly printed as $\xi_{\lambda(1,Z)}$]:

(7.1) Let k be any field which need not be algebraically closed and whose characteristic chc k need not be positive. Let $n > t > 1$ be integers such that $\mathrm{GCD}(n,t) = 1$ and $t \not\equiv 0(\mathrm{chc}\ k)$, and let $\Omega(Z)$ be the monic polynomial of degree $n-1$ in Z with coefficients in $k(Y)$ obtained by putting

$$\Omega(Z) = \frac{[(Z+Y)^n - Y^n] - [Y^{n-t} + Y^{-t}][(Z+Y)^t - Y^t]}{Z}.$$

Then $\Omega(Z)$ is irreducible in $k(Y)[Z]$.

Proof. Since $t \not\equiv 0(\text{chc } k)$, upon letting

$$\xi'_{\lambda'}(Y,Z) = \frac{[(Z+Y)^n - Y^n]}{Z} \text{ and } \eta_\mu(Y,Z) = \frac{-[(Z+Y)^t - Y^t]}{Z}$$

by the proof of the above cited lemma we see that: $\xi'_{\lambda'}(Y,Z)$ and $\eta_\mu(Y,Z)$ are homogeneous polynomials of degree $\lambda' = n-1$ and $\mu = t-1$ respectively, the polynomials $\xi'_{\lambda'}(1,Z)$ and $\eta_\mu(1,Z)$ have no nonconstant common factor in $k[Z]$, and the polynomial $\eta_\mu(1,Z)$ has a nonconstant irreducible factor in $k[Z]$ which does not divide $\xi'_{\lambda'}(1,Z)$ and whose square does not divide $\eta_\mu(1,Z)$. Upon letting

$$\xi_\lambda(Y,Z) = \frac{Y^t[(Z+Y)^n - Y^n] - Y^n[(Z+Y)^t - Y^t]}{Z}$$

we see that $\xi_\lambda(Y,Z)$ is a homogeneous polynomial of degree $\lambda = n+t-1$ and $\xi_\lambda(1,Z) = \xi'_{\lambda'}(1,Z) + \eta_\mu(1,Z)$ and therefore: the polynomials $\xi_\lambda(1,Z)$ and $\eta_\mu(1,Z)$ have no nonconstant common factor in $k[Z]$, and the polynomial $\eta_\mu(1,Z)$ has a nonconstant irreducible factor in $k[Z]$ which does not divide $\xi_\lambda(1,Z)$ and whose square does not divide $\eta_\mu(1,Z)$. The proof of the Second Irreducibility Lemma of Section 19 of [8] clearly remains valid if only one of the polynomials $\xi_\lambda(Y,Z)$ and $\eta_\mu(Y,Z)$ is assumed to be regular in Z, and in the present situation $\eta_\mu(Y,Z)$ is obviously regular in Z. Therefore by the said lemma, the polynomial $\xi_\lambda(Y,Z) + \eta_\mu(Y,Z)$ is irreducible in $k(Y)[Z]$. Obviously $\Omega(Z) = Y^{-t}[\xi_\lambda(Y,Z) + \eta_\mu(Y,Z)]$ and hence $\Omega(Z)$ is irreducible in $k(Y)[Z]$.

By using (7.1) we shall now prove:

(7.2) Let k be any field which need not be algebraically closed and whose characteristic chc k need not be positive. Let $0 \neq a \in k$ and let n, t, s be positive integers such that $1 < t \not\equiv 0(\text{chc } k)$ and $1 < n - t \not\equiv 0(\text{chc } k)$ and $\text{GCD}(n,t) = 1$. Then the polynomial $\Phi(Y) = Y^n - aX^sY^t + 1$ is irreducible in $k(X)[Y]$, its Y–discriminant is nonzero, and for its Galois group we have: $\text{Gal}(\Phi(Y), k(X)) = A_n$ or S_n. Similarly, the polynomial $\Psi(Y) = Y^n - aY^t + X^s$ is irreducible in $k(X)[Y]$, its Y–discriminant is nonzero, and for its Galois group we have: $\text{Gal}(\Psi(Y), k(X)) = A_n$ or S_n.

Proof. In view of the Basic Extension Principle and Corollaries (3.2) and (3.5) of the Substitutional Principle of Section 19 of [8], without loss of generality we may assume that k is algebraically closed and $a = 1 = s$. Since Φ and Ψ are linear in X, they are irreducible. By the discriminant calculation in Section 20 of [8] we see that their Y–discriminants are nonzero.

As in the beginning of Section 21 of [8] we see that the valuation $X = \infty$ of $k(X)/k$ splits into two valuations in the root field of $\Phi(Y)$ and their reduced ramification exponents are t and $n-t$. Now t and $n-t$ are both nondivisible by chc k and $\mathrm{GCD}(t, n-t) = 1$, and hence by the Cycle Lemma of Section 19 of [8] we conclude that $\mathrm{Gal}(\Phi(Y), k(X))$ contains a t–cycle and an $(n-t)$–cycle. By throwing away a root of $\Phi(Y)$ we get $[\Phi(Z + Y) - \Phi(Y)]/Z$ which equals $\Omega(Z)$ because by solving $\Phi(Y) = 0$ we get $X = Y^{n-t} + Y^{-t}$. Consequently by (7.1) we Conclude that $\mathrm{Gal}(\Phi(Y), k(X))$ is doubly transitive. Clearly either $1 < t < (n/2)$ or $1 < n - t < (n/2)$, and hence by Marggraff's Second Theorem as stated in Section 20 of [8] we get $\mathrm{Gal}(\Phi(Y), k(X)) = A_n$ or S_n. By Corollaries (3.2) and (3.5) of the Substitutional Principle of Section 19 of [8] it now follows that $\mathrm{Gal}(Y^n - X^{t-n}Y^t + 1, k(X)) = A_n$ or S_n. By multiplying the roots of $Y^n - X^{t-n}Y^t + 1$ by X, i.e., by writing Y/X for Y and multiplying throughout by X^n, we obtain the polynomial $Y^n - Y^t + X^n$ whose Galois group must be the same as the Galois group of $Y^n - X^{t-n}Y^t + 1$. Therefore $\mathrm{Gal}(Y^n - Y^t + X^n, k(X)) = A_n$ or S_n, and hence again by Corollaries (3.2) and (3.5) of the Substitutional Principle of Section 19 of [8] we conclude that $\mathrm{Gal}(\Psi(Y), k(X)) = A_n$ or S_n.

To get back to the polynomial $\widetilde{F}_{n,t,s,a}$, let us return to the assumption of k being an algebraically closed field of nonzero characteristic p. Let $0 \neq a \in k$ and let n, t, s be positive integers with

$$t < n \equiv 0(p) \text{ and } \mathrm{GCD}(n, t) = 1 \text{ and } s \equiv 0(t)$$

and recall that $\widetilde{F}_{n,t,s,a} = 0$ gives an unramified covering of L_k where

$$\widetilde{F}_{n,t,s,a} = Y^n - aY^t + X^s$$

and we want to consider the Galois group $\widetilde{G}_{n,t,s,a} = \mathrm{Gal}(\widetilde{F}_{n,t,s,a}, k(X))$. Since every member of $\pi_A(L_k)$ is a quasi p–group and since S_n is not a quasi p–group for $p \geq 3$, in view of (2.28) of [14], by the Ψ case of (7.2) we get the following sharper version of (6.1*) to (6.5*):

(7.1*) $1 < t < n - 1 \Rightarrow \widetilde{G}_{n,t,s,a} = A_n$.

Just as the small *border values* of t play a *special role* for the bar polynomial in (6.1) to (6.8), likewise the *condition* $1 \neq t \neq n-1$ in (7.1*) is *not accidental* as shown by the following four assertions:

(7.2*) $1 = t = n - 5$ and $p = 2 \Rightarrow \widetilde{G}_{n,t,s,a} = PSL(2,5) \approx A_5$.
(7.3*) $5 = t = n - 1$ and $p = 2 \Rightarrow \widetilde{G}_{n,t,s,a} = PSL(2,5) \approx A_5$.
(7.4*) $1 = t = n - 11$ and $p = 3 \Rightarrow \widetilde{G}_{n,t,s,a} = \widehat{M}_{11} \approx M_{11}$.
(7.5*) $1 = t$ and $n = p^m \Rightarrow \widetilde{G}_{n,t,s,a} = (Z_p)^m$.

Out of these four assertions, (7.2*) and (7.3*) may be found in my joint paper [17] with Yie, and (7.4*) may be found in my joint paper [15] with Popp and Seiler. It may be noted that $PSL(2,5)$ and A_5 are isomorphic as abstract groups but not as permutation groups. Likewise $\widehat{M}_{11}$ and M_{11} are isomorphic as abstract groups but not as permutation groups; here $\widehat{M}_{11}$ is the 3–transitive but not 4–transitive incarnation of M_{11} found by taking the image of the M_{11} under a noninner automorphism of M_{12}; see my paper [8] or volume III of the encyclopedic group theory book [36] of Huppert and Blackburn.

Concerning assertion (7.5*), multiplying the roots by a suitable nonzero element of k we can reduce to the case of $t = 1 = a$ and $n = p^m$ and then, remembering that $(Z_p)^m$ = the m–fold direct product of the cyclic group Z_p of order p = the underlying additive group of $\mathrm{GF}(p^m)$, our claim follows from the following remark:

(7.1**) If $Y^{p^m} - Y + x$ is irreducible over a field K of characteristic p, with $x \in K$ and $\mathrm{GF}(p^m) \subset K$, then by taking a root y of $Y^{p^m} - Y + x$ we have $Y^{p^m} - Y + x = \prod_{i \in \mathrm{GF}(p^m)} [Y - (y + i)]$ and hence exactly as in the p–cyclic case we get $\mathrm{Gal}(Y^{p^m} - Y + x, K) = (Z_p)^m$.

By throwing away a root of $\widetilde{F}_{n,t,s,a}$ and deforming things suitably we get the monic polynomial $\widetilde{F}'_{n,t,s,a}$ of degree $n-1$ in Y with coefficients in $k(X)$ given by

$$\widetilde{F}'_{n,t,s,a} = Y^{-1}[(Y+1)^n - 1] - aX^{-s}Y^{-1}[(Y+1)^t - 1]$$

and for its Galois group $\widetilde{G}'_{n,t,s,a} = \mathrm{Gal}(\widetilde{F}'_{n,t,s,a}, k(X))$, in my joint paper [15] with Popp and Seiler it is shown that:

(7.1′) $1 = t = n - 11$ and $p = 3$ and $s \equiv 0(n-1) \Rightarrow \widetilde{G}'_{n,t,s,a} = PSL(2,11)$, where, for the said values of the parameters, the equation $\widetilde{F}'_{n,t,s,a} = 0$ gives an unramified covering of L_k.

Finally let

$$\overline{F}^{(d)}_{n,q,s,a,b} = Y^{dn} - aX^sY^{dt} + b \text{ with positive integer } d \not\equiv 0(p)$$

where once again a, b are nonzero elements of k and n, t, s are positive integers with $t < n$ and $\mathrm{GCD}(n,t) = 1$ and $n - t = q$ = a positive power of p. For the Galois group $\overline{G}^{(d)}_{n,q,s,a,b} = \mathrm{Gal}(\overline{F}^{(d)}_{n,q,s,a,b}, k(X))$, in my joint paper [15] with Popp and Seiler it is shown that:

(7.2′) $d = t = 2 = n - 9$ and $q = 9$ and $p = 3 \Rightarrow \overline{G}^{(d)}_{n,q,s,a,b} = M^*_{11} \approx M_{11}$, where, for the said values of the parameters, the equation $\overline{F}^{(d)}_{n,q,s,a,b} = 0$ gives an unramified covering of L_k.

Again note that M_{11}^* and M_{11} are isomorphic as abstract groups but not as permutation groups; here M_{11}^* is the transitive but not 2–transitive incarnation of M_{11} obtained by considering its cosets according to the index 22 subgroup $PSL(2,9)$.

8 Grothendieck

Having dropped my prejudice against Statistics, it is high time to show my appreciation of Grothendieck.

For example by using the (very existential and highly nonequational) work of Grothendieck [33] on tame coverings of curves, in [7] and [11] I have shown that for any pairwise nonisomorphic nonabelian finite simple groups $D_1, \ldots, D_u$ with $|\text{Aut}D_1| \not\equiv 0(p), \ldots, |\text{Aut}D_u| \not\equiv 0(p)$ the *wreath product* $(D_1 \times \ldots \times D_u)$ Wr Z_p belongs to $\pi_A(L_k)$. For instance we may take $D_1 = A_{m_1}, \ldots, D_u = A_{m_u}$ with $4 < m_1 < m_2 < \ldots < m_u < p$.

Actually, using Grothendieck [33] I first prove an Enlargement Theorem and then from it deduce the above result about wreath products as a group theoretic consequence. The *Enlargement Theorem* asserts that if Θ is any finite group whose order is prime to p, and J is any nonidentity member of $\pi_A(L_k)$, then some enlargement of Θ by J belongs to $\pi_A(L_K)$.

Now enlargement is a generalization of group extensions. Namely, an *enlargement* of a group Θ by a group J is a group G together with an exact sequence $1 \to H \to G \to J \to 1$ and a normal subgroup Δ of $\lambda(H)$, where λ is the given map of H into G, such that $\lambda(H)/\Delta$ is isomorphic to Θ and no nonidentity normal subgroup of G is contained in Δ. Note that here G is an *extension* of H by J. The motivation behind enlargements is the fact that a Galois extension of a Galois extension need not be Galois and if we pass to the relevant least Galois extension then its Galois group is an enlargement of the second Galois group by the first.

Talking of group extensions, as a striking consequence of CT it can be seen that the direct product of two finite nonabelian simple groups is the only extension of one by the other. Here the relevant direct consequence of CT is the *Schreier Conjecture* which says that the outer automorphism group of any finite nonabelian simple group is solvable; see Abhyankar [11] and Gorenstein [32].

As another interesting result, in [12] I proved that $\pi_A(L_k)$ is closed with respect to direct products. It should also be noted that Nori [49] has shown that $\pi_A(L_k)$ contains $SL(n,p^m)$ and some other Lie type simple groups of characteristic p.

Returning to Grothendieckian techniques, Serre [56] proved that if $\pi_A(L_k)$ contains a group H then it contains every quasi p–group which is an extension of H by a solvable group J.

Indeed it appears that the ongoing work of Harbater and Raynaud using Grothendieckian techniques is likely to produce *existential* proofs of the quasi p–group conjecture.

But it seems worthwhile to march on with the equational concrete approach at least because it gives results over the prime field GF(p) and also because we still have no idea what the *complete algebraic fundamental group* $\pi_A^C(L_k)$ looks like where $\pi_A^C(L_k)$ is the Galois group over $k(X)$ of the compositum of all finite Galois extensions of $k(X)$ which are ramified only at infinity and which are contained in a fixed algebraic closure of $k(X)$.

9 Ramanujan

In the equational approach, "modular" things seem destined to play a significant role. For instance the Carlitz–Serre construction of $PSL(2,q)$ coverings and Serre's alternative proof [57] that $\overline{G}_{n,q,s,a} = PSL(2,8)$ for $q = p = 7$ and $n = 9$, are both modular. Similarly my joint paper [16] with Popp and Seiler which uses the Klein and Macbeath curves for writing down $PSL(2,7)$ and $PSL(2,8)$ coverings for small characteristic is also modular in nature.

Inspired by all this, I am undertaking the project of browsing in the 2 volume treatise of Klein and Fricke [43] on Elliptic Modular Functions to prepare myself for understanding Ramanujan himself who may be called the King of Things Modular, where Things = Functions, Equations, Mode of Thought or what have you; see Ramanujan's Collected Papers [50] and Ramanujan Revisited [51].

To explain what are moduli varieties and modular functions in a very naive but friendly manner: The discriminant $b^2 - 4ac$ of a quadratic $aY^2 + bY + c$ is the oldest known invariant. Coming to cubics or quartics $aY^4 + bY^3 + cY^2 + dY + e$ we can, as in books on theory of equations, consider *algebraic invariants*, i.e, polynomial functions of a, b, c, d, e which do not change (much) when we change Y by a fractional linear transformation (see my Invariant Theory Paper [13]), or we may consider *transcendental invariants* and then essentially we get elliptic modular functions. More generally we may consider several (homogeneous) polynomials in several (sets of) variables; when thought of as functions of the variables they give us algebraic varieties or multi–periodic functions or abelian varieties and so on; but as functions of the coefficients we get algebraic invariants or moduli varieties or modular functions. Modular functions and their transforms are related by *modular equations*; think of the expansion of $\sin n\theta$ in terms of $\sin\theta$!

But postponing this to another lecture on another day, let me end with a few equational questions suggested by the experimental data presented in this lecture.

Question (9.1) Which quasi p–groups can be obtained by *coverings of a line by a line*? In other words, which quasi p–groups are the Galois groups of f over $k(X)$ for some monic polynomial f in Y with coefficients in $k[X]$ such that f is linear in X, no valuation of $k(X)/k$ is ramified in the splitting field of f other than the valuations $X = 0$ and $X = \infty$, and the valuation $X = 0$ is at most tamely ramified in the said splitting field? We may even allow the given quasi p–group to equal $p(\text{Gal}(f, k(X))$; note that $\text{Gal}(f, k(X))/p(\text{Gal}(f, k(X))$ is necessarily a cyclic group of order prime to p. In any case this prime to p cyclic quotient as well as the tame branch point at $X = 0$ can be removed by Abhyankar's Lemma. [Hoped for Answer: many quasi p–groups if not all].

Question (9.2) Do *fewnomials* suffice for all simple quasi p–groups? Etymology: binomial, trinomial, ..., fewnomial. In other words, is there a positive integer d (hopefully small) such that every simple quasi p–group can be realized as the Galois group of a polynomial f containing at most d terms in Y (more precisely, at most d monomials in Y) with coefficients in $k[X]$ which gives an unramified covering of the X–axis L_k? Indeed, do fewnomials suffice for most (if not all) quasi p–groups (without requiring them to be simple)? If not, then do *sparnomials* suffice for most (if not all) quasi p–groups? Etymology: sparnomial = sparse polynomial = polynomial with many terms in Y missing; for example a fewnomial in Y plus a polynomial in Y^p.

Question (9.3) Which quasi p–groups can be realized as Galois groups of polynomials in Y whose coefficients are polynomials in X over the *prime field* $\text{GF}(p)$ such that no valuation of $\text{GF}(p)[X]$ is ramified in the relevant splitting field and such that $\text{GF}(p)$ is relatively algebraically closed in that splitting field? Same question where we drop the condition of $\text{GF}(p)$ being relatively algebraically closed but where we replace quasi p–groups by finite groups G for which $G/p(G)$ is cyclic. For instance, given any positive power q' of any prime p' such that the order of $PSL(2, q')$ is divisible by p, we may ask whether there exists an unramified covering of the affine line over $\text{GF}(p)$ whose Galois group is the *semidirect* product of $PSL(2, q')$ with $\text{Aut}(\text{GF}(q'))$, i.e., equivalently, whether there exists a polynomial in Y over $\text{GF}(p)[X]$, with Galois group the said semidirect product, such that no valuation of $\text{GF}(p)[X]$ is ramified in the relevant splitting field (without requiring $\text{GF}(p)$ to be relatively algebraically closed in that splitting field). Note that, in [9], this last question has been answered affirmatively for $p = 7$ and $q' = 8$. Also note that for even q' the said semidirect product is the *projective semilinear group* $P\Gamma L(2,q')$, whereas for odd q' it is an index 2 subgroup of $P\Gamma L(2,q')$; for definitions see [8].

Question (9.4) Do we get fewer Galois groups if we replace branch locus by *discriminant locus*? For instance, can every quasi p–group be realized as the Galois group of a monic polynomial in Y over $k[X]$ whose Y–discriminant is a nonzero element of k? Likewise which members of $\pi_A(L_{k,w})$, where $L_{k,w} = L_k$ minus w points, can be realized as Galois groups of monic polynomials in Y over $k[X]$ whose Y–discriminants have no roots other than the assigned w points. We may ask the same thing also for ground fields of characteristic zero.

Question (9.5) Concerning the bar and tilde equations discussed in (6.1) to (6.8) and (7.1*) to (7.5*) respectively, what further interesting groups do we get out of the *border values* of t?

(9.6) Can we describe the *complete algebraic fundamental group* $\pi_A^C(L_k)$? More generally, for a nonsingular projective curve C_g of genus g minus $w+1$ points, can we describe the complete algebraic fundamental group $\pi_A^C(C_{g,w})$?

(9.7) Descriptively speaking, can the "same" equation give unramified coverings of the affine line for all quasi p–groups in the "same family" of groups? For instance, $Y^{q+1} - XY + 1 = 0$ gives an unramified covering of the affine line, over a field of characteristic p, with Galois group $PSL(2,q)$ for every power q of every prime p. Now thinking of the larger family of groups $PSL(m,q)$, can we find a "single" equation with integer coefficients, "depending" on the parameters m and q, giving an unramified covering of the affine line, over a field of characteristic p, whose Galois group is $PSL(m,q)$ for every integer $m > 1$ and every power q of every prime p? Can we also arrange that the "same" equation gives an unramified covering of the affine line, over every field whose characteristic divides the order of $PSL(m,q)$, whose Galois group is $PSL(m,q)$? Even more, can we arrange that at the same time the Galois group of that equation over $\overline{\mathbf{Q}}(X)$ is $PSL(m,q)$ (but no condition on ramification) where $\overline{\mathbf{Q}}$ is the algebraic closure of $\mathbf{Q}$?

Note (9.8) Two or more of the above questions can be combined in an obvious manner to formulate more questions.

References

[1] S. S. Abhyankar, *On the ramification of algebraic functions,* Amer. Jour. Math. **77** (1955) 572–592.

[2] S. S. Abhyankar, *Local uniformization on algebraic surfaces over ground fields of characteristics* $p \neq 0$, Ann. Math. **63** (1956) 491–526.

[3] S. S. Abhyankar, *Coverings of algebraic curves,* Amer. Jour. Math. **79** (1957) 825–856.

[4] S. S. Abhyankar, *Tame coverings and fundamental groups of algebraic varieties, Parts I to VI,* Amer. Jour. Math. **81, 82** (1959-60).

[5] S. S. Abhyankar, *Historical ramblings in algebraic geometry and related algebra,* Amer. Math. Monthly **83** (1976) 409–448.

[6] S. S. Abhyankar, *Algebraic Geometry for Scientists and Engineers,* Mathematical Surveys and Monographs **35**, Amer. Math. Soc., Providence, 1990.

[7] S. S. Abhyankar, *Group enlargements,* C. R. Acad. Sci. Paris **312** (1991) 763–768.

[8] S. S. Abhyankar, *Galois theory on the line in nonzero characteristic,* Bull. Amer. Math. Soc. **27** (1992) 68–133.

[9] S. S. Abhyankar, *Square–root parametrization of plane curves,* Algebraic Geometry and Its Applications, Springer-Verlag (1994) 19–84.

[10] S. S. Abhyankar, *Alternating group coverings of the affine line in characteristic greater than two,* Math. Annalen **296** (1993) 63–68

[11] S. S. Abhyankar, *Wreath products and enlargements of groups,* Discrete Math. **120** (1993) 1–12.

[12] S. S. Abhyankar, *Linear disjointness of polynomials,* Proc. Amer. Math. Soc. **116** (1992) 7–12.

[13] S. S. Abhyankar, *Invariant theory and enumerative combinatorics of Young tableaux,* Geometric Invariance in Computer Vision, Edited by J. L. Mundy and A. Zisserman, MIT Press (1992) 45–76.

[14] S. S. Abhyankar, J. Ou and A. Sathaye, *Alternating group coverings of the affine line in characteristic two,* Discrete Mathematics, (To Appear).

[15] S. S. Abhyankar, H. Popp and W. K. Seiler, *Mathieu group coverings of the affine line,* Duke Math. Jour. **68** (1992) 301–311.

[16] S. S. Abhyankar, H. Popp and W. K. Seiler, *Construction techniques for Galois coverings of the affine line,* Proceedings of the Indian Academy of Sciences **103** (1993) 103–126.

[17] S. S. Abhyankar and I. Yie, *Small degree coverings of the affine line in characteristic two,* Discrete Mathematics (To Appear).

[18] C. Arf, *Untersuchungen über quadratische Formen in Körpern der Characteristik 2 (Teil I.),* Crelle Journal **183** (1941) 148–167.

[19] E. Berlekamp, *Algebraic Coding Theory* McGraw-Hill, 1968.

[20] E. Berlekamp, *An analog to the discriminant over fields of characteristic 2,* Journal of Algebra **38** (1976) 315–317.

[21] R. C. Bose, *On the construction of balanced incomplete block designs,* Ann. Eugenics **9** (1939) 353–399.

[22] P. J. Cameron, *Finite permutation groups and finite simple groups,* Bull. Lond. Math. Soc. **13** (1981) 1–22.

[23] P. J. Cameron. *Permutation groups,* Handbook of Combinatorics, Elsevier, Forthcoming.

[24] P. J. Cameron and J. Cannon, *Fast recognition of doubly transitive groups,* Journal of Symbolic Computations **12** (1991) 459–474.

[25] L. Carlitz, *Resolvents of certain linear groups in a finite field,* Canadian Jour. Math. **8** (1956) 568–579.

[26] C. Chevalley, *Theory of Lie Groups,* Princeton University Press, 1946.

[27] G. Chrystal, *Textbook of Algebra,* Parts I and II, Edinburgh, 1886.

[28] R. Dedekind and H. Weber, *Theorie der Algebraischen Funktionen einer Veränderlichen,* Crelle Journal **92** (1882) 181–290.

[29] P. Dembowski, *Finite Geometries,* Springer-Verlag, 1968.

[30] W. Feit, *On a class of doubly transitive permutation groups,* Illinois Jour. Math. **4** (1960) 170–186.

[31] A. R. Forsyth, *Theory of Functions of a Complex Variable,* Cambridge University Press, 1918.

[32] D. Gorenstein, *Classifying the finite simple groups,* Bull. Amer. Math. Soc. **14** (1986) 1–98.

[33] A. Grothendieck, *Revêtements Étales et Groupe Fondamental (SGA 1),* Lecture Notes in Mathematics **224** Springer-Verlag, 1971.

[34] D. G. Higman, *Finite permutation groups of rank 3,* Math. Z. **86** 145–156.

[35] D. Hilbert, *Über das Dirichletsche Prinzip,* Mathematische Annalen **59** (1904) 161–186.

[36] B. Huppert and N. Blackburn, *Finite Groups I, II, III,* Springer-Verlag, New York, 1982.

[37] N. Jacobson, *Lectures in Abstract Algebra, Vol III,* Van Nostrand, Princeton, 1964.

[38] N. Jacobson, *Basic Algebra, Vol I,* W. H. Freeman and Co. San Francisco, 1974.

[39] C. Jordan, *Sur la limite de transitivité des groupes non alternés,* Bull. Soc. Math. France **1** (1973) 40–71.

[40] H. W. E. Jung, *Darstellung der Funktionen eines algebraischen Körpers zweier unabhängigen Veränderlichen in der Umgebung einer Stelle,* Crelle Journal **133** (1908) 289–314.

[41] W. M. Kantor, *Homogeneous designs and geometric lattices,* Journal of Combinatorial Theory, Series A **38** (1985) 66–74.

[42] W. M. Kantor and R. L. Liebler, *The rank 3 permutation representations of the finite classical groups,* Trans. Amer. Math. Soc. **271** (1982) 1–71.

[43] F. Klein and R. Fricke, *Vorlesungen über die Theorie der Elliptischen Modulfunctionen,* 2 volumes, Teubner, Leipzig, 1890.

[44] M. W. Liebeck, *The affine permutation groups of rank three,* Proc. Lond. Math. Soc. **54** (1987) 477–516.

[45] B. Marggraff, *Über primitive Gruppen mit transitiven Untergruppen geringeren Grades,* Dissertation, Giessen, 1892.

[46] E. Mathieu, *Mémoire sur l'étude des fonctions de plusieurs quantités, sur la manière de les former, et sur les substitutions qui les laissent invariables,* J. Math Pures Appl. **18** (1861) 241–323.

[47] I. Newton, *Geometria Analytica* 1680.

[48] M. Noether, *Über einen Satz aus der Theorie der algebraischen Funktionen*, Math. Annalen **6** (1873) 351–359.

[49] M. V. Nori, *Unramified coverings of the affine line in positive characteristic,* Algebraic Geometry and Its Applications, Springer-Verlag (1994) 209–212.

[50] S. Ramanujan, *Collected Papers,* Cambridge University Press, 1927.

[51] S. Ramanujan, *Ramanujan Revisited,* Proceedings of the Centenary Conference at University of Illinois, Edited by G. E. Andrews, R. A. Askey, B. C. Berndt, K. G. Ramanathan and R. A. Rankin, Academic Press, 1987.

[52] B. Riemann, *Grundlagen für eine allgemeine Theorie der Functionen einer veränderlichen complexen Grösse,* Inauguraldissertation, Göttingen (1851). Collected Works, Dover (1953) 3–43.

[53] B. Riemann, *Theorie der Abelschen Funktionen*, Jour. für die reine und ang. Mathematik **64** (1865) 115–155.

[54] H. Seifert and W. Threlfall, *Lehrbuch der Topologie,* Teubner, 1934.

[55] J–P. Serre, *Géométrie algébrique et géométrie analytique,* Ann. Inst. Fourier **6** (1956) 1–42.

[56] J–P. Serre, *Construction de revêtments étales de la droite affine en charactéristic p,* C. R. Acad. Sci. Paris **311** (1990) 341–346.

[57] J–P. Serre, *A letter as an appendix to the square–root parametrization paper of Abhyankar,* Algebraic Geometry and Its Applications, Springer-Verlag (1994) 85–88.

[58] S. S. Shrikhande, *The impossibility of certain symmetrical balanced incomplete block designs,* Ann. Math. Statist. **21** (1950) 106–111.

[59] M. Suzuki. *On a class of doubly transitive groups,* Ann. Math. **75** (1962) 104–145.

[60] H. Wielandt. *Finite Permutation Groups,* Academic Press, New York, 1964.

[61] E. Witt. *Theorie der quadratischen Formen in beliebigen Körpern,* Crelle Journal **176** (1937) 31–41.

[62] H. Zassenhaus. *Kennzeichung endlicher linearen Gruppen als Permutationgruppen,* Abh. Math. Sem. Univ. Hamburg (**11**) (1936) 17–744.

Mathematics Department
Purdue University,
West Lafayette, IN 47907
U.S.A.

Impact of geometry of the boundary on the positive solutions of a semilinear Neumann problem with Critical nonlinearity

Adimurthi

Dedicated to M.S. Narasimhan and C.S. Seshadri on their 60th Birthdays

Let $n \geq 3$ and $\Omega \subset \mathbb{R}^n$ be a bounded domain with smooth boundary. We are concerned with the problem of existence of a function u satisfying the nonlinear equation

$$\begin{aligned} -\Delta u &= u^p - \lambda u \quad \text{in } \Omega \\ u &> 0 \\ \frac{\partial u}{\partial \nu} &= 0 \quad \text{on } \partial\Omega \end{aligned} \tag{1}$$

where $p = \frac{n+2}{n-2}, \lambda > 0$. Clearly $u = \lambda^{1/(p-1)}$ is a solution of (1) and we call it a trivial solution. The exponent $p = \frac{n+2}{n-2}$ is critical from the view point of Sobolev imbedding. Indeed the solution of (1) corresponds to critical points of the functional

$$Q_\lambda(u) = \frac{\int_\Omega |\nabla u|^2 dx + \lambda \int_\Omega u^2 dx}{\left(\int_\Omega |u|^{p+1} dx\right)^{2/p+1}} \tag{2}$$

on the manifold

$$M = \left\{ u \in H^1(\Omega) \; ; \; \int_\Omega |u|^{p+1} dx = 1 \right\}. \tag{3}$$

In fact, if $v \geq 0$ is a critical point of (2) on M, then $u = Q(v)^{1/(p-1)} v$ satisfies (1). Note that $p+1 = \frac{2n}{n-2}$ is the limiting exponent for the imbedding $H^1(\Omega) \mapsto L^{2n/(n-2)}(\Omega)$. Since this imbedding is not compact, the manifold M is not weakly closed and hence Q_λ need not satisfy the Palais Smale condition at all levels. Therefore there are serious difficulties when trying to find critical points by the standard variational methods. In fact there is a sharp contrast between the sub critical case $p < \frac{n+2}{n-2}$ and the critical case $p = \frac{n+2}{n-2}$.

Our motivation for investigation comes from a question of Brezis [11]. If we replace the Neumann condition by Dirichlet condition $u = 0$ on $\partial\Omega$ in (1), then the existence and non existence of solutions depends on topology and geometry of the domain (see Brezis-Nirenberg [14], Bhari-Coron [9], Brezis [10]). In view of this, Brezis raised the following problem

"Under what conditions on λ and Ω, (1) admits a solution ?"

The interest in this problem not only comes from a purely mathematical question, but it has applications in mathematical biology, population dynamics (see [16]) and geometry.

In order to answer the above question let us first look at the subcritical case where the compactness is assured.

Subcritical case $1 < p < \frac{n+2}{n-2}$.

This had been studied extensively in the recent past by Ni [18], and Lin-Ni-Takagi [16]. In [16], Lin-Ni-Takagi have proved the following

Theorem 1. *There exist two positive constants $\lambda_\star$ and $\lambda^\star$ such*

a) If $\lambda < \lambda_\star$, then (1) admits only trivial solutions.
a) If $\lambda > \lambda_\star$, then (1) admits non constant solutions.

Further Ni [18] and Lin-Ni [15] studied the radial case for all $1 < p < \infty$ and proved the following

Theorem 2. *Let $\Omega = \{x; |x| < 1\}$ is a ball. Then there exists two positive constants $\lambda_\star$ and $\lambda^\star$ such that $\lambda_\star \leq \lambda^\star$ and*

a) For $1 < p < \infty, \lambda > \lambda^\star$, (1) admits a radially increasing solution
b) if $p \neq \frac{n+2}{n-2}$, then for $0 < \lambda < \lambda_\star$, (1) does not admit a non constant radial solution.
c) Let $\Omega = \{x; 0 < \alpha < |x| < \beta\}$ be an annuluar domain and $1 < p < \infty$. Then there exist two positive constants $\lambda_\star \leq \lambda^\star$ such that for $\lambda \geq \lambda^\star$, (1) admits a non constant radial solution and if $\lambda \leq \lambda^\star$, then (1) does not admit a non constant radial solution.

In view of these results Lin and Ni [15] made the following

Conjecture : Let $p \geq \frac{n+2}{n-2}$, then there exist two positive constants $\lambda_\star \leq \lambda^\star$ such that

(A) For $0 < \lambda < \lambda_\star$, (1) does not admit non constant solutions
(B) For $\lambda > \lambda^\star$, (1) admits a non constant solution.

In this article we analyze this conjecture in the **critical case** $p = \frac{n+2}{n-2}$. Surprisingly enough, the critical case is totally different from the subcritical. In fact the part (A) of the conjecture in general is false. The following results of Adimurthi and Yadava [4] and Budd, Knaap and Peletier [12] gives a counter example to the Part (A) of the conjecture.

Theorem 3. *Let $n = 4, 5, 6$ and $\Omega = \{x : |x| < 1\}$. Then there exist a $\lambda_\star > 0$ such that for $0 < \lambda < \lambda_\star$, (1) admits a radially decreasing solution.*

Let us now turn our attention to Part (B) of the conjecture. Let S denote the best Sobolev constant for the imbedding $H^1(\mathbb{R}^n) \mapsto L^{2n/(n-2)}(\mathbb{R}^n)$ given by

$$S = \inf\left\{\int_{\mathbb{R}^n} |\nabla u|^2 dx : \int_{\mathbb{R}^n} |u|^{2n/n-2} dx = 1\right\} \tag{4}$$

Then S is achieved and any minimizer is given by U_{ε,x_0} for some $\varepsilon > 0$, $x_0 \in \mathbb{R}^n$ where

$$U(x) = \left[\frac{n(n-2)}{n(n-2) + |x|^2}\right]^{\frac{n-2}{2}} \tag{5}$$

$$U_{\varepsilon,x_0}(x) = \frac{1}{\varepsilon^{\frac{n-2}{2}}} U\left(\frac{x - x_0}{\varepsilon}\right) \tag{6}$$

In order to answer part (B) of the conjecture, geometry of the boundary play an important role. To see this, we look at a more general problem than (1), the mixed problem.

Let $\partial\Omega = \Gamma_0 \cup \Gamma_1, \Gamma_0 \cap \Gamma_1 = \phi, \Gamma_i$ are submanifolds of dimension $(n-1)$. The problem is to find function u satisfying

$$\begin{aligned} -\Delta u + \lambda u &= u^{\frac{n+2}{n-2}} \text{ in } \Omega \\ u &> 0 \\ u &= 0 \quad \text{on } \Gamma_0 \\ \frac{\partial u}{\partial \nu} &= 0 \quad \text{on } \Gamma_1 \end{aligned} \tag{7}$$

Let

$$H^1(\Gamma_0) = \{u \in H^1(\Omega) \; : \; u = 0 \quad \text{on } \Gamma_0\} \tag{8}$$

$$S(\lambda, \Gamma_0) = \inf \{Q_\lambda(u) \; ; \; u \in H^1(\Gamma_0) \cap M\} \tag{9}$$

Clearly, if $S(\lambda, \Gamma_0)$ is achieved by some v, then we can take $v \geq 0$ and $u = S(\lambda, \Gamma_0)^{\frac{n-2}{4}} v$ satisfies (7). u is called a *minimal energy* solution. Existence of a minimal energy solution is proved in Adimurthi and Mancini [1] (See also X.J. Wang [22]) and have the following

Theorem 4. *Assume that there exist an x_0 belonging to the interior of Γ_1 such that the mean curvature $H(x_0)$ at x_0 with respect to unit outward normal is positive. Then $S(\lambda, \Gamma_0)$ is achieved.*

Sketch of the Proof. The proof consists of two steps.

Step 1. Suppose $S(\lambda, \Gamma_0) < S/2^{2/n}$, then $S(\lambda, \Gamma_0)$ is achieved.

Let $v_k \in H^1(\Gamma_0) \cap M$ be a minimizing sequence. Clearly $\{v_k\}$ is bounded in $H^1(\Omega)$. Let for subsequence of $\{v_k\}$ still denoted by $\{v_k\}$, converges weakly to v_0 and almost everywhere in Ω. We first claim that $v_0 \not\equiv 0$. Suppose $v_0 \equiv 0$, then by Cherrier imbedding (See [8]) for every $\varepsilon > 0$, there exists $C(\varepsilon) > 0$ such that

$$1 = \left(\int_\Omega |v_k|^{p+1} dx \right)^{2/p+1} \leq \frac{2^{2/n}}{S}(1+\varepsilon) \int_\Omega |\nabla v_k|^2 dx + C(\varepsilon) \int v_k^2 dx \, .$$

By Rellich's compactness, $v_k \to 0$ in $L^2(\Omega)$ and hence in the above inequality letting $k \to \infty$ and $\varepsilon \to 0$ we obtain

$$\begin{aligned} 1 &\leq \lim_{\varepsilon \to 0} \frac{2^{2/n}}{S}(1+\varepsilon) \lim_{k\to\infty} Q_\lambda(v_k) \\ &= \frac{2^{2/n}}{S} S(\lambda, \Gamma_0) \\ &< 1 \end{aligned}$$

which is a contradiction. Hence $v_0 \not\equiv 0$. Let $h_k = v_k - v_0$, then $h_k \to 0$ weakly in $H^1(\Omega)$ and strongly in $L^2(\Omega)$. Hence

$$\begin{aligned} S(\lambda, \Gamma_0) &= Q_\lambda(v_k) + 0(1) \\ &= Q_\lambda(v_0) \left(\int_\Omega |v_0|^{p+1} dx \right)^{2/p+1} + \int_\Omega |\nabla h_k|^2 dx + 0(1) \end{aligned}$$

Now by Brezis-Lieb Lemma, Cherrier imbedding, from the above inequality, and by the hypothesis, we have for sufficiently small $\varepsilon > 0$,

$$S(\lambda, \Gamma_0) = S(\lambda, \Gamma_0) \left(\int_\Omega |v_k|^{p+1} dx \right)^{2/p+1}$$

$$
\begin{aligned}
&\le S(\lambda,\Gamma_0)\left\{\left(\int_\Omega |v_0|^{p+1}dx\right)^{2/p+1} + \left(\int_\Omega |h_k|^{p+1}dx\right)^{2/p+1}\right\} + 0(1)\\
&= S(\lambda,\Gamma_0)\left\{\left(\int_\Omega |v_0|^{p+1}dx\right)^{2/p+1} + \frac{2^{2/n}}{S}(1+\varepsilon)\int_\Omega |\nabla h_k|^2dx\right\} + 0(1)\\
&= S(\lambda,\Gamma_0)\left(\int_\Omega |v_0|^{p+1}dx\right)^{2/p+1} + \int_\Omega |\nabla h_k|^2dx + 0(1)\\
&= S(\lambda,\Gamma_0)\left(\int_\Omega |v_0|^{p+1}dx\right)^{2/p+1} + S(\lambda,\Gamma_0) - Q_\lambda(v_0)\left(\int_\Omega |v_0|^{p+1}dx\right)^{2/p+1}
\end{aligned}
$$

this implies that $Q_\lambda(v_0) \le S(\lambda,\Gamma_0)$. Hence v_0 is a minimizer.

Step 2. $S(\lambda,\Gamma_0) < S/2^{2/n}$

Let x_0 belong to the interior of Γ_1 at which $H(x_0) > 0$ and $r > 0$ such that $B(x_0,r)\cap\Gamma_0 = \phi$. Let $\varphi \in C_0^\infty(B(x_0,r))$ such that $\varphi = 1$ for $|x-x_0| < r/2$. Let $\varepsilon > 0$ and $v_\varepsilon = \varphi U_{\varepsilon,x_0}$. Then $v_\varepsilon \in H^1(\Gamma_0)$ and we can find positive constants A_n and a_n depending only on n such that

$$
Q_\lambda(v_\varepsilon) = \frac{S}{2^{2/n}} - A_nH(x_0)\beta_1(\varepsilon) + a_n\lambda\beta_2(\varepsilon) + 0(\beta_1(\varepsilon)+\beta_2(\varepsilon)) \tag{10}
$$

where

$$
\beta_1(\varepsilon) = \begin{cases} \varepsilon\log^{1/\varepsilon} & \text{if } n=3\\ \varepsilon & \text{if } n\ge 4\end{cases}
$$

$$
\beta_2(\varepsilon) = \begin{cases} \varepsilon & \text{if } n=3\\ \varepsilon^2\log 1/\varepsilon & \text{if } n=4\\ \varepsilon^2 & \text{if } n\ge 4\end{cases}
$$

Hence for ε small and since $H(x_0) > 0$ we obtain $Q_\lambda(v_\varepsilon) < S/2^{2/n}$ and this proves step (2) and hence the theorem.

Now it is to be noted that the curvature condition on Γ_1 is very essential. If the curvature conditions fails, then in general (7) may not admit any solution.

Example. Let $B = \{x : |x| < 1\}$ and
$\Omega = \{x\in B : x_n > 0\}$
$\Gamma_1 = \{x\in\partial\Omega : x_n = 0\}$
$\Gamma_0 = \{x\in\partial\Omega : x_n > 0\}$

Let $u\in H^1(\Gamma_0)$ be a solution of (7). Define w on B by

$$
w(x',x_n) = \begin{cases} w(x',x_n) & \text{if } x_n > 0\\ w(x',x_n) & \text{if } -x_n < 0\end{cases}
$$

Since $\frac{\partial u}{\partial \nu} = 0$ on Γ_1, w satisfies

$$\begin{aligned} -\Delta w + \lambda w &= w^{\frac{n+2}{n-2}} \text{ in } B \\ w &> 0 \\ w &= 0 \quad \text{on } \partial B\,. \end{aligned}$$

Hence by Pohozaev's identity we obtain

$$-\lambda \int_B w^2 dx = \int_{\partial B} |\nabla w|^2 \langle x.\nu \rangle d\xi$$

which is a contradiction. Notice that the mean curvature is zero on Γ_1.

Proof of Part (B) of the Conjecture

Let $\Gamma_1 = \partial\Omega$. Since $\partial\Omega$ is smooth, we can find an $x_0 \in \partial\Omega$ such that $H(x_0) > 0$. Hence from theorem (4), (1) admits a minimal energy solution u_λ. Let $u_0 = \lambda^{1/p-1}$ and $\lambda^\star = \frac{S}{(2|\Omega|)^{2/n}}$. Then for $\lambda > \lambda^\star$

$$Q_\lambda(u_\lambda) < \frac{S}{2^{2/n}} < \lambda|\Omega|^{2/n} = Q_\lambda(u_0)$$

Hence u_λ is a non constant solution of (1) and this proves part (b) of the conjecture.

Properties of the minimal energy solutions

1. By Theorem 3, Part (A) of the conjecture in general is false. Now we can ask whether this is true among minimal energy solutions ? In fact it is true. The following is proved in Adimurthi-Yadava [6].

Theorem 5. *There exist a* $\lambda_\star > 0$ *such that for all* $0 < \lambda < \lambda_\star$, *the minimal energy solutions are constant.*

Proof. By using the blow up technique [13], we can prove that for every $\varepsilon > 0$, there exist a $\lambda(\varepsilon) > 0$ such that for $0 < \lambda < \lambda(\varepsilon)$, if u_λ is a minimal energy solution, then

$$|u_\lambda|_\infty \le \varepsilon \tag{11}$$

where $|\cdot|_\infty$ denotes the L^∞ norm. Let μ_1 be the first non zero eigenvalue of

$$\begin{aligned} -\Delta\psi &= \mu\psi \quad \text{in } \Omega \\ \frac{\partial\psi}{\partial\nu} &= 0 \quad \text{on } \partial\Omega. \end{aligned}$$

Let $\bar{u}_\lambda = \frac{1}{|\Omega|}\int_\Omega u_\lambda dx$ and $\varphi_\lambda = u_\lambda - \bar{u}_\lambda$. Then φ_λ satisfies

$$-\Delta\varphi_\lambda + \lambda\varphi_\lambda = \bar{u}_\lambda^p + p\int_0^1 (\bar{u}_\lambda + t\varphi_\lambda)^{p-1}\varphi_\lambda^2 dt dx$$

From (11) we have $0 \le \bar{u}_\lambda + t\varphi_\lambda \le u_\lambda \le \varepsilon$ and $\int_\Omega \varphi_\lambda dx = 0$. Therefore we obtain

$$(\mu_1 + \lambda)\int_\Omega \varphi_\lambda^2 dx \le \int_\Omega (|\nabla\varphi_\lambda|^2 + \lambda\varphi_\lambda^2)dx \le p\varepsilon^{p-1}\int_\Omega \varphi_\lambda^2 dx$$

Now choose $\varepsilon^{p-1} = \frac{\mu_1}{2p}$ and $\lambda_* = \lambda(\varepsilon)$, then the above inequality implies that $\varphi_\lambda \equiv 0$ and hence u_λ is a constant. This proves the theorem.

2. Concentration and multiplicity results. From the concentration compactness results of P.L. Lions [17] if u_λ is a minimal energy solution of (1), then for any sequence $\lambda_k \to \infty$ with $|\nabla u_{\lambda_k}|^2 dx \to d\mu$, there exist a $x_0 \in \partial\Omega$ such that $d\mu = \frac{S^{n/2}}{2}\delta_{x_0}$. Now the natural question is "is it possible to characterize the concentration points x_0" ? One expects from the asymptotic formula (10) that x_0 must be a point of maximum mean curvature. This has been proved in Adimurthi, Pacella and Yadava [7] and we have the following

Theorem 6. *Let u_λ be a minimal energy solution of (1) and $P_\lambda \in \bar{\Omega}$ be such that*

$$u_\lambda(P_\lambda) = \max\{u_\lambda(x); x \in \bar{\Omega}\}$$

then there exist a $\lambda_0 > 0$ such that for all $\lambda > \lambda_0$

a) $P_\lambda \in \partial\Omega$ and is unique,
b) Let $n \ge 7$. The limit points of $\{P_\lambda\}$ are contained in the points of maximum mean curvature.

Part (a) of this Theorem is also proved in [19].

In view of the concentration at the boundary, it follows that the minimal energy solutions are not radial for λ sufficiently large and Ω being the ball. Hence in a ball, for large λ, we obtain at least two solutions one radial and the other non radial (see [5]). If Ω is not a ball then in Adimurthi and Mancini [2], they obtained that $\mathrm{Cat}_{\partial\Omega}(\partial\Omega^+)$ number of solutions for (1) where $\partial\Omega^+$ is the set of points in $\partial\Omega$ where the mean curvature is positive (here for $X \subset Y, Y$ a topological space, then $\mathrm{Cat}_Y(X)$ is the category of X in Y). Further if $\partial\Omega$ has rich geometry in the sense described below, then Adimurthi-Pacella and Yadava [7] have obtained more solutions of (1). They have proved the following

Theorem 7. *Let $n \geq 7$. Assume that $\partial\Omega$ has k-peaks, that is there exist k-points $\{x_1, \ldots x_k\} \subset \partial\Omega$ at which $H(x_i)$ is strictly local maxima. Then there exists a $\lambda_0 > 0$ such that for $\lambda > \lambda_0$, there are k distinct solutions $\{u_{i_\lambda}\}_{i=1}^k$ of (1) such that u_{i_λ} concentrates at x_i as $\lambda \to \infty$.*

Theorems 6 and 7 has been extended for the mixed boundary value problems.

Theorem 7 is not applicable in the case when Ω is a ball. On the otherhand, given a positive integer k, there exist a $\lambda(k)$ such that for $\lambda > \lambda(k)$, (1) admits at least k number of radial solutions (see [18]). Part (B) of the conjecture gives infinitely many rotationally equivalent solutions of minimal energy. In view of this it is not clear how to obtain more non radial solutions which are not rotationally equivalent.

References

[1] Adimurthi and G. Mancini, *The Neumann problem for elliptic equations with critical non-linearity*, A tribute in honour of G. Prodi, Ed by Ambrosetti and Marino, Scuola Norm. Sup. Pisa (1991) 9–25.

[2] Adimurthi and G. Mancini, *Effect of geometry and topology of the boundary in the critical Neumann problem*, J. Reine Angew. Math, to appear.

[3] Adimurthi and S.L. Yadava, *Critical Sobolev exponent problem in* $\mathbb{R}^n (n \geq 4)$ *with Neumann boundary condition*, Proc. Ind. Acad. Sci. **100** (1990) 275–284.

[4] Adimurthi and S.L. Yadava, *Existence and nonexistence of positive radial solutions of Neumann problems with Critical Sobolev exponents*, Arch. Rat. Mech. Anal. **115** (1991) 275–296.

[5] Adimurthi and S.L. Yadava, *Existence of nonradial positive solution for critical exponent problem with Neumann boundary condition*, J. Diff. Equations, **104** (1993) 298–306.

[6] Adimurthi and S.L. Yadava, *On a conjecture of Lin-Ni for semilinear Neumann problem*, Trans. Amer. Math. Soc. **336** (1993) 631–637.

[7] Adimurthi, F. Pacella and S.L. Yadava, *Interaction between the geometry of the boundary and positive solutions of a semilinear Neumann problem with Critical non linearity*, J. Funct. Anal. **113** (1993) 318–350.

[8] T. Aubin, *Nonlinear analysis of manifold : Monge-Ampere equations*, New York, Springer-Verlag (1982).

[9] A. Bahri and J.M. Coron, *On a nonlinear elliptic equation involving the critical Sobolev exponent : The effect of the topology of the domain*, Comm. Pure Appl. Math. **41** (1988) 253 –290.

[10] H. Brezis, *Elliptic equations with limiting Sobolev exponent. The impact of topology*, Comm. Pure Appl. Math. **39** (1986) S17–S39.

[11] H. Brezis, *Non linear elliptic equations involving the Critical Sobolev exponent in Survey and perspectives*, Direction in Partial differential equations, Ed by Crandall etc. (1987), 17–36.

[12] C. Budd, M.C. Knapp and L.A. Peletier, *Asymptotic behaviour of solution of elliptic equations with critical exponent and Neumann boundary conditions*, Proc. Roy. Soc. Edinburg **117A** (1991) 225–250.

[13] B. Gidas and J. Spruck, *A priori bounds for positive solutions of nonlinear elliptic equations*, Comm. Part. Diff. Equations **6** (1981) 883–901.

[14] H. Brezis and L. Nirenberg, *Positive solutions of nonlinear elliptic equations involving Critical exponents*, Comm. Pure Appl. Maths. Vol. **36** (1983) 437–477.

[15] C.S. Lin and W.M. Ni, *On the diffusion coefficient of a semilinear Neumann problem*, Springer Lecture Notes **1340** (1986).

[16] C.S. Lin, W.M. Ni and I. Takagi, *Large amplitude stationary solutions to a chemotaxis system*, J. Diff. Equations **72** (1988) 1–27.

[17] P.L. Lions, *The concentration compactness principle in the calcular of variations: The limit case* (Part 1 and part 2), Riv. Mat. Iberoamericana **1** (1985) 145–201; 45–121.

[18] W.M. Ni, *On the positive radial solutions of some semilinear elliptic equations on* $\mathbb{R}^n$, Appl. Math. Optim. **9** (1983) 373–380.

[19] W.M. Ni, X.B. Pan and I. Takagi, *Singular behaviour of least energy solutions of a semilinear Neumann problem involving critical Sobolev exponents*, Duke. Math. J. **67** (1992) 1–20.

[20] W.M. Ni and I. Takagi, *On the existence and the shape of solutions to a semilinear Neumann problem*, to appear in Proceedings of the conference on Nonlinear diffusion equations and their equilibrium states ; held at Gregynog, Wales, August 1989.

[21] W.M. Ni and I. Takagi, *On the shape of least-energy solutions to a semilinear Neumann problem*, Comm. Pure Appl. Math. **45** (1991) 819–851.

[22] X.J. Wang, *Neumann problem of semilinear elliptic equations involving critical Sobolev exponents*, J. Diff. Equations, **93** (1991) 283–310.

T.I.F.R. Centre
P.O. Box No. 1234
Bangalore 560 012.

Sur la cohomologie de certains espaces de modules de fibrés vectoriels

Arnaud Beauville *

Dédié à M.S. Narasimhan et C.S. Seshadri pour leur 60ème anniversaire

Soit X une surface de Riemann compacte. Fixons des entiers r et d *premiers entre eux*, avec $r \geq 1$, et notons $\mathcal{M}$ l'espace des modules $\mathcal{U}_X(r,d)$ des fibrés vectoriels stables sur X, de rang r et de degré d. C'est une variété projective et lisse, et il existe un *fibré de Poincaré* $\mathcal{E}$ sur $X \times \mathcal{M}$; cela signifie que pour tout point e de $\mathcal{M}$, correspondant à un fibré E sur X, la restriction de $\mathcal{E}$ à $X \times \{e\}$ est isomorphe à E .

Notons p, q les projections de $X \times \mathcal{M}$ sur X et $\mathcal{M}$ respectivement. Soit m un entier $\leq r$; la classe de Chern $c_m(\mathcal{E})$ admet une décomposition de Künneth

$$c_m(\mathcal{E}) = \sum_i p^*\xi_i \cdot q^*\mu_i \ ,$$

avec $\xi_i \in H^*(X, \mathbf{Z})$, $\mu_i \in H^*(\mathcal{M}, \mathbf{Z})$, $\deg(\xi_i) + \deg(\mu_i) = 2m$.

Nous dirons que les classes μ_i sont les *composantes de Künneth* de $c_m(\mathcal{E})$. Un des résultats essentiels de [A–B] est la détermination d'un ensemble de générateurs de l'algèbre de cohomologie $H^*(\mathcal{M}, \mathbf{Z})$; il a la conséquence suivante :

Théorème. *L'algèbre de cohomologie $H^*(\mathcal{M}, \mathbf{Q})$ est engendrée par les composantes de Künneth des classes de Chern de $\mathcal{E}$.*

Le but de cette note est de montrer comment la méthode de la diagonale utilisée dans [E–S] fournit une démonstration très simple de ce théorème. Celui-ci résulte de l'énoncé un peu plus général que voici :

*Avec le support partiel du projet européen Science "Geometry of Algebraic Varieties", Contrat n^o SCI-0398-C(A).

Proposition. *Soient X une variété complexe projective et lisse, et $\mathcal{M}$ un espace de modules de faisceaux stables sur X (par rapport à une polarisation fixée, cf.* [M]). *On fait les hypothèses suivantes :*

(i) La variété $\mathcal{M}$ est projective et lisse.

(ii) Il existe un faisceau de Poincaré $\mathcal{E}$ sur $\mathcal{M}$.

(iii) Pour E , F dans $\mathcal{M}$, on a $\mathrm{Ext}^i(E,F)=0$ pour $i\geq 2$.

Alors l'algèbre de cohomologie $H^*(\mathcal{M},\mathbf{Q})$ est engendrée par les composantes de Künneth des classes de Chern de $\mathcal{E}$.

La démonstration suit de près celle du th. 1 de [E–S]. Rappelons-en l'idée fondamentale ; soit δ la classe de cohomologie de la diagonale dans $H^*(\mathcal{M}\times\mathcal{M},\mathbf{Q})$; notons p et q les deux projections de $\mathcal{M}\times\mathcal{M}$ sur $\mathcal{M}$. Soit $\delta=\sum\limits_i p^*\mu_i\cdot q^*\nu_i$ la décomposition de Künneth de δ ; alors *l'espace $H^*(\mathcal{M},\mathbf{Q})$ est engendré par les ν_i* . En effet, pour λ dans $H^*(\mathcal{M},\mathbf{Q})$, on a

$$\lambda=q_*(\delta\cdot p^*\lambda)=\sum\deg(\lambda\cdot\mu_i)\,\nu_i$$

d'où notre assertion. Il s'agit donc d'exprimer la classe δ en fonction des classes de Chern du fibré universel.

Notons p_1,p_2 les deux projections de $C\times\mathcal{M}\times\mathcal{M}$ sur $C\times\mathcal{M}$, et π la projection sur $\mathcal{M}\times\mathcal{M}$; désignons par $\mathcal{H}$ le faisceau $\mathrm{Hom}(p_1^*\mathcal{E},p_2^*\mathcal{E})$. Vu l'hypothèse (iii), l'hypercohomologie $R\pi_*\mathcal{H}$ est représentée dans la catégorie dérivée par un complexe de fibrés $K^\bullet$, nul en degré différent de 0 et 1. Autrement dit, il existe un morphisme de fibrés $u:K^0\longrightarrow K^1$ tel qu'on ait, pour tout point $x=(E,F)$ de $\mathcal{M}$, une suite exacte

$$0\rightarrow\mathrm{Hom}(E,F)\longrightarrow K^0(x)\xrightarrow{u(x)}K^1(x)\longrightarrow\mathrm{Ext}^1(E,F)\longrightarrow 0\ .$$

Comme l'espace $\mathrm{Hom}(E,F)$ est non nul si et seulement si E et F sont isomorphes, on voit que la diagonale Δ de $\mathcal{M}\times\mathcal{M}$ coïncide ensemblistement avec le lieu de dégénérescence D de u (défini par l'annulation des mineurs de rang maximal de u). On peut prouver comme dans [E–S] l'égalité schématique, mais cela n'est pas nécessaire pour démontrer la proposition.

Soit E un élément de $\mathcal{M}$. On a

$$\mathrm{rg}(K^0)-\mathrm{rg}(K^1)=\dim\mathrm{Hom}(E,F)\ -\ \dim\mathrm{Ext}^1(E,F)$$

quel que soit le point (E,F) de $\mathcal{M}\times\mathcal{M}$. Puisque $\mathrm{Ext}^2(E,E)=0$, la dimension m de $\mathcal{M}$ est égale à dim $\mathrm{Ext}^1(E,E)$; ainsi la sous-variété déterminantale D de $\mathcal{M}\times\mathcal{M}$ a la codimension attendue $\mathrm{rg}(K^1)-\mathrm{rg}(K^0)+1$. Sa classe de cohomologie $\delta'\in H^m(\mathcal{M}\times\mathcal{M},\mathbf{Z})$ est alors donnée par la formule de Porteous

$$\delta'=c_m(K^1-K^0)=c_m(-\pi_!\mathcal{H}),$$

où $\pi_!$ désigne le foncteur image directe en K-théorie. Cette classe étant multiple de la classe δ de la diagonale, on conclut avec le lemme suivant:

Lemme. *Soit $\mathcal{A}$ la sous-$\mathbf{Q}$-algèbre de $H^*(\mathcal{M},\mathbf{Q})$ engendrée par les composantes de Künneth des classes de Chern de $\mathcal{E}$, et soient p et q les deux projections de $\mathcal{M}\times\mathcal{M}$ sur $\mathcal{M}$. Les classes de Chern de $\pi_!\mathcal{H}$ sont de la forme $\sum p^*\mu_i\cdot q^*\nu_i$, avec μ_i , $\nu_i\in\mathcal{A}$.*

Notons r la projection de $C\times\mathcal{M}\times\mathcal{M}$ sur C . Tout polynôme en les classes de Chern de $p_1^*\mathcal{E}$ et de $p_2^*\mathcal{E}$ est une somme de produits de la forme $r^*\gamma\cdot\pi^*p^*\mu\cdot\pi^*q^*\nu$, où μ et ν appartiennent à $\mathcal{A}$. Le lemme résulte alors de la formule de Riemann-Roch

$$ch(\pi_!\mathcal{H})=\pi_*\big(r^*\mathrm{Todd(C)}\,\mathrm{ch}(\mathcal{H})\big)\ .$$

Remarque. La condition (iii) de la proposition est évidemment très contraignante. Donnons deux exemples:

a) X est une surface rationnelle ou réglée, et la polarisation H vérifie $H\cdot K_X<0$. L'argument de [M, cor. 6.7.3] montre que la condition (iii) est satisfaite. Si de plus les coefficients a_i du polynôme de Hilbert des éléments de $\mathcal{M}$, écrit sous la forme $\chi(E\otimes H^m)=\sum_{i=0}^{2}a_i\binom{m+i}{i}$, sont premiers entre eux, les conditions (i) à (iii) sont satisfaites [M, §6].

Dans le cas d'une surface *rationnelle*, on obtient mieux. Pour toute variété T, désignons par $CH^*(T)$ l'anneau de Chow de T; grâce à l'isomorphisme $CH^*(X\times\mathcal{M})\cong CH^*(X)\otimes CH^*(\mathcal{M})$, on peut remplacer dans la démonstration de la proposition l'anneau de cohomologie par l'anneau de Chow. On en déduit que *la cohomologie rationnelle de $\mathcal{M}$ est algébrique,* c'est-à-dire que l'application "classe de cycles" $CH^*(\mathcal{M})\otimes\mathbf{Q}\longrightarrow H^*(\mathcal{M},\mathbf{Q})$ est un isomorphisme d'anneaux. Dans le cas $X=\mathbf{P}^2$, Ellingsrud et Strømme obtiennent le même résultat *sur* $\mathbf{Z}$, plus le fait que ces groupes sont sans torsion, grâce à l'outil supplémentaire de la suite spectrale de Beilinson.

b) X est une variété de Fano de dimension 3. Soit S une surface lisse appartenant au système linéaire $|-K_X|$ (de sorte que S est une surface $K3$).

Lorsqu'elle est satisfaite, la condition (iii) a des conséquences remarquables [T]: elle entraîne que l'application de restriction $E \longmapsto E_{|S}$ définit un isomorphisme de $\mathcal{M}$ sur une *sous-variété lagrangienne* d'un espace de modules $\mathcal{M}_S$ de fibrés sur S (muni de sa structure symplectique canonique). Il me semble intéressant de mettre en évidence des espaces de modules de fibrés sur une variété de Fano (et déjà sur $\mathbf{P}^3$) possédant la propriété (iii).

Références

[A-B] M. Atiyah et R. Bott, *The Yang-Mills equations over Riemann surfaces,* Phil. Trans. R. Soc. London A **308** (1982) 523–615.

[E–S] G. Ellingsrud et S.A. Strømme, *Towards the Chow ring of the Hilbert scheme of* $\mathbf{P}^2$, J. reine angew. Math. **441** (1993) 33–44.

[M] M. Maruyama, *Moduli of stable sheaves, II.* J. Math. Kyoto Univ. **18** (1978) 557–614.

[T] A. N. Tyurin, *The moduli space of vector bundles on threefolds, surfaces and curves I,* preprint (1990).

Arnaud Beauville
Université Paris-Sud
Mathématiques – Bât. 425
91 405 Orsay Cedex, France

Some quantum analogues of solvable Lie groups

C. De Concini, V.G. Kac and C. Procesi

Introduction

In the papers [DK1–2],[DKP1–2] the quantized enveloping algebras introduced by Drinfeld and Jimbo have been studied in the case $q = \varepsilon$, a primitive l–th root of 1 with l odd (cf. §2 for basic definitions). Let us only recall for the moment that such algebras are canonically constructed starting from a Cartan matrix of finite type and in particular we can talk of the associated classical objects (the root system, the simply connected algebraic group G, etc.). For such an algebra the generic (resp. any) irreducible representation has dimension equal to (resp. bounded by) l^N where N is the number of positive roots and the set of irreducible representations has a canonical map to the big cell of the corresponding group G.

In this paper we analyze the structure of some subalgebras of quantized enveloping algebras corresponding to unipotent and solvable subgroups of G. These algebras have the non–commutative structure of iterated algebras of twisted polynomials with a derivation, an object which has often appeared in the general theory of non–commutative rings (see e.g. [KP], [GL] and references there). In particular, we find maximal dimensions of their irreducible representations. Our results confirm the validity of the general philosophy that the representation theory is intimately connected to the Poisson geometry.

1 Twisted polynomial rings

1.1 In this section we will collect some well known definitions and properties of twisted derivations.

Let A be an algebra and let σ be an automorphism of A. A *twisted derivation* of A relative to σ is a linear map $D: A \to A$ such that:

$$D(ab) = D(a)b + \sigma(a)D(b).$$

Example. An element $a \in A$ induces an inner twisted derivation $ad_\sigma a$ relative to σ defined by the formula:

$$(ad_\sigma a)b = ab - \sigma(b)a.$$

The following well–known fact is very useful in calculations with twisted derivations. (Here and further we use the usual "box" notation:

$$[n] = \frac{q^n - q^{-n}}{q - q^{-1}},\ [n]! = [1][2]\dots[n],\ \begin{bmatrix} m \\ n \end{bmatrix} = \frac{[m][m-1]\dots[m-n+1]}{[n]!}$$

One also writes $[n]_d$, etc. if q is replaced by q^d.)

Proposition. *Let $a \in A$ and let σ be an automorphism of A such that $\sigma(a) = q^2 a$, where q is a scalar. Then*

$$(ad_\sigma a)^m(x) = \sum_{j=0}^{m} (-1)^j q^{j(m-1)} \begin{bmatrix} m \\ j \end{bmatrix} a^{m-j}\sigma^j(x)a^j.$$

Proof. Let L_a and R_a denote the operators of left and right multiplications by a in A. Then

$$ad_\sigma a = L_a - R_a\sigma.$$

Since L_a and R_a commute, due to the assumption $\sigma(a) = q^2 a$ we have

$$L_a(R_a\sigma) = q^{-2}(R_a\sigma)L_a.$$

Now the proposition is immediate from the following well–known binomial formula applied to the algebra End A.

Lemma. *Suppose that x and y are elements of an algebra such that $yx = q^2 xy$ for some scalar q. Then*

$$(x+y)^m = \sum_{j=0}^{m} \begin{bmatrix} m \\ j \end{bmatrix} q^{j(m-j)} x^j y^{m-j}.$$

Proof. is by induction on m using

$$\begin{bmatrix} m \\ j-1 \end{bmatrix} q^{m+1} + \begin{bmatrix} m \\ j \end{bmatrix} = \begin{bmatrix} m+1 \\ j \end{bmatrix} q^j,$$

which follows from

$$q^b[a] + q^{-a}[b] = [a+b]. \quad \square$$

Let ℓ be a positive integer and let q be a primitive ℓ–th root of 1. Let $\ell' = \ell$ if ℓ is odd and $= \frac{1}{2}\ell$ if ℓ is even. Then, by definition, we have

$$\begin{bmatrix} \ell' \\ j \end{bmatrix} = 0 \text{ for all } j \text{ such that } 0 < j < \ell'.$$

This together with Proposition 1.1 implies

Corollary. *Under the hypothesis of Proposition 1.1 we have:*

$$(ad_\sigma a)^{\ell'}(x) = a^{\ell'}x - \sigma^{\ell'}(x)a^{\ell'} \text{ if q is a primitive } \ell - \text{th root of } 1.$$

Remark. Let D be a twisted derivation associated to an automorphism σ such that $\sigma D = q^2 D\sigma$. Then by induction on m one obtains the following well–known q–analogue of the Leibniz formula:

$$D^m(xy) = \sum_{j=0}^{m} \begin{bmatrix} m \\ j \end{bmatrix} q^{j(m-j)} D^{m-j}(\sigma^j x) D^j(y).$$

It follows that if q is a primitive ℓ–th root of 1, then $D^{\ell'}$ is a twisted derivation associated to $\sigma^{\ell'}$.

1.2 Given an automorphism σ of A and a twisted derivation D of A relative to σ we define the *twisted polynomial algebra* $A_{\sigma,D}[x]$ in the indeterminate x to be the $\mathbb{F}$–module $A \otimes_{\mathbb{F}} \mathbb{F}[x]$ thought as formal polynomials with multiplication defined by the rule:

$$xa = \sigma(a)x + D(a).$$

When $D = 0$ we will also denote this ring by $A_\sigma[x]$. Notice that the definition has been chosen in such a way that in the new ring the given twisted derivation becomes the inner derivation $ad_\sigma x$.

Let us notice that if $a, b \in A$ and a is invertible we can perform the change of variables $y := ax + b$ and we see that $A_{\sigma,D}[x] = A_{\sigma',D'}[y]$. It is better to make the formulas explicit separately when $b = 0$ and when $a = 1$. In the first case $yc = axc = a(\sigma(c)x + D(c)) = a(\sigma(c))a^{-1}y + aD(c)$ and we see that the new automorphism σ' is the composition $(Ada)\sigma$, so that $D' := aD$ is a twisted derivation relative to σ'. Here and further Ada stands for the inner automorphism:

$$(Ada)x = axa^{-1}.$$

In the case $a = 1$ we have $yc = (x + b)c = \sigma(c)x + D(b) + bc = \sigma(c)y + D(b) + bc - \sigma(c)b$, so that $D' = D + ad_\sigma b$. Summarizing we have

Proposition. *Changing σ, D to $(Ada)\sigma, aD$ (resp. to $\sigma, D + D_b$) does not change the twisted polynomial ring up to isomorphism.* □

We may express the previous fact with a definition: For a ring A two pairs (σ, D) and (σ', D') are *equivalent* if they are obtained one from the other by the above moves.

If $D = 0$ we can also consider the twisted Laurent polynomial algebra $A_\sigma[x, x^{-1}]$. It is clear that if A has no zero divisors, then the algebras $A_{\sigma,D}[x]$ and $A_\sigma[x, x^{-1}]$ also have no zero divisors.

The importance for us of twisted polynomial algebras will be clear in the section on quantum groups.

1.3 We want to study special cases of the previous construction.

Let us first consider a finite dimensional semisimple algebra A over an algebraically closed field $\mathbb{F}$, let $\bigoplus_i \mathbb{F}e_i$ be the fixed points of the center of A under σ where the e_i are central idempotents. We have $D(e_i) = D(e_i^2) = 2D(e_i)e_i$ hence $D(e_i) = 0$ and, if $x = xe_i$, then $D(x) = D(x)e_i$. It follows that, decomposing $A = \bigoplus_i Ae_i$, each component Ae_i is stable under σ and D and thus we have

$$A_{\sigma,D}[x] = \bigoplus_i (Ae_i)_{\sigma,D}[x].$$

This allows us to restrict our analysis to the case in which 1 is the only fixed central idempotent.

The second special case is described by the following:

Lemma. *Consider the algebra $A = \mathbb{F}^{\oplus k}$ with σ the cyclic permutation of the summands, and let D be a twisted derivation of this algebra relative to σ. Then D is an inner twisted derivation.*

Proof. Compute D on the idempotents: $D(e_i) = D(e_i^2) = D(e_i)(e_i + e_{i+1})$. Hence we must have $D(e_i) = a_i e_i - b_i e_{i+1}$ and from $0 = D(e_i e_{i+1}) = D(e_i)e_{i+1} + e_{i+1}D(e_{i+1})$ we deduce $b_i = a_{i+1}$. Let now $a = (a_1, a_2, \ldots, a_k)$; an easy computation shows that $D = ad_\sigma a$. □

Proposition. *Let σ be the cyclic permutation of the summands of the algebra $\mathbb{F}^{\oplus k}$. Then*

(a) $\mathbb{F}_\sigma^{\oplus k}[x, x^{-1}]$ *is an Azumaya algebra of degree k over its center* $\mathbb{F}[x^k, x^{-k}]$.

(b) $R := \mathbb{F}_\sigma^{\oplus k}[x, x^{-1}] \otimes_{\mathbb{F}[x^k, x^{-k}]} \mathbb{F}[x, x^{-1}]$ *is the algebra of $k \times k$ matrices over* $\mathbb{F}[x, x^{-1}]$.

Proof. It is enough to prove (b). Let $u := x \otimes x^{-1}$, $e_i := e_i \otimes 1$; we have $u^k = x^k \otimes x^{-k} = 1$ and $ue_i = e_{i+1}u$. From these formulas it easily follows that the elements $e_i u^j$ $(i, j = 1, \ldots, k)$ span a subalgebra A and that there exists an isomorphism $A \xrightarrow{\sim} M_k(\mathbb{F})$ mapping $\mathbb{F}^{\oplus k}$ to the diagonal matrices and u to the matrix of the cyclic permutation. Then $R = A \otimes_{\mathbb{F}} \mathbb{F}[x, x^{-1}]$. □

1.4 Assume now that A is semisimple and that σ induces a cyclic permutation of the central idempotents.

Lemma.

(a) $A = M_d(\mathbb{F})^{\oplus k}$.

(b) Let D be a twisted derivation of A relative to σ. Then the pair (σ, D) is equivalent to the pair $(\sigma', 0)$ where

$$\sigma'(a_1, a_2, \ldots, a_k) = (a_k, a_1, a_2, \ldots, a_{k-1}). \tag{1.4.1}$$

Proof. Since σ permutes transitively the simple blocks they must all have the same degree d so that $A = M_d(F)^{\oplus k}$. Furthermore we can arrange the identifications of the simple blocks with matrices so that:

$$\sigma(a_1, a_2, \ldots, a_k) = (\tau(a_k), a_1, a_2, \ldots, a_{k-1}),$$

where τ is an automorphism of $M_d(\mathbb{F})$. Any such automorphism is inner, hence after composing σ with an inner automorphism, we may assume in the previous formula that $\tau = 1$. Then we think of A as $M_d(\mathbb{F}) \otimes \mathbb{F}^{\oplus k}$, the new automorphism being of the form $1 \otimes \sigma'$ where $\sigma' : \mathbb{F}^{\oplus k} \to \mathbb{F}^{\oplus k}$ is given by (1.4.1).

We also have that $M_d(\mathbb{F}) = A^\sigma$ and $\mathbb{F}^{\oplus k}$ is the centralizer of A^σ. Next observe that D restricted to A^σ is a derivation of $M_d(\mathbb{F})$ with values in $\oplus_{i=1}^k M_d(\mathbb{F})$, i.e., $D(a) = (D_1(a), D_2(a), \dots, D_k(a))$ where each D_i is a derivation of $M_d(\mathbb{F})$. Since for $M_d(\mathbb{F})$ all derivations are inner we can find an element $u \in A$ such that $D(a) = [u, a]$ for all $a \in M_d(\mathbb{F})$. So $(D - ad_\sigma u)(a) = [u, a] - (ua - \sigma(a)u) = 0$ for $a \in A^\sigma$. Thus, changing D by adding $-ad_\sigma u$ we may assume that $D = 0$ on $M_d(\mathbb{F})$.

Now consider $b \in \mathbb{F}^{\oplus k}$ and $a \in M_d(\mathbb{F})$; we have $D(b)a = D(ba) = D(ab) = aD(b)$. Since $\mathbb{F}^{\oplus k}$ is the centralizer of $M_d(\mathbb{F})$ we have $D(b) \in \mathbb{F}^{\oplus k}$ and D induces a twisted derivation of $\mathbb{F}^{\oplus k}$. By Lemma 1.3 this last derivation is inner and the claim is proved. □

Summarizing we have

Proposition. *Let A be a finite-dimensional semisimple algebra over an algebraically closed field $\mathbb{F}$. Let σ be an automorphism of A which induces a cyclic permutation of order k of the central idempotents of A. Let D be a twisted derivation of A relative to σ. Then:*

$$A_{\sigma,D}[x] \cong M_d(\mathbb{F}) \otimes \mathbb{F}_\sigma^{\oplus k}[x],$$

$$A_{\sigma,D}[x, x^{-1}] \cong M_d(\mathbb{F}) \otimes \mathbb{F}_\sigma^{\oplus k}[x, x^{-1}].$$

This last algebra is Azumaya of degree dk.

1.5 We can now globalize the previous construction. Let A be a prime algebra (i.e. $aAb = 0$, $a, b \in A$, implies that $a = 0$ or $b = 0$) over a field $\mathbb{F}$ and let Z be the center of A. Then Z is a domain and A is a torsion free module over Z. Assume that A is a finite module over Z. Then A embeds in a finite-dimensional central simple algebra $Q(A) = A \otimes_Z Q(Z)$, where $Q(Z)$ is the ring of fractions of Z. If $\overline{Q(Z)}$ denotes the algebraic closure of $Q(Z)$ we have that $A \otimes_Z \overline{Q(Z)}$ is the full ring $M_d(\overline{Q(Z)})$ of $d \times d$ matrices over $\overline{Q(Z)}$. Then d is called the *degree* of A.

Let σ be an automorphism of the algebra A and let D be a twisted derivation of A relative to σ. Assume that

(a) There is a subalgebra Z_0 of Z, such that Z is finite over Z_0.

(b) D vanishes on Z_0 and σ restricted to Z_0 is the identity.

These assumptions imply that σ restricted to Z is an automorphism of finite order. Let d be the degree of A and let k be the order of σ on the center Z.

Assume that the field $\mathbb{F}$ has characteristic 0. The main result of this section is:

Theorem. *Under the above assumptions the twisted polynomial algebra $A_{\sigma,D}[x]$ is an order in a central simple algebra of degree kd.*

Proof. Let Z^σ be the fixed points in Z of σ. By the definition, it is clear that D restricted to Z^σ is a derivation. Since it vanishes on a subalgebra over which it is finite hence algebraic and since we are in characteristic zero it follows that D vanishes on Z^σ. Let us embed Z^σ in an algebraically closed field $\mathbb{L}$ and let us consider the algebra $A \otimes_{Z^\sigma} \mathbb{L}$. This algebra of course equals $A \otimes_Z Z \otimes_{Z^\sigma} \mathbb{L}$, but clearly $Z \otimes_{Z^\sigma} \mathbb{L} = \mathbb{L}^{\oplus k}$ and $A \otimes_Z \mathbb{L} = M_d(\mathbb{L})$. Thus we get that $A \otimes_{Z^\sigma} \mathbb{L} = \oplus_{i=1}^k M_d(\mathbb{L})$. The pair σ, D extends to $A \otimes_{Z^\sigma} \mathbb{L}$ and using the same notations we have that $(A \otimes_{Z^\sigma} \mathbb{L})_{\sigma,D}[x] = (A_{\sigma,D}[x]) \otimes_{Z^\sigma} \mathbb{L}$. We are now in the situation of a semisimple algebra which we have already studied and the claim follows. □

Corollary. *Under the above assumptions, $A_{\sigma,D}[x]$ and $A_\sigma[x]$ have the same degree.*

Remark. The previous analysis yields in fact a stronger result. Consider the open set of Spec Z where A is an Azumaya algebra; it is clearly σ–stable. In it we consider the open part where σ has order exactly k. Every orbit of k elements of the group generated by σ gives a point $F(p)$ in Spec Z^σ and $A \otimes_Z Z \otimes_{Z^\sigma} F(p) = \oplus_{i=1}^k M_d(F(p))$. Thus we can apply the previous theory which allows us to describe the fiber over $F(p)$ of the spectrum of $A_{\sigma,D}[x]$.

1.6 Let A be a prime algebra over a field $\mathbb{F}$ of characteristic 0, let $x_1, \ldots, x_n$ be a set of generators of A and let Z_0 be a central subalgebra of A. For each $i = 1, \ldots, k$, denote by A^i the subalgebra of A generated by $x_1, \ldots, x_i$, and let $Z_0^i = Z_0 \cap A^i$. We assume that the following three conditions hold for each $i = 1, \ldots, k$:

(a) $x_i x_j = b_{ij} x_j x_i + P_{ij}$ if $i > j$, where $b_{ij} \in \mathbb{F}$, $P_{ij} \in A^{i-1}$.

(b) A^i is a finite module over Z_0^i.

(c) Formulas $\sigma_i(x_j) = b_{ij} x_j$ for $j < i$ define an automorphism of A^{i-1} which is the identity on Z_0^{i-1}.

Note that letting $D_i(x_j) = P_{ij}$ for $j < i$, we obtain $A^i = A^{i-1}_{\sigma_i, D_i}[x_i]$, so that A is an iteratated twisted polynomial algebra. Note also that each triple (A^{i-1}, σ_i, D_i) satisfies assumptions 1.5(a) and (b).

We may consider the twisted polynomial algebras $\overline{A}^i$ with zero derivations, so that the relations are $x_i x_j = b_{ij} x_j x_i$ for $j < i$. We call this the *associated quasipolynomial algebra* (as in [DK1]).

We can prove now the main theorem of this section.

Theorem. *Under the above assumptions, the degree of A is equal to the degree of the associated quasipolynomial algebra $\overline{A}$.*

Proof. We use the following remark. If there is an index h such that the elements $P_{ij} = 0$ for all $i > h$ and all j, then the monomials in the variables different from x_h form a subalgebra B and the algebra A is a twisted polynomial ring $B_{\sigma,D}[x_h]$. The associated ring $B_\sigma[x_h]$ is obtained by setting $P_{hj} = 0$ for all j. Having made this remark we see that we can inductively modify the relations 1.6(a) so that at the h-th step we have an algebra A_h^n with the same type of relations but $P_{ij} = 0$ for all $i > n - h$ and all j. Since A_h^n and A_{h-1}^n are of type $B_{\sigma,D}[x]$ and $B_\sigma[x]$ respectively we see, by Corollary 1.5, that they have all the same degree. □

2 Quantum groups

2.1 Let (a_{ij}) be an indecomposable $n \times n$ Cartan matrix and let $d_1, \ldots, d_n$ be relatively prime positive integers such that $d_i a_{ij} = d_j a_{ji}$. Recall the associated notions of the weight, coroot and root lattices $P, Q^\vee$ and Q, of the root and coroot systems R and $R^\vee$, of the Weyl group W, the W-invariant bilinear form $(.|.)$, etc.:

Let P be a lattice over $\mathbb{Z}$ with basis $\omega_1, \ldots, \omega_n$ and let $Q^\vee = \mathrm{Hom}_{\mathbb{Z}}(P, \mathbb{Z})$ be the dual lattice with the dual basis $\alpha_1^\vee, \ldots, \alpha_n^\vee$, i.e. $\langle \omega_i, \alpha_j^\vee \rangle = \delta_{ij}$. Let $P_+ = \sum_{i=1}^n \mathbb{Z}_+ \omega_i$. Let

$$\rho = \sum_{i=1}^n \omega_i, \ \alpha_j = \sum_{i=1}^n a_{ij} \omega_i \ (j = 1, \ldots, n),$$

and let $Q = \sum_{j=1}^n \mathbb{Z} \alpha_j \subset P$, and $Q_+ = \sum_{j=1}^n \mathbb{Z}_+ \alpha_j$.

Define automorphisms s_i of P by $s_i(\omega_j) = \omega_j - \delta_{ij} \alpha_i$ $(i, j = 1, \ldots, n)$. Then $s_i(\alpha_j) = \alpha_j - a_{ij} \alpha_i$. Let W be the subgroup of $GL(P)$ generated by $s_1, \ldots, s_n$. Let

$$\begin{aligned} \Pi &= \{\alpha_1, \ldots, \alpha_n\}, \ \Pi^\vee = \{\alpha_1^\vee, \ldots, \alpha_n^\vee\}, \\ R &= W\Pi, \ R^+ = R \cap Q_+, \ R^\vee = W\Pi^\vee. \end{aligned}$$

The map $\alpha_i \longmapsto \alpha_i^\vee$ extends uniquely to a bijective W–equivariant map $\alpha \longmapsto \alpha^\vee$ between R and $R^\vee$. The reflection s_α defined by $s_\alpha(\lambda) = \lambda - \langle\lambda, \alpha^\vee\rangle\alpha$ lies in W for each $\alpha \in R$, so that $s_{\alpha_i} = s_i$.

Define a bilinear pairing $P \times Q \to \mathbb{Z}$ by $(\omega_i|\alpha_j) = \delta_{ij}d_j$. Then $(\alpha_i|\alpha_j) = d_i a_{ij}$, giving a symmetric $\mathbb{Z}$–valued W–invariant bilinear form on Q such that $(\alpha|\alpha) \in 2\mathbb{Z}$. We may identify $Q^\vee$ with a sublattice of the $\mathbf{Q}$–span of P (containing Q) using this form. Then:

$$\alpha_i^\vee = d_i^{-1}\alpha_i,\ \alpha^\vee = 2\alpha/(\alpha|\alpha).$$

One defines the *simply connected quantum group* $\mathcal{U}$ associated to the matrix (a_{ij}) as an algebra over the ring $\mathcal{A} := \mathbb{C}[q, q^{-1}, (q^{d_i} - q^{-d_i})^{-1}]$ with generators E_i, F_i $(i = 1, \ldots, n)$, K_α $(\alpha \in P)$ subject to the following relations (this is a simple variation of the construction of Drinfeld and Jimbo):

$$\begin{aligned}
&K_\alpha K_\beta = K_{\alpha+\beta},\ K_0 = 1,\\
&\sigma_\alpha(E_i) = q^{(\alpha|\alpha_i)}E_i, \sigma_\alpha(F_i) = q^{-(\alpha|\alpha_i)}F_i,\\
&[E_i, F_j] = \delta_{ij}\tfrac{K_{\alpha_i} - K_{-\alpha_i}}{q^{d_i} - q^{-d_i}},\\
&(ad_{\sigma_{-\alpha_i}}E_i)^{1-a_{ij}}E_j = 0, (ad_{\sigma_{\alpha_i}}F_i)^{1-a_{ij}}F_j = 0\ (i \neq j),
\end{aligned}$$

where $\sigma_\alpha = Ad\ K_\alpha$. Recall that $\mathcal{U}$ has a Hopf algebra structure with comultiplication Δ, antipode S and counit η defined by:

$$\begin{aligned}
&\Delta E_i = E_i \otimes 1 + K_{\alpha_i} \otimes E_i,\ \Delta F_i = F_i \otimes K_{-\alpha_i} + 1 \otimes F_i,\ \Delta K_\alpha = K_\alpha \otimes K_\alpha,\\
&SE_i = -K_{-\alpha_i}E_i,\ SF_i = -F_iK_i,\ SK_\alpha = K_{-\alpha},\\
&\eta E_i = 0,\ \eta F_i = 0,\ \eta K_\alpha = 1.
\end{aligned}$$

Recall that the braid group $\mathcal{B}_W$ (associated to W), whose canonical generators one denotes by T_i, acts as a group of automorphisms of the algebra $\mathcal{U}$ ([L]):

$$\begin{aligned}
T_iK_\alpha &= K_{s_i(\alpha)},\ T_iE_i = -F_iK_{\alpha_i},\\
T_iE_j &= \tfrac{1}{[-a_{ij}]_{d_i}!}(ad_{\sigma_{\alpha_i}}(-E_i))^{-a_{ij}}E_j,\\
T_i\kappa &= \kappa T_i,
\end{aligned}$$

where κ is a conjugate–linear anti–automorphism of $\mathcal{U}$, viewed as an algebra over $\mathbb{C}$, defined by:

$$\kappa E_i = F_i,\ \kappa F_i = E_i,\ \kappa K_\alpha = K_{-\alpha},\ \kappa q = q^{-1}.$$

2.2 Fix a reduced expression $w_0 = s_{i_1} \ldots s_{i_N}$ of the longest element of W, and let

$$\beta_1 = \alpha_{i_1},\ \beta_2 = s_{i_1}(\alpha_{i_2}), \ldots, \beta_N = s_{i_1} \ldots s_{i_{N-1}}(\alpha_{i_N})$$

be the corresponding convex ordering of R^+. Introduce the corresponding *root vectors* $(m = 1, \ldots, N)$ ([L]):

$$E_{\beta_m} = T_{i_1} \ldots T_{i_{m-1}} E_{i_m},\ E_{-\beta_m} = T_{i_1} \ldots T_{i_{m-1}} F_{i_m} = \kappa E_\beta$$

(they depend on the choice of the reduced expression).

For $k = (k_1, \ldots, k_N) \in \mathbb{Z}_+^N$ we let

$$E^k = E_{\beta_1}^{k_1} \ldots E_{\beta_N}^{k_N},\ F^k = \kappa E^k.$$

Lemma.

(a) [L] The elements $F^k K_\alpha E^r$, where $k, r \in \mathbb{Z}_+^N$, $\alpha \in P$, form a basis of $\mathcal{U}$ over $\mathcal{A}$.

(b) [LS] For $i < j$ one has:

(2.2.1) $$E_{\beta_i} E_{\beta_j} - q^{(\beta_i|\beta_j)} E_{\beta_j} E_{\beta_i} = \sum_{k \in \mathbb{Z}_+^N} c_k E^k,$$

where $c_k \in \mathbb{C}[q, q^{-1}]$ and $c_k \neq 0$ only when $k = (k_1, \ldots, k_N)$ is such that $k_s = 0$ for $s \leq i$ and $s \geq j$. □

An immediate corollary is the following:

Let w be any element of the Weyl group. We can choose for it a reduced expression $w = s_{i_1} \ldots s_{i_k}$ which we complete to a reduced expression $w_0 = s_{i_1} \ldots s_{i_N}$ of the longest element of W. Consider the elements E_{β_j}, $j = 1, \ldots, k$. Then we have:

Proposition. *(a) The elements E_{β_j}, $j = 1, \ldots, k$, generate a subalgebra $\mathcal{U}^w$ which is independent of the choice of the reduced expression of w.*

(b) If $w' = ws$ with s a simple reflection and $l(w') = l(w) + 1 = k + 1$, then $\mathcal{U}^{w'}$ is a twisted polynomial algebra of type $\mathcal{U}^w_{\sigma,D}[E_{\beta_{k+1}}]$, where the formulas for σ and D are implicitly given in the formulas (2.2.1).

Proof. (a) Using the fact that one can pass from one reduced expression of w to another by braid relations one reduces to the case of rank 2 where one repeats the analysis made by Lusztig ([L]). (b) is clear by Lemma 2.2. □

The elements K_α clearly normalize the algebras $\mathcal{U}^w$ and when we add them to these algebras we are performing an iterated construction of Laurent twisted polynomials. The resulting algebras will be called $\mathcal{B}^w$.

Since the algebras $\mathcal{U}^w$ and $\mathcal{B}^w$ are iterated twisted polynomial rings with relations of the type 1.6(a) we can consider the associated quasipolynomial algebras, and we will denote them by $\overline{\mathcal{U}}^w$ and $\overline{\mathcal{B}}^w$. Notice that the latter algebras depend on the reduced expression chosen for w. Of course the defining relations for these algebras are obtained from (2.2.1) by replacing the right–hand side by zero. We could of course also perform the same construction with the negative roots but this is not strictly necessary since we can simply apply the anti–automorphism κ to define the analogous negative objects.

3 Degrees of algebras $\mathcal{U}_\varepsilon^w$ and $\mathcal{B}_\varepsilon^w$

3.1 We specialize now the previous sections to the case $q = \varepsilon$, a primitive ℓ–th root of 1. Assuming that $\ell' > \max_i d_i$, we may consider the specialized algebras:

$$\mathcal{U}_\varepsilon = \mathcal{U}/(q-\varepsilon),\ \mathcal{U}_\varepsilon^w = \mathcal{U}^w/(q-\varepsilon),\ \mathcal{B}_\varepsilon^w = \mathcal{B}^w/(q-\varepsilon), \text{etc.}$$

We have obvious subalgebra inclusions $\mathcal{U}_\varepsilon^w \subset \mathcal{B}_\varepsilon^w \subset \mathcal{U}_\varepsilon$.

First, let us recall and give a simple proof of the following crucial fact [DK1]:

Proposition. *Elements E_α^ℓ ($\alpha \in R$) and K_β^ℓ ($\beta \in P$) lie in the center Z_ε of $\mathcal{U}_\varepsilon$ if $\ell' > \max_{i,j} |a_{ij}|$ (for any generalized Cartan matrix (a_{ij})).*

Proof. The only non–trivial thing to check is that $[E_i^\ell, E_j] = 0$ for $i \neq j$. From the "Serre relations" it is immediate that $(ad_{\sigma_{-\alpha_i}} E_i)^{\ell'} E_j = 0$. Due to Corollary 1.1, this can be rewritten as

$$E_i^{\ell'} E_j = \varepsilon^{-\ell'(\alpha_i|\alpha_j)} E_j E_i^{\ell'},$$

proving the claim. □

As has been already remarked, the algebras $\mathcal{U}_\varepsilon^w$ and $\mathcal{B}_\varepsilon^w$ are iterated twisted polynomial algebras with relations of the type 1.6(a). Proposition 3.1

shows that they satisfy conditions 1.6(b) and (c). Hence Theorem 1.6 implies

Corollary. *Algebras* $\mathcal{U}_\varepsilon^w$ *and* $\overline{\mathcal{U}}_\varepsilon^w$ *(resp.* $\mathcal{B}_\varepsilon^w$ *and* $\overline{\mathcal{B}}_\varepsilon^w$*) have the same degree.* □

3.2 We proceed to calculate the degrees of algebras $\overline{\mathcal{U}}_\varepsilon^w$ and $\overline{\mathcal{B}}_\varepsilon^w$. Recall that these algebras are, up to inverting some variables, quasipolynomial algebras whose generators satisfy relations of type $x_i x_j = b_{ij} x_j x_i$, $i, j = 1, \ldots, s$, where the elements b_{ij} have the special form $b_{ij} = \varepsilon^{c_{ij}}$, the c_{ij} being entries of a skew–symmetric integral $s \times s$ matrix H. As we have shown in [DKP2, Proposition 2.2], considering H as the matrix of a linear map $\mathbb{Z}^s \to (\mathbb{Z}/(\ell))^s$, the degree of the corresponding twisted polynomial algebra is $\sqrt{h}$, where h is the number of elements of the image of this map.

Fix $w \in W$ and its reduced expression $w = s_{i_1} \ldots s_{i_k}$. We shall denote the matrix H for the algebras $\overline{\mathcal{U}}_\varepsilon^w$ and $\overline{\mathcal{B}}_\varepsilon^w$ by A and S respectively. First we describe explicitly these matrices.

Let $d = 2$ unless (a_{ij}) is of type G_2 in which case $d = 6$, and let $\mathbb{Z}' = \mathbb{Z}[d^{-1}]$. Consider the roots $\beta_1, \ldots, \beta_k$ as in Section 2.2, and consider the free $\mathbb{Z}'$–module V with basis $u_1, \ldots, u_k$. Define on V a skew–symmetric bilinear form by

$$\langle u_i | u_j \rangle = (\beta_i | \beta_j) \text{ if } i < j.$$

Then A is the matrix of this bilinear form in the basis $\{u_i\}$. Identifying V with its dual V^* using the given basis, we may think of A as a linear operator from V to itself.

Furthermore,

$$S = \begin{pmatrix} A & -{}^tC \\ C & 0 \end{pmatrix},$$

where C is the $n \times k$ matrix $((\omega_i|\beta_j))_{1\le i\le n, 1\le j\le k}$. We may think of the matrix C as a linear map from the module V with the basis $u_1, \ldots, u_k$ to the module $Q^\vee \otimes_{\mathbb{Z}} \mathbb{Z}'$ with the basis $\alpha_1^\vee, \ldots, \alpha_n^\vee$. Then we have:

(3.2.1) $$C(u_i) = \beta_i, \ i = 1, \ldots, k.$$

To study the matrices A and S we need the following

Lemma. *Given* $\omega = \sum_{i=1}^{n} \delta_i \omega_i$ *with* $\delta_i = 0$ *or* 1, *set*

$$I_\omega = \{t \in \{1, \ldots, k\} | s_{i_t}(\omega) \neq \omega\}.$$

Then

$$w(\omega) = \omega - \sum_{j \in I_\omega} \beta_j.$$

Proof. by induction on the length of w. Write $w = w' s_{i_k}$. If $k \notin I_\omega$ then $w(\omega) = w'(\omega)$ and we are done by induction. Otherwise $w(\omega) = w'(\omega - \alpha_{i_k}) = w'(\omega) - \beta_k$ and again we are done by induction. □

Note that $\{1, 2, \ldots, k\} = \coprod_{i=1}^{n} I_{\omega_i}$.

3.3 Consider the operators: $M = (A \ \ -{}^tC)$ and $N = (C \ 0)$ so that $S = M \oplus N$.

Lemma.

(a) The operator M *is surjective.*

(b) The vectors $v_\omega := (\sum_{t \in I_\omega} u_t) - \omega - w(\omega)$, *as* ω *runs through the fundamental weights, form a basis of the kernel of* M.

(c) $N(v_\omega) = \omega - w(\omega) = \sum_{t \in I_\omega} \beta_t$.

Proof. (a) We have by a straightforward computation:

$$S(u_i + \beta_i) = -(\beta_i|\beta_i) u_i - 2 \sum_{j>i} (\beta_i|\beta_j) u_j - \beta_i,$$

and

$$M(u_i + \beta_i) = -(\beta_i|\beta_i) u_i - 2 \sum_{j>i} (\beta_i|\beta_j) u_j.$$

Since $(\beta_i|\beta_i)$ is invertible in $\mathbb{Z}'$ the claim follows.

(b) Since the vectors v_ω are part of a basis and, by (a), the kernel of M is a direct summand, it is enough to show that these vectors lie in the kernel. Now to check that $M(v_{\omega_i}) = 0$ is equivalent to seeing that v_{ω_i} lies in the kernel of the corresponding skew-symmetric form, i.e. $\langle u_j | v_{\omega_i} \rangle = 0$ for all $j = 1, \ldots, k$:

Using Lemma 3.2, we have

(3.3.1) $$\langle u_j | v_{\omega_i} \rangle = -2 \sum_{t>j} (\beta_j | \beta_t) + 2(\beta_j | w(\omega_i)) + a_j,$$

where $a_j = 0$ if $j \notin I_{\omega_i}$ and $a_j = (\beta_j|\beta_j)$ otherwise.

We proceed by induction on $k = l(w)$. Let us write $v_{\omega_i}(w)$ to stress the dependence on w. For $k = 0$ there is nothing to prove. Let $w = w's_{i_k}$ with $l(w') = l(w) - 1$. We distinguish two cases according to whether $i = i_k$ or not. If $i \neq i_k$, i.e. $k \notin I_{\omega_i}$, we have that $v_{\omega_i}(w) = v_{\omega_i}(w')$ hence the claim follows by induction if $j < k$. For $j = k$ we obtain from (3.3.1):

$$\langle u_k | v_{\omega_i} \rangle = -2(\beta_k | w(\omega_i)) = -2(w'(\alpha_{i_k}) | w'(\omega_i)) = -2(\alpha_{i_k} | \omega_i) = 0.$$

Assume now that $i_k = i$ so that $w = w's_i$. Then $v_{\omega_i}(w) = v_{\omega_i}(w') + u_k - \beta_k$. For $j < k$ by induction we get:

$$\langle u_j | v_{\omega_i} \rangle = \langle u_j | u_k \rangle - \langle u_j | \beta_k \rangle = -(\beta_j | \beta_k) + (\beta_j | \beta_k) = 0.$$

Finally if $j = k$ we have:

$$2(\beta_k | w(\omega_i)) + (\beta_k | \beta_k) = 2(w'\alpha_i | w'(\omega_i - \alpha_i)) + (\alpha_i | \alpha_i) = 2(\alpha_i | \omega_i) - (\alpha_i | \alpha_i) = 0.$$

Finally using (3.2.1), we have: $N(v_{\omega_i}) = \sum_{t \in I_{\omega_i}} \beta_t$, hence (c) follows from Lemma 3.2. □

3.4 In order to compute the kernel of S we need to compute the kernel of N on the submodule spanned by the vectors v_{ω_i}. Let us identify this module with the weight lattice P by identifying v_{ω_i} with ω_i. By Lemma 3.3(c), we see that N is identified with the map $1 - w : P \to Q$. At this point we need the following fact:

Lemma. *Consider the highest root* $\theta = \sum_{i=1}^n a_i \alpha_i$ *of the root system* R. *Let* $\mathbb{Z}'' = \mathbb{Z}''[a_1^{-1}, \ldots, a_n^{-1}]$, *and let* $M' = M \otimes_{\mathbb{Z}} \mathbb{Z}'$, $M'' = M \otimes_{\mathbb{Z}} \mathbb{Z}''$ *for* $M = P$ *or* Q. *Then for any* $w \in W$, *the* $\mathbb{Z}''$*-submodule* $(1 - w)P''$ *of* Q'' *is a direct summand.*

Proof. Recall that one can represent w in the form $w = s_{\gamma_1} \ldots s_{\gamma_m}$ where $\gamma_1, \ldots, \gamma_m$ is a linearly independent set of roots (see e.g. [C]). Since in the decomposition $\gamma^\vee = \sum_i r_i \alpha_i^\vee$ one of the r_i is 1 or 2, it follows that $(1 - s_\gamma)P' =$

$\mathbb{Z}'\gamma$. Since $1-w=(1-s_{\gamma_1}\dots s_{\gamma_{m-1}})s_{\gamma_m}+(1-s_{\gamma_m})$, we deduce by induction that

$$(3.4.1)\qquad (1-w)P'=\sum_{i=1}^{m}\mathbb{Z}'\gamma_m$$

Recall now that any sublattice of Q spanned over $\mathbb{Z}$ by some roots is a $\mathbb{Z}$–span of a set of roots obtained from Π by iterating the following procedure: add a highest root to the set of simple roots, then remove several other roots from this set. The index of the lattice M thus obtained in $M\otimes_{\mathbb{Z}}\mathbf{Q}\cap Q$ is equal to the product of coefficients of removed roots in the added highest root. Hence it follows from (3.4.1) that

$$((1-w)P'')\otimes_{\mathbb{Z}}\mathbf{Q}\cap Q''=(1-w)P'',$$

proving the claim. □

We call $\ell>1$ a *good* integer if it is relatively prime to d and to all the a_i.

Theorem. *If ℓ is a good integer, then*

$$\deg\mathcal{B}^w_\varepsilon=\deg\overline{\mathcal{B}}^w_\varepsilon=ell^{\frac{1}{2}(\ell(w)+\textbf{rank }(1-w))}.$$

Proof. From the above discussion we see that $\deg\overline{\mathcal{B}}^w_\varepsilon=\ell^s$, where $s=(\ell(w)+n)-(n-\text{rank}(1-w))$, which together with Corollary 3.1 proves the claim. □

3.5 We pass now to $\mathcal{U}^w_\varepsilon$. For this we need to compute the image of the matrix A. Computing first its kernel, we have that $KerA$ is identified with the set of linear combinations $\sum_i c_i v_{\omega_i}$ for which $\sum_i c_i(\omega_i+w(\omega_i))=0$ i.e. $\sum_i c_i\omega_i\in\ker(1+w)$. This requires a case by case analysis. A simple case is when $w_0=-1$, so that $1+w=w_0(-1+w_0w)$ and one reduces to the previous case. Thus we get

Proposition. *If $w_0=-1$ (i.e. for types different from A_n, D_{2n+1} and E_6) and if ℓ is a good integer, we have:*

$$\deg\mathcal{U}^w_\varepsilon=\deg\overline{\mathcal{U}}^w_\varepsilon=\ell^{\frac{1}{2}(\ell(w)+\textbf{rank}(1+w)-n)}.$$

Let us note the special case $w=w_0$. Remark that defining ${}^t\omega:=-w_0(\omega)$ we have an involution $\omega\rightarrow{}^t\omega$ on the set of fundamental weights. Let us denote by s the number of orbits of this involution.

Theorem. *If ε is a primitive ℓ–th root of* 1*, where ℓ is an integer greater than* 1 *and relatively prime to d, then the algebras $\mathcal{U}_\varepsilon^{w_0}$ and $\mathcal{B}_\varepsilon^{w_0}$ have degrees $\ell^{\frac{N-s}{2}}$ and $\ell^{\frac{N+s}{2}}$ respectively.*

Proof. In this case $l(w_0) = N$ and the maps $\omega \to \omega + w_0(\omega)$ and $\omega \to \omega - w_0(\omega)$ are $\omega \to \omega - {}^t\omega$ and $\omega \to \omega + {}^t\omega$ and so their ranks are clearly $n - s$ and s respectively. □

4 Poisson structure

4.1 Before we revert to the discussion of our algebras we want to make a general remark. Assume that we have a manifold M and a vector bundle V of algebras with 1 (i.e. 1 and the multiplication map are smooth sections). We identify the functions on M with the sections of V which are multiples of 1. Let D be a derivation of V, i.e. a derivation of the algebra of sections which maps the algebra of functions on M into itself and let X be the corresponding vector field on M.

Proposition. *For each point $p \in M$ there exists a neighborhood U_p and a map φ_t defined for $|t|$ sufficiently small on $V|U_p$ which is a morphism of vector bundles covering the germ of the* 1*–parameter group generated by X and is also an isomorphism of algebras.*

Proof. The hypotheses on D imply that it is a vector field on V linear on the fibers, hence we have the existence of a local lift of the 1–parameter group as a morphism of vector bundles. The condition of being a derivation implies that the lift preserves the multiplication section i.e. it is a morphism of algebras. □

We will have to consider a variation of this: suppose M is a Poisson manifold and assume furthermore that the Poisson structure lifts to V i.e. for each (local) function f and section s we have a Poisson bracket which is a derivation. This means that we have a lift of the Hamiltonian vector fields as in the previous proposition. We deduce:

Corollary. *Under the previous hypotheses, the fibers of V over points of a given symplectic leaf of M are all isomorphic as algebras.*

Proof. The proposition implies that in a neighborhood of a point in a leaf the algebras are isomorphic but since the notion of isomorphism is transitive this implies the claim. □

4.2 Let us recall some basic facts on Poisson groups (we refer to [D], [STS], [LW] for basic definitions and properties). Since a Poisson structure on a manifold M is a special type of a section of $\Lambda^2 T(M)$ it can be viewed as a linear map from $T^*(M)$ to $T(M)$. The image of this map is thus a distribution on the manifold M. It can be integrated so that we have a decomposition of M into symplectic leaves which are connected locally closed submanifolds whose tangent spaces are the spaces of the distribution. In fact in our case the leaves will turn out to be Zariski open sets of algebraic subvarieties.

For a group H the tangent space at each point can be identified to the Lie algebra $\hbar$ by left translation and thus a Poisson structure on H can be given as a family of maps $\gamma_h : \hbar^* \to \hbar$ as $h \in H$. Let G be an algebraic group and $H, K \subset G$ algebraic subgroups. We shall say that (G, H, K) form a Manin triple of algebraic groups if their corresponding Lie algebras $(g, \hbar, k)$ form a Manin triple, i.e. (cf. [D], [LW]) if g has a non–degenerate, symmetric invariant bilinear form with respect to which the Lie subalgebras $\hbar$ and k are isotropic and $g = \hbar \oplus k$ (as vector spaces). Then it follows that we have a canonical isomorphism $\hbar^* = k$. Having identified $\hbar^*$ with k, the Poisson structure on H is thus described by giving for every $h \in H$ a linear map $\gamma_h : k \to \hbar$.

Let $x \in k$, consider x as an element of g, set $\pi : g \to \hbar$ to be the projection with kernel k. Set:

$$\gamma_h(x) = (Adh)\pi(Adh)^{-1}(x).$$

Then one can verify (as in [LW]) that the corresponding tensor satisfies the required properties of a Poisson structure. (In fact any Poisson structure on H can be obtained in this way.)

Notice now that the (restriction of the) canonical map:

$$\delta : H \to G/K$$

is an étale covering of some open set in G/K. Thus for every point $h \in H$ we can identify the tangent space to H in h with the tangent space to G/K at $\delta(h)$. By using right translation we can then identify the tangent space to G/K at $\delta(h)$ with $g/(Adh)k$, the tangent space at $h \in H$ with $\hbar$ by right translation and the isomorphism between them with the projection $\hbar \to g/(Adh)k$.

Using all these identifications one verifies that the map γ_h previously considered is the map induced by differentiating the left K–action on G/K. From this it follows:

Proposition. *The symplectic leaves for the symplectic structure on H coincide with the connected components of the preimages under δ of K–orbits* *under the left multiplications on G/K.*

Consider now a quotient Poisson group S of H, that is S is a quotient group of H and the ring $\mathbb{C}[S] \subset \mathbb{C}[H]$ is a Poisson subalgebra. Let U be the kernel of the quotient homomorphism $\varphi: H \to S$, let $s = Lie\ S$, $u = Lie\ U$ and $d\varphi: \hbar \to s$ the Lie algebra quotient map. Then u is an ideal in $\hbar$ and we identify s^* with a subspace of $\hbar^* = k$ by taking $u^\perp \subset g$ under the invariant form and intersecting it with k. Then for $p \in S$ the linear map: $\overline{\gamma}_p : s^* \to s$ giving rise to the Poisson structure is given by:

$$\overline{\gamma}_p = (d\varphi) \cdot (\gamma_{\tilde{p}}|_{s^*})$$

where $\tilde{p} \in H$ is any representative of p ($\overline{\gamma}_p$ is independent of the choice of $\tilde{p}$).

The construction of the Manin triple corresponding to the Poisson manifold S is obtained from the following simple fact:

Lemma. *Let $(g, \hbar, k)$ be a Manin triple of Lie algebras, and let $u \subset \hbar$ be an ideal such that $u^\perp$ (in g) intersected with k is a subalgebra of the Lie algebra k. Then*

(a) $u^\perp$ is a subalgebra of g and u is an ideal of $u^\perp$.

(b) $(u^\perp/u,\ \hbar/u,\ k \cap u^\perp)$ is a Manin triple, where the bilinear form on $u^\perp$ is induced by that on g.

Proof. Straightforward. □

4.3 In the remaining sections we will apply the above remarks to the Poisson groups associated to the Hopf algebra $\mathcal{U}_\varepsilon$ and its Hopf subalgebra $\mathcal{B}_\varepsilon := \mathcal{B}_\varepsilon^{w_0}$, and will derive some results on representations of the algebra $\mathcal{B}_\varepsilon$. From now on ε is a primitive ℓ–th root of 1 where $\ell > 1$ is relatively prime to d.

Let Z_0 (resp. Z_0^+) be the subalgebra of $\mathcal{U}_\varepsilon$ (resp. $\mathcal{B}_\varepsilon$) generated by the elements E_α^ℓ with $\alpha \in R$ (resp. $\alpha \in R^+$) and K_β with $\beta \in P$. (We assume fixed a reduced expression of w_0; Z_0 and Z_0^+ are independent of this choice [DK1].) Recall that they are central subalgebras (Proposition 3.1).

It was shown in [DK1] that Z_0 and Z_0^+ are Hopf subalgebras, hence Spec Z_0 and Spec Z_0^+ have a canonical structure of an affine algebraic group. Furthermore, since $\mathcal{U}_\varepsilon$ is a specialization of the algebra $\mathcal{U}$ at $q = \varepsilon$, the center Z_ε of $\mathcal{U}_\varepsilon$ possesses a canonical Poisson bracket given by the formula:

$$\{a, b\} = \frac{[\hat{a}, \hat{b}]}{2\ell^2(q-\varepsilon)} \bmod (q-\varepsilon), \quad a, b \in Z_\varepsilon,$$

where $\hat{a}$ denotes the preimage of a under the canonical homomorphism $\mathcal{U} \to \mathcal{U}_\varepsilon$. The algebras Z_0 and Z_0^+ are Poisson subalgebras of Z_ε, thus Spec Z_0 and Spec Z_0^+ have a canonical structure of Poisson algebraic groups, Spec Z_0^+ being a quotient Poisson group of Spec Z_0.

In [DKP1] an explicit isomorphism was constructed between the Poisson group Spec Z_0 and a Poisson group H which is described below. We shall identify these Poisson groups.

Let G be the connected simply connected algebraic group associated to the Cartan matrix (a_{ij}) and let g be its (complex) Lie algebra. We fix the triangular decomposition $g = u_- + t + u_+$, let $b_\pm = t + u_\pm$, and denote by $(.|.)$ the invariant bilinear form on g which on the set of roots $R \subset t^*$ coincides with that defined in Section 2.1. Let $U_\pm$, $B_\pm$ and T be the algebraic subgroups of G corresponding to Lie algebras $u_\pm$, $b_\pm$ and t. Then as an algebraic group, H is the following subgroup of $G \times G$:

$$H = \{(tu_+, t^{-1}u_-) | t \in T,\ u_\pm \in \mathcal{U}_\pm\}.$$

The Poisson structure on H is given by the Manin triple $(g \oplus g,\ \hbar,\ k)$, where

$$\hbar = \{(t + u_+,\ -t + u_-) | t \in t, u_\pm \in u_\pm\},$$
$$k = \{(g, g) | g \in g\},$$

and the invariant bilinear form on $g \oplus g$ is

$$((x_1, x_2)|(y_1, y_2)) = -(x_1|y_1) + (x_2|y_2).$$

We identify the group $B_+ = H/\{(1, u_-)|u_- \in U_-\}$. The Manin triple generating Poisson structure on B_+ is obtained from $(g \oplus g, \hbar, k)$ by taking the ideal $u = \{(0, u_-), u_- \in u_-\}$ and applying the construction given by Lemma 4.2. We clearly obtain the triple $(g \oplus t, b_+, b_-)$, where we used identifications

$$b_\pm = \{(u_\pm - t, \pm t) | u_\pm \in u_\pm, t \in t\}.$$

According to the general recipe of Proposition 4.2, the symplectic leaves of the Poisson group B_+ are obtained as follows. We identify the groups $B_\pm$ with the following subgroups of $G \times T$:

$$B_\pm = \{(t^{-1}u_\pm, t^{\pm 1}) | t \in T, u_\pm \in U_\pm\}.$$

The inclusion $B_+ \subset G \times T$ induces an étale morphism

$$\delta :\ B_+ \to (G \times T)/B_-.$$

Then the symplectic leaves of B_+ are the connected components of the preimages under the map δ of B_-–orbits on $G\times T/B_-$ under the left multiplication.

In order to analyze the B_-–orbits on $G\times T/B_-$, let $\mu_\pm : B_\pm \to T$ denote the canonical homomorphisms with kernels $U_\pm$ and consider the equivariant isomorphism of B_-–varieties $\gamma : G/U_- \to (G\times T)/B_-$ given by $\gamma(g\mathcal{U}_-) = (g,1)B_-$, where B_- acts on G/U_- by

$$b(gU_-) = bg\mu_-(b)U_-. \tag{4.3.1}$$

Then the map δ gets identified with the map $\delta : B_+ \to G/U_-$ given by

$$\delta(b) = b\mu_+(b)U_-.$$

We want to study the orbits of the action (4.3.1) of B_- on G/U_-. Consider the action of B_- on G/B_- by left multiplication. Then the canonical map $\pi : G/U_- \to G/B_-$ is B_-–equivariant, hence π maps every B_-–orbit $\mathcal{O}$ in G/U_- to a B_-–orbit in G/B_-, i.e. a Schubert cell $C_w = B_-wB_-/B_-$ for some $w\in W$. We shall say that the orbit $\mathcal{O}$ is associated to w.

Remark We have a sequence of maps:

$$B_+ \xrightarrow{\delta} (G\times T)/B_- \xrightarrow{\gamma^{-1}} G/U_- \xrightarrow{\pi} G/B_+.$$

Let $\psi = \pi\circ\gamma^{-1}\circ\delta$ and $X_w = B_+\cap B_-wB_-$. Then:

$$\pi^{-1}(C_w) = B_-wB_-/U_- \text{ and } \psi^{-1}(C_w) = X_w.$$

We can prove now the following

Proposition. *Let $\mathcal{O}$ be a B_-–orbit in G/U_- under the action (4.3.1) associated to $w\in W$. Then the morphism:*

$$\pi|_{\mathcal{O}} : \mathcal{O}\to C_w$$

is a principal torus bundle with structure group:

$$T^w := \{w^{-1}(t)t^{-1}, \text{ where } t\in T\}.$$

In particular:

$$\dim\mathcal{O} = \dim C_w + \dim T^w = l(w) + \operatorname{rank}(I-w).$$

Proof. For $g \in G$ we shall write $[g]$ for the coset gU_-. The mor phism π is clearly a principal T–bundle with T acting on the right by $[g]t := [gt]$. The action (4.3.1) of B_- commutes with the right T–action so that T permutes the B_-–orbits. Each B_-–orbit is a principal bundle whose structure group is the subtorus of T which stabilizes the orbit. This subtorus is independent of the orbit since T is commutative. In order to compute it we proceed as follows. Let $[g_1], [g_2]$ be two elements in $\mathcal{O}$ mapping to $w \in C_w$. We may assume that $g_1 = nh$, $g_2 = nk$ with $h, k \in T$ uniquely determined, where $n \in N_G(T)$ is representative of w. Suppose that $b[nh] = b[nk]$, $b \in B_-$. We can first reduce to the case $b = t \in T$; indeed, writing $b = ut$ we see that u must fix $w \in C_w$ hence $un = nu'$ with $u' \in U_-$ and hence u acts trivially on $t[nh]$. Next we have that, by definition of the T–action (4.3.1),

$$[nk] = [tnht^{-1}] = [n(n^{-1}tnht^{-1})]$$

hence $k = n^{-1}tnht^{-1}$ or $k = h(h^{-1}n^{-1}tnht^{-1}) = h(n^{-1}tnt^{-1})$ as required. □

Lemma. *Let $\mathcal{O} \subset B_+$ be a symplectic leaf associated to w. Then $\mathcal{O}T = X_w$.*

Proof. From our proof we know that the map δ is a principal T–bundle and T permutes transitively the leaves lying over C_w. □

We thus have a canonical stratification of B_+, indexed by the Weyl group, by the subsets X_w. Each such subset is a union of leaves permuted transitively by the right multiplications of the group T.

We say that a point $a \in \text{Spec} Z_0^+ = B_+$ lies over w if $\psi(a) \in C_w$.

4.4 Recall that $T = \mathbb{Z}^\times \otimes_{\mathbb{Z}} Q^\vee$ and therefore any $\lambda \in P = \text{Hom}_{\mathbb{Z}}(Q^\vee, \mathbb{Z}^\times)$ defines a homomorphism (again denoted by) $\lambda : T \to \mathbb{Z}^\times$ For each $t \in T$ we define an automorphism β_t of the algebra $\mathcal{B}_\varepsilon$ by:

$$\beta_t(K_\alpha) = \alpha(t)K_\alpha, \; \beta_t(E_\alpha) = \alpha(t)E_\alpha.$$

Note that the automorphisms β_t leave Z_0^+ invariant and permute transitively the leaves of each set $\psi^{-1}(C_w) \subset B_+$.

Given $a \in B_+ = \text{Spec } Z_0^+$, denote by m_a the corresponding maximal ideal of Z_0^+ and let

$$A_a = \mathcal{B}_\varepsilon / m_a \mathcal{B}_\varepsilon.$$

These are finite–dimensional algebras and we may also consider these algebras as algebras with trace in order to use the techniques of [DKP2].

Theorem. *If $a, b \in \operatorname{Spec} Z_0^+$ lie over the same element $w \in W$, then the algebras A_a and A_b are isomorphic (as algebras with trace).*

Proof. We just apply Proposition 4.1 to the vector bundle of algebras A_a over a symplectic leaf and the group T of algebra automorphisms which permutes the leaves in $\psi^{-1}(C_w)$ transitively. □

4.5 Let $B^w := B_+ \cap wB_-w^{-1}$ and $U^w := U_+ \cap wU_-w^{-1}$ so that $B^w = U^wT$. Set also $U_w := U_+ \cap wU_+w^{-1}$. One knows that $\dim B^w = n + l(w)$ and that the multiplication map:

$$\sigma : U_w \times B^w \to B_+$$

is an isomorphism of algebraic varieties. We define the map

$$p_w : B_+ \to B^w$$

to be the inverse of σ followed by the projection on the second factor.

Proposition. *The map*

$$p_w|_{X_w} : X_w \to B^w$$

is birational.

Proof. We need to exhibit a Zariski open set $\Omega \subset B^w$ such that for any $b \in \Omega$ there is a unique $u \in U_w$ with $ub \in X_w$.

Let $n \in N_G(T)$ be as above a representative for w so that:

$$X_w = \{b \in B_+ | b = b_1 n b_2, \text{ where } b_1, b_2 \in B_-\}.$$

Consider the Bruhat cell $B_-n^{-1}B_- \subset G$. Every element in $B_-n^{-1}B_-$ can be written uniquely in the form:

$$bn^{-1}u, \text{ where } b \in n^{-1}B^w n,\ u \in U_-.$$

The set $B_+U_- = B_+B_-$ is open dense and so it intersects $B_-n^{-1}B_- = n^{-1}B^wU_-$ in a non–empty open set which is clearly B_-–stable for the right multiplication, hence
$B_+B_- \cap B_-n^{-1}B_- = n^{-1}\Omega U_-$ for some non empty open set $\Omega \subset B^w$. In particular $\Omega \subset nB_+B_- = nU_wB^wU_-$. Take $b \in \Omega$ and write it as $b = nxcv$ with $x \in U_w$, $c \in B^w$, $v \in U_-$. By the remarks made above this decomposition is unique; furthermore, $nxn^{-1} \in U_w$, $ncn^{-1} \in B_-$. For the

element $n^{-1}b = xcv$ we have by construction that $xcv \in B_-n^{-1}B_-$ and $nx^{-1}n^{-1}b = (ncn^{-1})nv \in B_-nB_-$ and $nx^{-1}n^{-1} \in U_w$. Thus setting $u := nx^{-1}n^{-1}$ we have found $u \in U_w$ such that $ub \in X_w$. This u is unique since the element x is unique. □

We are ready now for the concluding theorem which is in the spirit of the conjecture formulated in [DKP1].

Theorem. *Let $p \in X_w$ be a point over $w \in W$ and let A_p be the corresponding algebra. Assume that l is a good integer. Then the dimension of each irreducible representation of A_p is divisible by $= l^{\frac{1}{2}(l(w)+\mathrm{rank}(1-w))}$.*

Proof. Consider the algebra $\mathcal{B}_\varepsilon^w$ for which we know by Theorem 3.5 that

$$\deg \mathcal{B}^w = l^{\frac{1}{2}(l(w)+\mathrm{rank}(1-w))}.$$

The subalgebra $Z_{0,w}$ of Z_0 generated by the elements K_λ^l and E_α^l, where $\lambda \in P$ and $\alpha \in R^+$ is such that $-w^{-1}\alpha \in R^+$, is isomorphic to the coordinate ring of B^w, and $\mathcal{B}_\varepsilon^w$ is a finite free module over $Z_{0,w}$. Thus by [DKP2] there is a non empty open set $\mathcal{A}$ of B^w such that for $p \in \mathcal{A}$ any irreducible representation of $\mathcal{B}^w$ lying over p is of maximal dimension, equal to the degree of $\mathcal{B}_\varepsilon^w$. Now the ideal I defining X_w has intersection 0 with $Z_{0,w}$ and so when we restrict a generic representation of $\mathcal{B}_\varepsilon$ lying over points of X_w to the algebra $\mathcal{B}_\varepsilon^w$ we have, as a central character of $Z_{0,w}$, a point in $\mathcal{A}$. Thus the irreducible representation restricted to $\mathcal{B}^w$ has all its composition factors irreducible of dimension equal to $\deg \mathcal{B}^w$. This proves the claim. □

It is possible that the dimension of any irreducible representation of $\mathcal{B}_\varepsilon$ whose central character restricted to Z_0^+ is a point of X_w is exactly $\ell^{\frac{1}{2}(\ell(w)+\mathrm{rank}(1-w))}$. This fact if true would require a more detailed analysis in the spirit of Section 1.3.

We would like, in conclusion, to propose a more general conjecture, similar to one of the results of [WK] on solvable Lie algebras of characteristic p.

Let A be an algebra over $\mathbb{C}[q, q^{-1}]$ on generators $x_1, \ldots, x_n$ satisfying the following relations:

$$x_i x_j = q^{h_{ij}} x_j x_i + P_{ij} \text{ if } i > j,$$

where (h_{ij}) is a skew–symmetric matrix over $\mathbb{Z}$ and $P_{ij} \in \mathbb{C}[q, q^{-1}][x_1, \ldots, x_{i-1}]$. Let $\ell > 1$ be an integer relatively prime to all elementary divisors of the matrix (h_{ij}) and let ε be a primitive ℓ–th root of 1. Let $A_\varepsilon = A/(q - \varepsilon)$ and assume that all elements x_i^ℓ are central. Let $Z_0 = \mathbb{C}[x_1^\ell, \ldots, x_n^\ell]$; this algebra has a canonical Poisson structure.

Conjecture. Let π be an irreducible representation of the algebra A_ε and let $\mathcal{O}_\pi \subset \text{Spec } Z_0$ be the symplectic leaf containing the restriction of the central character of π to Z_0. Then the dimension of this representation is equal to $\ell^{\frac{1}{2}\dim \mathcal{O}_\pi}$.

This conjecture of course holds if all P_{ij} are 0, and it is in complete agreement with Theorems 1.6, 3.5 and 4.6.

References

[C] R.W. Carter, *Conjugacy classes in the Weyl group,* Compositio Math. **25** (1972) 1–59.

[D] V.G. Drinfeld, *Quantum groups, Proc.* ICM Berkeley **1** (1986) 789–820.

[DK1] C. De Concini, V.G. Kac, *Representations of quantum groups at roots of* 1, in Progress in Math. **92**, Birkhäuser, (1990) 471–506.

[DK2] C. De Concini, V.G. Kac, *Representations of quantum groups at roots of* 1*: reduction to the exceptional case*, in Infinite Analysis, World Scientific, (1992) 141-150.

[DKP1] C. De Concini, V.G. Kac, C. Procesi, *Quantum coadjoint action,* Journal of AMS, **5** (1992) 151–190.

[DKP2] C. De Concini, V.G. Kac, C. Procesi, *Some remarkable degenerations of quantum groups,* preprint, (1991).

[GL] K.R. Goodearl, E.S. Letzter, *Prime ideals in skew and q–skew polynomial rings,* preprint, (1991).

[KP] V.G. Kac, D.H. Peterson, *Generalized invariants of groups generated by reflections*, in Progress in Math. **60**, Birkhäuser, (1985) 231–250.

[L] G. Lusztig, *Quantum groups at roots of* 1, Geom. Ded. **35** (1990) 89–114.

[LS] S.Z. Levendorskii, Ya. S. Soibelman,*Algebras of functions on compact quantum groups, Schubert cells and quantum tori*, Comm. Math. Physics **139** (1991) 141–170.

[LW] J.–H. Lu, A Weinstein, *Poisson Lie groups, dressing transformations and Bruhat decompositions*, J. Diff. Geom. **31** (1990) 501–526.

[STS] M.A. Semenov–Tian–Shansky, *Dressing transformations and Poisson group actions*, Publ. RIMS **21** (1985) 1237–1260.

[WK] B. Yu, Weisfeiler, V.G. Kac, *On irreducible representations of Lie p–algebras*, Funct. Anal. Appl. **5:2** (1971) 28–36.

Classe di Scienze
Scuda Normale Superiore
56100 Pisa
Piazza dei Cavalieri
Italy

Department di Mathematics
Instituto Guide Gastelnuovo
Universita di Ronala Sapienza
Piazzale Aldo Moro, 5
1-00185 Rome, Italy

Compact complex manifolds whose tangent bundles satisfy numerical effectivity properties

Jean-Pierre Demailly

(joint work with Thomas Peternell and Michael Schneider)

Dedicated to M.S. Narasimhan and C.S. Seshadri on their sixtieth birthdays

0 Introduction

A compact Riemann surface always has a hermitian metric with constant curvature, in particular the curvature sign can be taken to be constant: the negative sign corresponds to curves of general type (genus ≥ 2), while the case of zero curvature corresponds to elliptic curves (genus 1), positive curvature being obtained only for $\mathbb{P}^1$ (genus 0). In higher dimensions the situation is much more subtle and it has been a long standing conjecture due to Frankel to characterize $\mathbb{P}_n$ as the only compact Kähler manifold with positive holomorphic bisectional curvature. Hartshorne strengthened Frankel's conjecture and asserted that $\mathbb{P}_n$ is the only compact complex manifold with ample tangent bundle. In his famous paper [Mo79], Mori solved Hartshorne's conjecture by using characteristic p methods. Around the same time Siu and Yau [SY80] gave an analytic proof of the Frankel conjecture. Combining algebraic and analytic tools Mok [Mk88] classified all compact Kähler manifolds with semi-positive holomorphic bisectional curvature.

From the point of view of algebraic geometry, it is natural to consider the class of projective manifolds X whose tangent bundle is numerically effective (nef). This has been done by Campana and Peternell [CP91] and – in case of dimension 3 – by Zheng [Zh90]. In particular, a complete classification is obtained for dimension at most three.

The main purpose of this work is to investigate compact (most often Kähler) manifolds with nef tangent or anticanonical bundles in arbitrary dimension. We first discuss some basic properties of nef vector bundles which will be needed in the sequel in the general context of compact complex manifolds. We refer to [DPS91] and [DPS92] for detailed proofs. Instead, we put here the emphasis on some unsolved questions.

1 Numerically effective vector bundles

In algebraic geometry a powerful and flexible notion of semi-positivity is *numerical effectivity* ("nefness"). We will explain here how to extend this notion to arbitrary compact complex manifolds.

1.1 Definition. A line bundle L on a projective manifold X is said to be *numerically effective* (nef for short) if $L \cdot C \geq 0$ for all compact curves $C \subset X$.

It is clear that a line bundle with semi-positive curvature is nef. The converse had been conjectured by Fujita [Fu83]. Unfortunately this is not true; a simple counterexample can be obtained as follows:

1.2 Example. Let Γ be an elliptic curve and let E be a rank 2 vector bundle over Γ which is a non-split extension of $\mathcal{O}$ by $\mathcal{O}$; such a bundle E can be described as the locally constant vector bundle over Γ whose monodromy is given by the matrices

$$\begin{pmatrix} 1 & 0 \\ 0 & 1 \end{pmatrix}, \quad \begin{pmatrix} 1 & 1 \\ 0 & 1 \end{pmatrix}$$

associated to a pair of generators of $\pi_1(\Gamma)$. We take $L = \mathcal{O}_E(1)$ over the ruled surface $X = \mathbb{P}(E)$. Then L is nef and it can be checked that, up to a positive constant factor, there is only one (possibly singular) hermitian metric on L with semi-positive curvature; this metric is unfortunately singular and has logarithmic poles along a curve. Thus L cannot be semi-positive for any smooth hermitian metric. □

1.3 Definition. A vector bundle E is called *nef* if the line bundle $\mathcal{O}_E(1)$ is nef on $\mathbb{P}(E)$ (= projectivized bundle of hyperplanes in the fibres of E).

Again it is clear that a vector bundle E which admits a metric with semi-positive curvature (in the sense of Griffiths) is nef. A compact Kähler manifold X having semi-positive holomorphic bisectional curvature has by definition a tangent bundle TX with semi-positive curvature. Again the converse does not hold. One difficulty in carrying over the algebraic definition of nefness to the Kahler case is the possible lack of curves. This is overcome by the following:

1.4 Definition. Let X be a compact complex manifold with a fixed hermitian metric ω. A line bundle L over X is *nef* if for every $\varepsilon > 0$ there exists a smooth hermitian metric h_ε on L such that the curvature satisfies

$$\Theta_{h_\varepsilon} \geq -\varepsilon\,\omega\,.$$

This means that the curvature of L can have an arbitrarily small negative part. Clearly a nef line bundle L satisfies $L \cdot C \geq 0$ for all curves $C \subset X$, but the converse is not true (X may have no curves at all, as is the case for instance for generic complex tori). For projective algebraic X both notions coincide; this is an easy consequence of Seshadri's ampleness criterion: take L to be a nef line bundle in the sense of Definition 1.1 and let A be an ample line bundle; then $L^{\otimes k} \otimes A$ is ample for every integer k and thus L has a smooth hermitian metric with curvature form $\Theta(L) \geq -\frac{1}{k}\Theta(A)$.

Definition 1.3 can still be used to define the notion of nef vector bundles over arbitrary compact manifolds. If (E,h) is a hermitian vector bundle, recall that the Chern curvature tensor

$$\Theta_h(E) = \frac{i}{2\pi}D^2_{E,h} = i\sum_{\substack{1\leq j,k\leq n\\ 1\leq\lambda,\mu\leq r}} a_{jk\lambda\mu}dz_j \wedge d\overline{z}_k \otimes e^\star_\lambda \otimes e_\mu$$

is a hermitian (1,1)-form with values in $\mathrm{Hom}(E,E)$. We say that (E,h) is *semi-positive in Griffiths' sense* [Gr69] and write $\Theta_h(E) \geq 0$ if $\Theta_h(E)(\xi\otimes t) = \sum a_{jk\lambda\mu}\xi_j\overline{\xi}_k v_\lambda \overline{v}_\mu \geq 0$ for every $\xi \in T_xX$, $v \in E_x$, $x \in X$. We write $\Theta_h(E) > 0$ in case there is strict inequality for $\xi \neq 0, v \neq 0$. Numerical effectivity can then be characterized by the following differential geometric criterion (see [De91]).

1.5 Criterion. Let ω be a fixed hermitian metric on X. A vector bundle E on X is nef if and only if there is a sequence of hermitian metrics h_m on S^mE and a sequence ε_m of positive numbers decreasing to 0 such that

$$\Theta_{h_m}(S^mE) \geq -m\varepsilon_m\omega \otimes \mathrm{Id}_{S^mE}$$

in the sense of Griffiths.

The main functorial properties of nef vector bundles are summarized in the following proposition.

1.6 Proposition. *Let X be an arbitrary compact complex manifold and let E be a holomorphic vector bundle over X.*

(i) [1] *If* $f : Y \to X$ *is a holomorphic map with equidimensional fibres, then* E *is nef if and only if* $f^\star E$ *is nef.*

(ii) *Let* $\Gamma^a E$ *be the irreducible tensor representation of* $Gl(E)$ *of highest weight* $a = (a_1, \ldots, a_r) \in \mathbb{Z}^r$, *with* $a_1 \geq \ldots \geq a_r \geq 0$. *Then* $\Gamma^a E$ *is nef. In particular, all symmetric and exterior powers of* E *are nef.*

(iii) *let* F *be a holomorphic vector bundle over* X. *If* E *and* F *are nef, then* $E \otimes F$ *is nef.*

(iv) *If some symmetric power* $S^m E$ *is nef* $(m > 0)$, *then* E *is nef.*

(v) *Let* $0 \to F \to E \to Q \to 0$ *be an exact sequence of holomorphic vector bundles over* X. *Then*

(α) E *nef* $\Rightarrow$ Q *nef.*

(β) F, Q *nef* $\Rightarrow$ E *nef.*

(γ) E *nef,* $(\det Q)^{-1}$ *nef* $\Rightarrow$ F *nef.*

The proof of these properties in the general analytic context can be easily obtained by curvature computations. The arguments are parallel to those of the algebraic case and will therefore be omitted (see [Ha66] and [CP91] for that case). Another useful result which will be used over and over in the sequel is

1.7 Proposition. *Let* E *be a nef vector bundle over a connected compact n-fold* X *and let* $\sigma \in H^0(X, E^\star)$ *be a non zero section. Then* σ *does not vanish anywhere.*

Proof. We merely observe that if h_m is a sequence of hermitian metrics on $S^m E$ as in criterion 5, then

$$T_m = \frac{i}{\pi}\partial\overline{\partial}\frac{1}{m}\log\|\sigma^m\|_{h_m}$$

has zero $\partial\overline{\partial}$-cohomology class and satisfies $T_m \geq -\varepsilon_m\omega$. It follows that T_m converges to a weak limit $T \geq 0$ with zero cohomology class. Thus $T = i\partial\overline{\partial}\varphi$ for some global plurisubharmonic function φ on X. By the maximum principle this implies $T = 0$. However, if σ vanishes at some point x, then all T_m have Lelong number ≥ 1 at x. Therefore so has T, contradiction. □

One of our key results is a characterization of vector bundles E which are numerically flat, i.e. such that both E and $E^\star$ are nef.

[1] We expect (i) to hold whenever f is surjective, but there are serious technical difficulties to overcome in the nonalgebraic case.

1.8 Theorem. *Suppose that X is Kähler. Then a holomorphic vector bundle E over X is numerically flat iff E admits a filtration*

$$\{0\} = E_0 \subset E_1 \subset \ldots \subset E_p = E$$

by vector subbundles such that the quotients E_k/E_{k-1} are hermitian flat, i.e. given by unitary representations $\pi_1(X) \to U(r_k)$.

Sketch of Proof. It is clear by 1.6 (v) that every vector bundle which is filtrated with hermitian flat quotients is nef as well as its dual. Conversely, suppose that E is numerically flat. This assumption implies $c_1(E) = 0$. Fix a Kähler metric ω. If E is ω-stable, then E is Hermite-Einstein by the Uhlenbeck-Yau theorem [UY86]. Moreover we have $0 \le c_2(E) \le c_1(E)^2$ by Theorem 1.9 below, so $c_2(E) = 0$. Kobayashi's flatness criterion derived from Lübke's inequality on Chern classes then implies that E is hermitian flat. Now suppose that E is unstable and take $\mathcal{F} \subset \mathcal{O}(E)$ to be a destabilizing subsheaf of minimal rank p. We then have by definition $c_1(\mathcal{F}) = c_1(\det \mathcal{F}) = 0$ and the morphism $\det \mathcal{F} \to \Lambda^p E$ cannot have any zero (otherwise the zeroes would contribute to give a semi-negative non zero curvature current on the line bundle $\det \mathcal{F}$, contradiction). This implies easily that $\mathcal{F}$ is locally free, and we infer that $\mathcal{F}$ is also numerically flat. Since $\mathcal{F}$ is stable by definition, $\mathcal{F}$ must be hermitian flat. We set $E_1 = \mathcal{F}$, observe that $E' = E/E_1$ is again numerically flat and proceed by induction on the rank. □

Another key point, which has been indeed used in the above proof, is the fact that the Fulton-Lazarsfeld inequalities [FL83] for Chern classes of ample vector bundles still hold for nef vector bundles over compact Kähler manifolds:

1.9 Theorem. *Let (X, ω) be a compact Kähler manifold and let E be a nef vector bundle on X. Then for all positive polynomials P the cohomology class $P(c(E))$ is numerically positive, that is, $\int_Y P(c(E)) \wedge \omega^k \ge 0$ for any k and any subvariety Y of X.*

By a positive polynomial in the Chern classes, we mean as usual a homogeneous weighted polynomial $P(c_1, \ldots, c_r)$ with $\deg c_i = 2i$, such that P is a positive integral combination of Schur polynomials:

$$P_a(c) = \det(c_{a_i - i + j})_{1 \le i,j \le r}, \qquad r \ge a_1 \ge a_2 \ge \ldots \ge a_r \ge 0$$

(by convention $c_0 = 1$ and $c_i = 0$ if $i \ne [0, r]$, $r = \operatorname{rank} E$). The proof of Theorem 1.9 is based essentially on the same arguments as the original proof of [FL83] for the ample case: the starting point is the nonnegativity of all Chern classes $c_k(E)$ (Bloch-Gieseker [BG71]); the general case then follows from a formula of Schubert calculus known as the Kempf-Laksov formula [KL74], which expresses any Schur polynomial $P_a(c(E))$ as a Chern class $c_k(F_a)$ of some related vector bundle F_a. The only change occurs in the proof of Gieseker's result, where the Hard Lefschetz theorem is needed for arbitrary Kähler metrics instead of hyperplane sections (fortunately enough, the technique then gets simplified, covering tricks being eliminated). Since $c_1 c_{k-1} - c_k$ is a Schur polynomial, we get by induction

$$0 \leq c_k(E) \leq c_1(E)^k \quad \text{for all } k.$$

Therefore all Chern monomials are bounded above by corresponding powers $c_1(E)^k$ of the same degree, and we infer:

1.10 Corollary. *If E is nef and $c_1(E)^n = 0$, $n = \dim X$, then all Chern polynomials $P(c(E))$ of degree $2n$ vanish.*

2 Compact Kähler manifolds with nef anti-canonical line bundle

Compact Kähler manifolds with zero or semi-positive Ricci curvature have been investigated by various authors (cf. [Ca57], [Ko61], [Li67, 71, 72], [Bo74a, b], [Be83], [Ko81] and [Kr86]). The purpose of this section is to discuss the following two conjectures.

2.1 Conjecture. Let X be a compact Kähler manifold with numerically effective anticanonical bundle K_X^{-1}. Then the fundamental group $\pi_1(X)$ has polynomial growth.

2.2 Conjecture. Let X be a compact Kähler manifold with K_X^{-1} numerically effective. Then the Albanese map $\alpha : X \to \mathrm{Alb}(X)$ is a smooth fibration onto the Albanese torus. If this holds, one can infer that there is a finite étale cover $\widetilde{X}$ of X such that the Albanese map of $\widetilde{X}$ has simply connected fibres. In particular, $\pi_i(X)$ would almost abelian (namely an extension of a finite group by a free abelian group).

These conjectures are known to be true if K_X^{-1} is semi-positive. In both cases, the proof is based on differential geometric techniques (see e.g. [Bi63], [HK78] for Conjecture 2.1 and [Li71] for Conjecture 2.2). However, the methods of proof are not so easy to carry over to the nef case. Our main contributions to these conjectures are derived from Theorem 2.3 below.

2.3 Theorem. *Let X be a compact Kähler manifold with K_X^{-1} nef. Then $\pi_1(X)$ is a group of subexponential growth.*

The proof actually gives the following additional fact (this was already known before, see [Bi63]).

2.4 Corollary. *If moreover $-K_X$ is hermitian semi-positive, then $\pi_1(X)$ has polynomial growth of degree $\leq 2\dim X$, in particular $h^1(X, \mathcal{O}_X) \leq \dim X$.*

As noticed by F. Campana (private communication), Theorem 2.3 also implies the following consequences.

2.5 Corollary. *Let X be a compact Kähler manifold with K_X^{-1} nef. Let $\alpha : X \to \mathrm{Alb}(X)$ be the Albanese map and set $n = \dim X$, $d = \dim \alpha(X)$. If $d = 0, 1$ or n, then α is surjective. The same is true if $d = n - 1$ and if X is projective algebraic.*

2.6 Corollary. *Let X be a Kähler surface or a projective 3-fold with K_X^{-1} nef. Then the Albanese map $\alpha : X \to \mathrm{Alb}(X)$ is surjective.*

We now explain the main ideas required in the proof of Theorem 2.3. If G is a finitely generated group with generators $g_1, \ldots, g_p$, we denote by $N(k)$ the number of elements $\gamma \in G$ which can be written as words

$$\gamma = g_{i_1}^{\varepsilon_1} \cdots g_{i_k}^{\varepsilon_k}, \qquad \varepsilon_j = 0,\ 1 \text{ or } -1$$

of length $\leq k$ in terms of the generators. The group G is said to have *subexponential growth* if for every $\varepsilon > 0$ there is a constant $C(\varepsilon)$ such that

$$N(k) \leq C(\varepsilon) e^{\varepsilon k} \quad \text{for } k \geq 0.$$

This notion is independent of the choice of generators. In the free group with two generators, we have $N(k) = 1 + 4(1 + 3 + 3^2 + \ldots + 3^{k-1}) = 2 \cdot 3^k - 1$, thus a group with subexponential growth cannot contain a non abelian free subgroup.

The first step consists in the construction of suitable Kähler metrics by means of the Calabi-Yau theorem [Y77]. Let ω be a fixed Kähler metric on X. Since K_X^{-1} is nef, for every $\varepsilon > 0$ there exists a smooth hermitian metric h_ε on K_X^{-1} such that

$$u_\varepsilon = \Theta_{h_\varepsilon}(K_X^{-1}) \geq -\varepsilon\omega.$$

By [Y77] and [Y78] there exists a unique Kähler metric ω_ε in the cohomology class $\{\omega\}$ such that

$$(+) \qquad \mathrm{Ricci}(\omega_\varepsilon) = -\varepsilon\omega_\varepsilon + \varepsilon\omega + u_\varepsilon.$$

In fact u_ε belongs to the Ricci class $c_1(K_X^{-1}) = c_1(X)$, hence so does the right hand side $-\varepsilon\omega_\varepsilon + \varepsilon\omega + u_\varepsilon$. In particular there exists a function f_ε such that

$$u_\varepsilon = \mathrm{Ricci}(\omega) + i\partial\overline{\partial} f_\varepsilon.$$

If we set $\omega_\varepsilon = \omega + i\partial\overline{\partial}\varphi$ (where φ depends on ε), equation (+) is equivalent to the Monge-Ampère equation

$$(++) \qquad \frac{(\omega + i\partial\overline{\partial}\varphi)^n}{\omega^n} = e^{\varepsilon\varphi - f_\varepsilon}$$

because

$$\begin{aligned} i\partial\overline{\partial}\log(\omega + i\partial\overline{\partial}\varphi)^n/\omega^n &= \mathrm{Ricci}(\omega) - \mathrm{Ricci}(\omega_\varepsilon) \\ &= \varepsilon(\omega_\varepsilon - \omega) + \mathrm{Ricci}(\omega) - u_\varepsilon \\ &= i\partial\overline{\partial}(\varepsilon\varphi - f_\varepsilon). \end{aligned}$$

It follows from the general results of [Y78] that (++) has a unique solution φ, thanks to the fact the right hand side of (++) is increasing in φ. Since $u_\varepsilon \geq -\varepsilon\omega$, equation (+) implies in particular that $\mathrm{Ricci}(\omega_\varepsilon) \geq -\varepsilon\omega$.

Now, recall the well-known differential geometric technique for bounding $N(k)$ (this technique has been explained to us in a very efficient way by S. Gallot). Let (M, g) be a compact Riemannian m-fold and let $E \subset \widetilde{M}$ be a fundamental domain for the action of $\pi_1(M)$ on the universal covering $\widetilde{M}$. Fix $a \in E$ and set $\beta = \operatorname{diam} E$. Since $\pi_1(M)$ acts isometrically on $\widetilde{M}$ with respect to the pull-back metric $\widetilde{g}$, we infer that

$$E_k = \bigcup_{\gamma\in\pi_1(M),\ \mathrm{length}(\gamma)\leq k} \gamma(E)$$

has volume equal to $N(k)\,\mathrm{Vol}(M)$ and is contained in the geodesic ball $B(a, \alpha k + \beta)$, where α is the maximum of the length of loops representing the generators g_j. Therefore

$$(*) \qquad N(k) \leq \frac{\mathrm{Vol}\big(B(a, \alpha k + \beta)\big)}{\mathrm{Vol}(M)}.$$

and it is enough to bound the volume of geodesic balls in $\widetilde{M}$. For this we use the following fundamental inequality due to R. Bishop [Bi63], Heintze-Karcher [HK78] and M. Gage [Ga80].

2.7 Lemma. *Let*

$$\Phi : T_a\widetilde{M} \to \widetilde{M}, \quad \Phi(\zeta) = \exp_a(\zeta)$$

be the (geodesic) exponential map. Denote by

$$\Phi^* dV_g = a(t,\zeta)\, dt\, d\sigma(\zeta)$$

the expression of the volume element in spherical coordinates with $t \in \mathbb{R}_+$ *and* $\zeta \in S_a(1) =$ *unit sphere in* $T_a\widetilde{M}$. *Suppose that* $a(t,\zeta)$ *does not vanish for* $t \in]0, \tau(\zeta)[$, *with* $\tau(\zeta) = +\infty$ *or* $a(\tau(\zeta),\zeta) = 0$. *Then* $b(t,\zeta) = a(t,\zeta)^{1/(m-1)}$ *satisfies on* $]0, \tau(\zeta)[$ *the inequality*

$$\frac{\partial^2}{\partial t^2} b(t,\zeta) + \frac{1}{m-1} \operatorname{Ricci}_g(v(t,\zeta), v(t,\zeta)) b(t,\zeta) \leq 0$$

where

$$v(t,\zeta) = \frac{d}{dt} \exp_a(t\zeta) \in S_{\Phi(t\zeta)}(1) \subset T_{\Phi(t\zeta)}\widetilde{M}.$$

If $\operatorname{Ricci}_g \geq -\varepsilon g$, we infer in particular

$$\frac{\partial^2 b}{\partial t^2} - \frac{\varepsilon}{m-1} b \leq 0$$

and therefore $b(t,\zeta) \leq \alpha^{-1} \sinh(\alpha t)$ with $\alpha = \sqrt{\varepsilon/(m-1)}$ (to check this, observe that $b(t,\zeta) = t + o(t)$ at 0 and that $\sinh(\alpha t)\, \partial b/\partial t - \alpha \cosh(\alpha t)\, b$ has a negative derivative). Now, every point $x \in B(a,r)$ can be joined to a by a minimal geodesic arc of length $< r$. Such a geodesic arc cannot contain any focal point (i.e. any critical value of Φ), except possibly at the end point x. It follows that $B(a,r)$ is the image by Φ of the open set

$$\Omega(r) = \big\{(t,\zeta) \in [0, r[\times S_a(1)\,;\, t < \tau(\zeta)\big\}.$$

Therefore

$$\operatorname{Vol}_g(B(a,r)) \leq \int_{\Omega(r)} \Phi^* dV_g = \int_{\Omega(r)} b(t,\zeta)^{m-1} dt\, d\sigma(\zeta).$$

As $\alpha^{-1} \sinh(\alpha t) \leq t\, e^{\alpha t}$, we get

$$(**) \qquad \operatorname{Vol}_g(B(a,r)) \leq \int_{S_a(1)} d\sigma(\zeta) \int_0^r t^{m-1} e^{(m-1)\alpha t} dt \leq v_m r^m e^{\sqrt{(m-1)\varepsilon}\, r}$$

where v_m is the volume of the unit ball in $\mathbb{R}^m$. □

In our application, the difficulty is that the metric $g = \omega_\varepsilon$ varies with ε as well as the constants $\alpha = \alpha_\varepsilon$, $\beta = \beta_\varepsilon$ in $(*)$, and $\alpha_\varepsilon \sqrt{(m-1)\varepsilon}$ need not converge to 0 as ε tends to 0. We overcome this difficulty by the following lemma.

2.8 Lemma. *Let U_1, U_2 be compact subsets of $\widetilde{X}$. Then for every $\delta > 0$, there are closed subsets $U_{1,\varepsilon,\delta} \subset U_1$ and $U_{2,\varepsilon,\delta} \subset U_2$ with $\mathrm{Vol}_\omega(U_j U_{j,\varepsilon,\delta}) < \delta$, such that any two points $x_1 \in U_{1,\varepsilon,\delta}$, $x_2 \in U_{2,\varepsilon,\delta}$ can be joined by a path of length $\leq C\,\delta^{-1/2}$ with respect to ω_ε, where C is a constant independent of ε and δ.*

We will not explain the details. The basic observation is that

$$\int_X \omega_\varepsilon \wedge \omega^{n-1} = \int_X \omega^n$$

does not depend on ε, therefore $\|\omega_\varepsilon\|_{L^1(X)}$ is uniformly bounded. This is enough to imply the existence of sufficiently many paths of bounded length between random points taken in X (this is done for example by computing the average length of piecewise linear paths). □

We let U be a compact set containing the fundamental domain E, so large that $U^\circ \cap g_j(U^\circ) \neq \emptyset$ for each generator g_j. We apply Lemma 2.8 with $U_1 = U_2 = U$ and $\delta > 0$ fixed such that

$$\delta < \frac{1}{2}\mathrm{Vol}_\omega(E), \quad \delta < \frac{1}{2}\mathrm{Vol}_\omega\big(U \cap g_j(U)\big).$$

We get $U_{\varepsilon,\delta} \subset U$ with $\mathrm{Vol}_\omega(UU_{\varepsilon,\delta}) < \delta$ and $\mathrm{diam}_{\omega_\varepsilon}(U_{\varepsilon,\delta}) \leq C\delta^{-1/2}$. The inequalities on volumes imply that $\mathrm{Vol}_\omega(U_{\varepsilon,\delta} \cap E) \geq \frac{1}{2}\mathrm{Vol}_\omega(E)$ and $U_{\varepsilon,\delta} \cap g_j(U_{\varepsilon,\delta}) \neq \emptyset$ for every j (note that all g_j preserve volumes). It is then clear that

$$W_{k,\varepsilon,\delta} := \bigcup_{\gamma\in\pi_1(X),\ \mathrm{length}(\gamma)\leq k} \gamma(U_{\varepsilon,\delta})$$

satisfies

$$\begin{aligned}\mathrm{Vol}_\omega(W_{k,\varepsilon,\delta}) &\geq N(k)\,\mathrm{Vol}_\omega(U_{\varepsilon,\delta}\cap E) \geq N(k)\,\tfrac{1}{2}\,\mathrm{Vol}_\omega(E) \quad \text{and}\\ \mathrm{diam}_{\omega_\varepsilon}(W_{k,\varepsilon,\delta}) &\leq k\,\mathrm{diam}_{\omega_\varepsilon} U_{\varepsilon,\delta} \leq kC\delta^{-1/2}.\end{aligned}$$

Since $m = \dim_{\mathbb{R}} X = 2n$, inequality (**) implies

$$\mathrm{Vol}_{\omega_\varepsilon}(W_{k,\varepsilon,\delta}) \leq \mathrm{Vol}_{\omega_\varepsilon}\big(B(a, kC\delta^{-1/2})\big) \leq C_4 k^{2n} e^{C_5\sqrt{\varepsilon}\,k}.$$

Now X is compact, so there is a constant $C(\varepsilon) > 0$ such that $\omega^n \leq C(\varepsilon)\omega_\varepsilon^n$. We conclude that

$$N(k) \leq \frac{2\,\mathrm{Vol}_\omega(W_{k,\varepsilon,\delta})}{\mathrm{Vol}_\omega(E)} \leq C_6 C(\varepsilon) k^{2n} e^{C_5\sqrt{\varepsilon}\,k}.$$

The proof of Theorem 2.3 is complete. □

2.9 Remark. It is well known and easy to check that equation (++) implies

$$C(\varepsilon) \leq \exp\left(\max_X f_\varepsilon - \min_X f_\varepsilon\right).$$

Therefore it is reasonable to expect that $C(\varepsilon)$ has polynomial growth in ε^{-1}; this would imply that $\pi_1(X)$ has polynomial growth by taking $\varepsilon = k^{-2}$. When K_X^{-1} has a semipositive metric, we can even take $\varepsilon = 0$ and find a metric ω_0 with $\mathrm{Ricci}(\omega_0) = u_0 \geq 0$. This implies Corollary 2.4.

Proof of Corollary 2.5. If $d = 0$, then by definition $H^0(X, \Omega_X^1) = 0$ and $\mathrm{Alb}(X) = \{0\}$.

If $d = n$, the Albanese map has generic rank n, so there exist holomorphic 1-forms $u_1, \ldots, u_n$ such that $u_1 \wedge \ldots \wedge u_n \not\equiv 0$. However $u_1 \wedge \ldots \wedge u_n$ is a section of K_X which has nef dual, so $u_1 \wedge \ldots \wedge u_n$ cannot vanish by Proposition 1.7 and K_X is trivial. Therefore $u_1 \wedge \ldots \hat{u}_k \ldots \wedge u_n \wedge v$ must be a constant for every holomorphic 1-form v and $(u_1, \ldots, u_n)$ is a basis of $H^0(X, \Omega_X^1)$. This implies $\dim A(X) = n$, hence α is surjective.

If $d = 1$, the image $C = \alpha(X)$ is a smooth curve. The genus g of C cannot be ≥ 2, otherwise $\pi_1(X)$ would be mapped onto a subgroup of finite index in $\pi_1(C)$, and thus would be of exponential growth, contradicting Theorem 2.3. Therefore C is an elliptic curve and is a subtorus of $\mathrm{Alb}(X)$. By the universal property of the Albanese map, this is possible only if $C = \mathrm{Alb}(X)$.

The case $d = n - 1$ is more subtle and uses Mori theory (this is why we have to assume X projective algebraic). We refer to [DPS92] for the details. □

3 Compact complex manifolds with nef tangent bundles

Several interesting classes of such manifolds are produced by the following simple observation.

3.1 Proposition. *Every homogeneous compact complex manifold has a nef tangent bundle.*

Indeed, if X is homogeneous, the Killing vector fields generate TX, so TX is a quotient of a trivial vector bundle. In particular, we get the following

3.2 Examples (homogeneous case)

(i) Rational homogeneous manifolds: $\mathbb{P}_n$, flag manifolds, quadrics Q_n (all are Fano manifolds, i.e. projective algebraic with K_X^{-1} ample).

(ii) Tori $\mathbb{C}^n/\Lambda$
(Kähler, possibly non algebraic).

(iii) Hopf manifolds $\mathbb{C}^n\{0\}/H$ where H is a discrete group of homotheties (non Kähler for $n \geq 2$).

(iv) Iwasawa manifolds G/Λ where G is the group of unipotent upper triangular $p \times p$ matrices and Λ the subgroup of matrices with entries in the ring of integers of some imaginary quadratic field, e.g. $\mathbb{Z}[i]$ (non Kähler for $p \geq$, although TX is trivial). □

We must remark at this point that not all manifolds X with nef tangent bundles are homogeneous, the automorphism group may even be discrete:

3.3 Example. Let $\Gamma = \mathbb{C}/(\mathbb{Z}+\mathbb{Z}\tau)$, $Im\,\tau > 0$, be an elliptic curve. Consider the quotient space $X = (\Gamma\times\Gamma\times\Gamma)/G$ where $G = \{1, g_1, g_2, g_1g_2\} \simeq \mathbb{Z}_2\times\mathbb{Z}_2$ is given by

$$\begin{aligned} g_1(z_1,z_2,z_3) &= (z_1+\tfrac{1}{2}, -z_2, -z_3),\\ g_2(z_1,z_2,z_3) &= (-z_1, z_2+\tfrac{1}{2}, -z_3+\tfrac{1}{2}),\\ g_1g_2(z_1,z_2,z_3) &= (-z_1+\tfrac{1}{2}, -z_2+\tfrac{1}{2}, z_3+\tfrac{1}{2}).\end{aligned}$$

Then G acts freely, so X is smooth. It is clear also that TX is nef (in fact TX is unitary flat). Since the pull-back of TX to $\Gamma\times\Gamma\times\Gamma$ is trivial, we easily conclude that TX has no sections, thanks to the change of signs in g_1, g_2, g_1g_2. Therefore the automorphism group $\mathrm{Aut}(X)$ is discrete. The same argument shows that $H^0(X, \Omega_X^1) = 0$.

3.4 Example. Let X be the ruled surface $\mathbb{P}(E)$ over the elliptic curve $\Gamma = \mathbb{C}/(\mathbb{Z}+\mathbb{Z}\tau)$ defined in Example 1.2. Then the relative tangent bundle of $\mathbb{P}(E) \to \Gamma$ (=relative anticanonical line bundle) is $\pi^\star(\det E^\star)\otimes\mathcal{O}_E(2) \simeq \mathcal{O}_E(2)$ and $T\Gamma$ is trivial, so TX is nef. Moreover X is almost homogeneous, with automorphisms induced by

$$(x, z_1, z_2) \mapsto (x+a, z_1+b, z_2), \quad (a,b)\in\mathbb{C}^2$$

and a single closed orbit equal to the curve $\{z_2 = 0\}$. Here, no finite étale cover of X can be homogeneous, otherwise $K_X^{-1} = \mathcal{O}_E(2)$ would be semi-positive. Observe that no power of K_X^{-1} is generated by sections, although K_X^{-1} is nef. □

Our main result is a structure theorem on the Albanese map of compact Kähler manifolds with nef tangent bundles.

3.5 Main Theorem. *Let X be a compact Kähler manifold with nef tangent bundle TX. Let $\widetilde{X}$ be a finite étale cover of X of maximum irregularity $q = q(\widetilde{X}) = h^1(\widetilde{X}, \mathcal{O}_{\widetilde{X}})$. Then*

(i) $\pi_1(\widetilde{X}) \simeq \mathbb{Z}^{2q}$.

(ii) *The Albanese map $\alpha : \widetilde{X} \to A(\widetilde{X})$ is a smooth fibration over a q-dimensional torus with nef relative tangent bundle.*

(iii) *The fibres F of α are Fano manifolds with nef tangent bundles.*

Recall that a Fano manifold is by definition a compact complex manifold with ample anticanonical bundle K_X^{-1}. It is well known that Fano manifolds are always simply connected (Kobayashi [Ko61]). As a consequence we get

3.6 Corollary. *With the assumtions of 3.5, the fundamental group $\pi_1(X)$ is an extension of a finite group by $\mathbb{Z}^{2q}$.*

In order to complete the classification of compact Kähler manifolds with nef tangent bundles (up to finite étale cover), a solution of the following two conjectures would be needed.

3.7 Conjecture. (Campana-Peternell [CP91]) Let X be a Fano manifold. Then X has a nef tangent bundle if and only if X is rational homogeneous.

The evidence we have for Conjecture 3.7 is that it is true up to dimension 3. In dimension 3 there are more than 100 different types of Fano manifolds, but only five types have a nef tangent bundle, namely $\mathbb{P}_3$, Q_3 (quadric), $\mathbb{P}_1 \times \mathbb{P}_2$, $\mathbb{P}_1 \times \mathbb{P}_1 \times \mathbb{P}_1$ and the flag manifold $F_{1,2}$ of lines and planes in $\mathbb{C}^3$; all five are homogeneous.

A positive solution to Conjecture 3.7 would clarify the structure of fibers in the Albanese map of Theorem 3.5. To get a complete picture of the situation, one still needs to know how the fibers are deformed and glued together to yield a holomorphic family over the Albanese torus. We note that $K_{\widetilde{X}}^{-1}$ is relatively ample, thus for m large the fibres can be embedded in the projectivized bundle of the direct image bundle $\alpha_*(K_{\widetilde{X}}^{-m}$. The structure of the deformation is described by the following theorem.

3.8. Theorem. *In the situation of Theorem 3.5, all direct image bundles $E_m = \alpha_*(K_{\widetilde{X}}^{-m})$ are numerically flat over the Albanese torus. Moreover, for $p \gg m \gg 0$, the fibers of the Albanese map can be described as Fano submanifolds of the fibers of $IP(E_m)$ defined by polynomial equations of degree p, in such a way that the bundle of equations $V_{m,p} \subset S^p(E_m)$ is itself numerically flat.*

Theorem 3.8 is proved in [DPS91] in case X is projective algebraic. The extension to the Kähler case has been obtained by Ch. Mourougane in his PhD Thesis work (Grenoble, still unpublished). We now explain the main steps in the proof of Theorem 3.5. One of the key points is the following

3.9 Proposition. *Let X be a compact Kähler n-fold with TX nef. Then*

(i) *If $c_1(X)^n > 0$, then X is a Fano manifold.*

(ii) *If $c_1(X)^n = 0$, then $\chi(\mathcal{O}_X) = 0$ and there exists a non zero holomorphic p-form, p suitable odd and a finite étale cover $\widetilde{X} \to X$ such that $q(\widetilde{X}) > 0$.*

Proof. We first check that every effective divisor D of X is nef. In fact, let $\sigma \in H^0(X, \mathcal{O}(D))$ be a section with divisor D. Then for k larger than the maximum vanishing order of σ on X, the k-jet section $J^k\sigma \in H^0(X, J^k\mathcal{O}(D))$ has no zeroes. Therefore, there is an injection $\mathcal{O} \to J^k\mathcal{O}(D)$ and a dual surjection

$$(J^k\mathcal{O}(D))^\star \otimes \mathcal{O}(D) \to \mathcal{O}(D).$$

Now, $J^k\mathcal{O}(D)$ has a filtration whose graded bundle is $\bigoplus_{0\le p\le k} S^pT^\star X \otimes \mathcal{O}(D)$, so $(J^k\mathcal{O}(D))^\star \otimes \mathcal{O}(D)$ has a dual filtration with graded bundle $\bigoplus_{0\le p\le k} S^pTX$. By 1.6 (ii) and 1.6 (v)(β), we conclude that $(J^k\mathcal{O}(D))^\star \otimes \mathcal{O}(D)$ is nef, so its quotient $\mathcal{O}(D)$ is nef by 1.6 (v)(α).

Part (i) is based on the solution of the Grauert-Riemenschneider conjecture as proved in [De85]. Namely, $L = K_X^{-1} = \Lambda^n TX$ is nef and satisfies $c_1(L)^n > 0$, so L has Kodaira dimension n (holomorphic Morse inequalities are needed at that point because X is not supposed a priori to be algebraic). It follows that X is Moishezon, thus projective algebraic, and for $m > 0$ large we have $L^m = \mathcal{O}(D + A)$ with divisors D, A such that D is effective and A ample. Since D must be in fact nef, it follows that $L = K_X^{-1}$ is ample, as desired.

The most difficult part is (ii). Since $c_1(X)^n = 0$, Corollary 1.10 implies $\chi(\mathcal{O}_X) = 0$. By Hodge symmetry, we get $h^0(X, \Omega_X^p) = h^p(X, \mathcal{O}_X)$ and

$$\chi(\mathcal{O}_X) = \sum_{0\le p\le n} (-1)^p h^0(X, \Omega_X^p) = 0.$$

From this and the fact that $h^0(X, \mathcal{O}_X) = 1$, we infer the existence of a non zero p-form u for some suitable odd number p. Let

$$\sigma : \Lambda^{p-1}TX \longrightarrow \Omega_X^1$$

be the bundle morphism obtained by contracting $(p-1)$-vectors with u. For every $k > 0$, the morphism $\Lambda^k\sigma$ can be viewed as a section of the bundle $\Lambda^k(\Lambda^{p-1}T^\star X) \otimes \Lambda^k T^\star X$ which has nef dual. Hence by Proposition 1.7 we know that $\Lambda^k\sigma$ is either identically zero or does not vanish. This means that σ has constant rank. Let E be the image of σ. Then E is a quotient bundle of $\Lambda^{p-1}TX$, so E is nef, and E is a subbundle of $\Omega^1_X = T^\star X$, so $E^\star$ is likewise nef. Theorem 1.8 implies the existence of a hermitian flat subbundle $E_1 \subset E$. If E_1 would be trivial after pulling-back to some finite etale cover $\widetilde{X}$, we would get a trivial subbundle of $\Omega^1_{\widetilde{X}}$, hence $q(\widetilde{X}) > 0$ and the proposition would be proved. Otherwise E_1 is given by some infinite representation of $\pi_1(X)$ into some unitary group. Let Γ be the monodromy group (i.e. the image of $\pi_1(X)$ by the representation). We use a result of Tits [Ti72] according to which every subgroup of a linear group contains either a non abelian free subgroup or a solvable subgroup of finite index. The first case cannot occur by Theorem 2.?. In the second case, we may assume Γ solvable by taking some finite étale cover. We consider the series of derived groups

$$\Gamma \supset \Gamma_1 \supset \ldots \supset \Gamma_N = \{0\}$$

and the largest index k such that Γ_k has finite index in Γ. Then the inverse image of Γ_k in $\pi_1(X)$ defines a finite étale cover $\widetilde{X}$ of X with infinite first homology group (the representation maps this group onto Γ_k/Γ_{k+1} which is infinite). Hence $q(\widetilde{X}) > 0$, as desired. □

Proof of the Main Theorem. Let X be compact Kähler manifold with nef tangent bundle. Since a non zero holomorphic form $u \in H^0(X, \Omega^1_X)$ can never vanish by Proposition 1.7, it follows immediately that the Albanese map α has rank equal to $q(X)$ at every point, hence α is a submersion and $q(X) \leq n$. Let $\widetilde{X}$ be a finite étale cover with maximum irregularity $q = q(\widetilde{X})$ (note that $\widetilde{X}$ also has a nef tangent bundle, so $q(\widetilde{X}) \leq n$). Let F denote the fibers of the Albanese map $\alpha : \widetilde{X} \to A(\widetilde{X})$. The relative tangent bundle exact sequence

$$0 \to TF \to TX \xrightarrow{d\alpha} \alpha^\star TA(X) \to 0$$

in which $TA(X)$ is trivial shows by 1.6 (v)(γ) that TF is nef. Lemma 3.10 (iii) below implies that all finite étale covers $\widetilde{F}$ of F satisfy $q(\widetilde{F}) = O$. Hence the fibers F must be Fano by Proposition 3.9 and the Main Theorem follows. □

3.10 Lemma. *Let X, Y be compact Kähler manifolds and let $g : X \to Y$ be a smooth fibration with connected fibres. We let $q(X)$ be the irregularity of X and $\widetilde{q}(X)$ be the sup of the irregularity of all finite étale covers. If F denotes any fibre of g, then*

(i) $q(X) \le q(Y) + q(F)$,

(ii) $\widetilde{q}(X) \le \widetilde{q}(Y) + \widetilde{q}(F)$.

(iii) *Suppose that the boundary map $\pi_2(Y) \to \pi_1(F)$ is zero, that $\pi_1(F)$ contains an abelian subgroup of finite index and that Y contains a subvariety S with $\pi_1(S) \simeq \pi_1(Y)$, such that any two generic points in the universal covering $\widehat{S}$ can be joined through a chain of holomorphic images $\mathbb{C} \to \widehat{S}$. Then*

$$\widetilde{q}(X) = \widetilde{q}(Y) + \widetilde{q}(F).$$

The proof is based on a use of the Leray spectral sequence and a study of the resulting monodromy on $H^1(F, \mathbb{C})$. Triviality of the monodromy is achieved in case (iii) because all Kähler deformations of tori over Y must be trivial. We refer the reader to [DPS91] for the details. In our application, Y is taken to be the Albanese torus, so assumption (iii) is satisfied with $S = Y$ ($\pi_1(F)$ contains an abelian subgroup of finite index thanks to Corollary 3.6, by using an induction on dimension).

4 Classification in dimension 2 and 3

By using the Kodaira classification of surfaces and the structure theorems of Section 3, it is not difficult to classify all Kähler surfaces with nef tangent bundles; except for tori, the Kähler classification is identical to the projective one. The projective case was already mentioned in [CP91] and [Zh90].

4.1 Theorem. *Let X be a smooth Kähler surface such that TX is nef. Then X is minimal and is exactly one of the surfaces in the following list:*

(i) *X is a torus;*

(ii) *X is hyperelliptic;*

(iii) $X = \mathbb{P}_2$;

(iv) $X = \mathbb{P}_1 \times \mathbb{P}_1$;

(v) $X = \mathbb{P}(E)$, *where E is a rank 2-vector bundle on an elliptic curve C with either*

[(α)] $E = \mathcal{O} \oplus L$, $L \in \mathrm{Pic}^0(C)$, or

[(β)] *E is given by a non split extension* $0 \to \mathcal{O} \to E \to L \to 0$ *with* $L = \mathcal{O}$ *or* $\deg L = 1$.

The list of non-Kähler surfaces in the Kodaira classification is much smaller. It is then rather easy to check nefness in each case:

4.2 Theorem. *The smooth non Kähler compact complex surfaces with nef tangent bundles are precisely:*

(i) *Kodaira surfaces (that is, surfaces of Kodaira dimension* 0 *with* $b_1(X)$ *odd)*;

(ii) *Hopf-surfaces (that is, surfaces whose universal cover is* $\mathbb{C}^2\{0\}$).

A similar classification can be obtained for 3-dimensional compact Kähler manifolds.

7.1 Theorem. *Let X be a Kähler* 3*-fold. Then TX is nef if and only if X is up to finite étale cover one of the manifolds in the following list:*

(i) $X = \mathbb{P}_3$;

(ii) $X = Q_3$, *the 3-dimensional quadric*;

(iii) $X = \mathbb{P}_1 \times \mathbb{P}_2$;

(iv) $X = \mathbb{P}_1 \times \mathbb{P}_1 \times \mathbb{P}_1$;

(v) $X = F_{1,2}$, *the flag manifold of subspaces of* $\mathbb{C}^3$;

(vi) $X = \mathbb{P}(E)$, *with E a numerically flat rank 3-bundle on an elliptic curve C*;

(vii) $X = \mathbb{P}(E) \times_C \mathbb{P}(F)$, *with E, F numerically flat rank 2-bundles over an elliptic curve C*;

(viii) $X = \mathbb{P}(E)$, *with E a numerically flat rank 2-bundle over a 2-dimensional complex torus*;

(ix) $X =$ *3-dimensional complex torus.*

The only non-algebraic manifolds appear in classes (viii) and (ix) when the Albanese torus is not algebraic. Let us mention that the classification of projective 3-folds with nef tangent bundles was already carried out in [CP91] and [Zh90]. In addition to Theorem 3.5, the main ingredient is the classification of Fano 3-folds by Shokurov and Mori-Mukai. An inspection of the list yields the five first classes (i)-(v).

References

[Bi63] Bishop. R, *A relation between volume, mean curvature and diameter,* Amer. Math. Soc. Not. **10** (1963) 364.

[BG71] Bloch. S, Gieseker. D, *The positivity of the Chern classes of an ample vector bundle,* Inv. Math. **12** (1971) 112–117.

[Bo74a] Bogomolov. F. A, *On the decomposition of Kähler manifolds with trivial canonical class,* Math. USSR Sbornik **22** (1974) 580–583.

[Bo74b] Bogomolov. F.A, *Kähler manifolds with trivial canonical class,* Izvestija Akad. Nauk **38** (1974) 11—21 and Math. USSR Izvestija **8** (1974) 9–20.

[BPV84] Barth. W, Peters. C, Van de Ven. A, *Compact Complex Surfaces,* Erg. der Math. 3. Folge, Band 4, Springer, Berlin (1984)

[Ca57] Calabi. E, *On Kähler manifolds with vanishing canonical class,* Alg. geometry and topology, Symposium in honor of S. Lefschetz, Princeton Univ. Press, Princeton (1957) 78–89.

[CP91] Campana. F, Peternell. Th, *Projective manifolds whose tangent bundles are numerically effective,* Math. Ann. **289** (1991) 169–187.

[De85] Demailly. J.P, *Champs magnétiques et inégalités de Morse pour la d''-cohomologie,* Ann. Inst. Fourier **35** (1985) 185–229.

[De91] Demailly. J.P, *Regularization of closed positive currents and intersection theory,* Preprint (1991)

[DPS91] Demailly. J.P, Peternell. Th, Schneider M., *Compact complex manifolds with numerically effective tangent bundles,* Preprint (1991) To appear in J.Alg. Geom.

[DPS92] Demailly. J.P, Peternell. Th, Schneider. M, *Kähler manifolds with numerically effective Ricci class*, Compositio Math. **89** (1993) 217–240.

[FL83] Fulton. W, Lazarsfeld. R, *Positive polynomials for ample vector bundles,* Ann. Math. **118** (1983) 35–60.

[Fu83] Fujita. T, *Semi-positive line bundles,* J. Fac. Sci. Univ. Tokyo. **30** (1983) 353–378.

[Ga80] Gage. M. E, *Upper bounds for the first eigenvalue of the Laplace- Beltrami operator,* Indiana Univ. J. **29** (1980) 897–912.

[Gr69] Griffiths. P, *Hermitian differential geometry, Chern classes, and positive vector bundles,* In: Global Analysis, Princeton Math. Series **29** (1969) 185–251.

[Há66] Hartshorne. R, *Ample vector bundles,* Publ. Math. Inst. Hautes Etud. Sci. **29** (1966) 319–394.

[HK78] Heintze. E, Karcher. H, *A general comparison theorem with applications to volume estimates for submanifolds,* Ann. Scient. Ec.e **11** (1978) 451–470.

[KL74] Kempf. G, Laksov. D, *The determinantal formula of Schubert calculus,* Acta Math. **132** (974) 153–162.

[Ko61] Kobayashi. S, *On compact Kähler manifolds with positive definite Ricci tensor,* Ann. Math **74** (1961) 570–574.

[Ko81] Kobayashi. S, *Recent results in complex differential geometry,* Jahresber. Deutsche Math. Ver. **83** (1981) 147–158.

[Ko87] Kobayashi. S, *Differential geometry of complex vector bundles,* Princeton Univ. Press (1987)

[Kd66] Kodaira. K, *On the structure of compact complex surfaces II,* Amer. J. Math. **88** (1966) 682–721.

[Kr86] Kollár. J, *Higher direct images of dualizing sheaves,* Ann. Math. **123** (1986) 11–42.

[Li67] Lichnerowicz. A, *Variétés Kählériennes et première classe de Chern,* J. Diff. Geom. **1** (1967) 195–224.

[Li71] Lichnerowicz. A, *Variétés Kählériennes à première classe de Chern non négative et variétés riemanniennes à courbure de Ricci généralisée non négative,* J. Diff. Geom. **6** (1971) 47–94.

[Li72] Lichnerowicz. A, *Variétés Kählériennes à première classe de Chern non négative et situation analogue dans le cas riemannien,* Ist. Naz. Alta Mat., Rome, Symposia Math., vol. 10, Academic Press, New-York (1972) 3–18.

[Mk88] Mok. N, *The uniformization theorem for compact Kähler manifolds of non negative holomorphic bisectional curvature,* J. Diff. Geom. **27** (1988) 179–214.

[Mo79] Mori. S, *Projective manifolds with ample tangent bundles,* Ann. Math. **110** (1979) 593-606.

[My41] Myers. S. B, *Riemannian manifolds with positive mean curvature,* Duke Math. J. **8** (1941) 401–404.

[SY80] Siu. Y.T, Yau. S.T, *Compact Kähler manifolds with positive bisectional curvature,* Inv. Math. **59** (1980) 189–204.

[Ti72] Tits. J, *Free subgroups in linear groups,* J. of Algebra **20** (1972) 250–270.

[UY86] Uhlenbeck. K, Yau. S.T, *On the existence of Hermitian-Yang-Mills connections in stable vector bundles,* Comm. Pure and Appl. Math. **39** (1986) 258–293.

[Y77] Yau. S.T, *Calabi's conjecture and some new results in algebraic geometry,* Proc. Nat. Acad. Sci. USA **74** (1977) 1789–1790.

[Y78] Yau. S.T, *On the Ricci curvature of a complex Kähler manifold and the complex Monge-Ampère equation I,* Comm. Pure and Appl. Math. **31** (1978) 339–411.

[Zh90] Zheng. F, *On semi-positive threefolds* Thesis, Harvard (1990)

Université de Grenoble
Institut Fourier, BP 74
U.R.A. 188 du C.N.R.S.
38402 Saint-Martin d'Hères, France

Universität Bayreuth
Mathematisches Institut
Postfach 10 12 51
D-8580 Bayreuth, Deutschland

Algebraic Representations of Reductive Groups over Local Fields

William J. Haboush *

Introduction

This paper is an extended study of the behaviour of simplicial co-sheaves on the buildings associated to algebraic groups, both finite and infinite dimensional. Recently the theory of simplicial sheaves and co-sheaves has found a number of important applications to the representation theory and cohomology theory of finite groups (see [T], [RS]), the computation of the cohomology of arithmetic groups and the problem of admissible representations of p-adic groups and the Langlands classification ([CW], [BW]). My interest has been, for the most part, the representation theory of semi-simple groups over fields of positive characteristic. In this area, of course, the driving force of much recent work has been the so-called Lusztig characteristic p conjecture [L1] (so called to distinguish it from a number of other equally interesting Lusztig conjectures). In contemplating this conjecture one is struck by certain resonances with work on admissible representations etc.

The line of argument I am hoping to achieve is something like this. One should attempt to use the homological algebra of simplicial co-sheaves to construct a category of representations of something like the loop group associated to the semisimple group, G, which have a computable character theory. Then one should attempt to express the finite dimensional representations of G as virtual representations in that category. Then presumably the "generic decomposition patterns" should be formulae expressing the character of a dual Weyl module in terms of these computable characters. The hope of constructing such a theory has led me to conduct the rather extended exploration below.

One is immediately tempted to replace harmonic analysis with a purely algebraic theory and to use this theory to do the representation-theoretic

*This research was funded in part by the National Science Foundation

computations necessary. My replacement for harmonic analysis is this. Let G be algebraic of simple type over $\mathbb{Z}$. Let K be a field complete with respect to a discrete rank one valuation, let $\mathcal{O}$ be the valuation ring and let L be its residue field. Then consider the Bruhat-Tits building of $G(K)$. It is a simplicial complex. Let $\mathbf{G} = G(K)$. Let $\mathcal{I}$ be its Bruhat-Tits building. Then $\mathcal{I}$ is $\mathbf{G}$ equivariant. Let k be another field. My idea is to consider the category of $\mathbf{G}$ equivariant co-sheaves of k-vector spaces which are, in some sense made precise within, locally algebraic. Then in a manner entirely analogous to the classical notion of rational representative functions, this category admits an injective co-generator. The endomorphism ring of this canonical co-generator is a certain algebra. Let it be denoted $\mathcal{H}$. [T] shows that for suitably finite $\mathbf{G}$-co-sheaves the co-limit is a finite $\mathcal{H}$-module. Then one may send the class of a finite dimensional G-representation to the alternating sum of the left derived functors of the co-limit of its induced $\mathbf{G}$-co-sheaf in the Grothendieck ring of $\mathcal{H}$. In this context that one would hope to obtain interesting identities relating finite dimensional representation theory to the representation theory of $\mathcal{H}$.

I have made certain choices in this discussion. As I am discussing sheaves and co-sheaves on simplicial complexes, I have decided to use the word carapaces for co-sheaves. There are three reasons: the first is that the word, co-sheaves, seems rather cobbled together, the second is that a carapace really would look like a lobster shell or such if one were to draw one and the third is that Leray used that word for sheaves and I don't like to see such a nice word go to waste. I am working for the most part with carapaces rather than sheaves because I am working on an infinite simplicial complex and in moving from limit to co-limit, one is moving from an infinite direct product sort of thing to an infinite co-product sort of thing. Thus, in using limits rather than co-limits, one loses structure just as one does in taking an adic completion of a commutative ring. I have also decided to include a discussion of the homological algebra of carapaces. Now all of this material is some sort of special case of certain kinds of sheaves on sites or the homological algebra of abelian group valued functors, but with all due apologies to those who have worked on those topics, I would prefer to formulate this material in a way which anticipates my intentions. A number of mathematicians have done this sort of thing. Tits and Solomon [S], [T], Stephen Smith and Mark Ronan [RS] immediately come to mind. But again, working on an infinite complex has dictated that I reformulate many of the elements for this situation.

1 Carapaces and their Homology

Basic references for this section are [Mac] and [Gr]. Basic notions and definitions all follow those two sources. A simplicial complex, X, will be a set $ver(X)$ together with a collection of finite subsets of $ver(X)$ such that whenever $\sigma \in X$ every subset of σ is in X. These finite subsets of $ver(X)$ are the simplices of X. The dimension of σ is its cardinal less one and the dimension of X, if it exists, is the maximum of the dimensions of simplices in X. We shall view X as a category, the morphisms being the inclusions of simplices. We identify $ver(X)$ with the zero simplices of X. If X and Y are simplicial complexes a morphism of complexes from X to Y is just a covariant functor from X to Y; a simplicial morphism is a morphism of complexes taking vertices to vertices.

Let R be a commutative ring with unit fixed for the remainder of this work and let $Mod(R)$ denote the category of R-modules. Let X be a simplicial complex.

1.1 Definition. *An R-carapace* on X is a covariant functor from X to $Mod(R)$. If A and B are two R-carapaces on X, *a morphism* of R-carapaces from A to B is a natural transformation of functors.

If A is an R-carapace on X and $\sigma \in X$ is a simplex, then $A(\sigma)$ is called the segment of A along σ. A sequence of morphisms of R-carapaces will be called exact if and only if the corresponding sequence of segments is exact for each simplex in X. The product (respectively co-product) of a family of R-carapaces is the R-carapace whose segments and morphisms are the products (respectively co-products) of those in the family of carapaces. If $\sigma \subseteq \tau$ is a pair of simplices in X write $e^{\tau}_{A,\sigma}$ or e^{τ}_{σ} when there is no possibility of confusion for the map from $A(\sigma)$ to $A(\tau)$. I will call it the expansion of A from σ to τ. Finally, if A and B are two R-carapaces on X, write $\mathbf{Hom}_{R,X}(A, B)$ for the R-module of morphisms of R-carapaces from A to B.

Let M be an R-module. Then let M_X denote the constant carapace with value M. That is, its segment along any simplex is M and its expansions are all the identity map. In addition for any given simplex, σ, there are two dually defined carapaces, $M \uparrow_{\sigma}$ and $M \downarrow^{\sigma}$ defined by:

$$
\begin{aligned}
M \uparrow_{\sigma} (\tau) &= M (\sigma \subseteq \tau) \\
M \uparrow_{\sigma} (\tau) &= (0) (\sigma \not\subseteq \tau)
\end{aligned}
\tag{1.2}
$$

$$
\begin{array}{lrcll}
(1.3) & M\downarrow^{\sigma}(\tau) & = & M & (\sigma \supseteq \tau) \\
 & M\downarrow^{\sigma}(\tau) & = & (0) & (\sigma \not\supseteq \tau)
\end{array}
$$

In $M\uparrow_\sigma$, the expansions are the identity for pairs τ, γ such that $\sigma \subseteq \tau \subseteq \gamma$ and 0 otherwise. In $M\downarrow^\sigma$ they are the identity for τ, γ such that $\tau \subseteq \gamma \subseteq \sigma$ and 0 otherwise.

1.4 Lemma. *Let X be a simplicial complex and let M be an R-module and let A be an R-carapace on X. Then,*

1. $\mathbf{Hom}_{R,X}(M\uparrow_\sigma, A) = Hom_R(M, A(\sigma))$
2. $\mathbf{Hom}_{R,X}(A, M\downarrow^\sigma) = Hom_R(A(\sigma), M)$
3. *If M is R-projective, then $M\uparrow_\sigma$ is projective in the category of R-carapaces on X.*
4. *If M is R-injective, then $M\downarrow^\sigma$ is injective in the category of R-carapaces on X.*

Proof. Statements (1) and (2) require no proof. Note that the functor which assigns to an R-carapace on X its segment along σ is an exact functor. This observation together with (1) and (2) implies (3) and (4). □

1.5 Proposition. *Let X be a simplicial complex. Then the category of R-carapaces on X has enough projectives and enough injectives.*

Proof. Let A be any R-carapace on X. We must show that there is a surjective map from a projective R-carapace to X and an injective map from A into an injective R-carapace. For each simplex, σ in X choose a projective R-module, P_σ, with a surjective map π_σ mapping P_σ onto $A(\sigma)$. Let

$$P = \coprod_{\sigma \in X} P_\sigma \uparrow_\sigma$$

For each σ, let $\overline{\pi}_\sigma$ be the morphism of carapaces corresponding to π_σ given by 1.4, 1). Define π by:

$$\pi = \coprod_{\sigma \in X} \overline{\pi}_\sigma.$$

Then π is a surjective map from a projective to A.

To construct an injective, choose an injective module and an inclusion, $j_\sigma : A(\sigma) \to I_\sigma$. Then define an injective carapace, I, and an inclusion, j, as products of the carapaces $I_\sigma \downarrow^\sigma$ and inclusions $\bar{j}_\sigma$ defined dually to the corresponding objects in the projective case. □

1.6 Proposition. *Let X be a simplicial complex. Then*

(1) If P is a projective R-carapace on X, then $P(\sigma)$ is R projective for each $\sigma \in X$.

(2) If I is an injective R-carapace on X, then $I(\sigma)$ is R injective for each $\sigma \in X$.

Proof. Suppose P is projective. For each σ let $\pi_\sigma : F_\sigma \to P(\sigma)$ be a surjective map from a projective R-module onto $P(\sigma)$. Then $Q = \coprod_{\sigma \in X} F_\sigma \uparrow_\sigma$ is projective by 1.4 and $\coprod_{\sigma \in X} \pi_\sigma$ maps Q onto P. Since P is projective, it is a direct summand of Q. But then $P(\sigma)$ is a direct summand of $Q(\sigma) = \coprod_{\tau \subseteq \sigma} F_\tau$ which is clearly projective. This establishes the first statement.

To prove the second statement, for each σ choose an embedding, $j_\sigma : I(\sigma) \to J_\sigma$ where J_σ is R-injective. Then use $\prod_{\sigma \in X}$ to embed I in $\prod_{\sigma \in X} J_\sigma$ and reason dually to the previous argument replacing, at each point where it occurs, the co-product with the product. □

It is clear that there are natural homology and cohomology theories on the category of R-carapaces on X. The most natural functors to consider are the limit and the co-limit over X. To simplify the discussion, *let $Car_R(X)$ denote the category of R-carapaces on X.*

1.7 Definition. Let U denote any sub-category of X, and let A be an R-carapace on X. Then

$$\Sigma(U, A) = \varinjlim_{\sigma \in U} A(\sigma)$$

$$\Gamma(U, A) = \varprojlim_{\sigma \in U} A(\sigma).$$

We will refer to $\Sigma(U, A)$ as the *segment* of A over U and to $\Gamma(U, A)$ as the *sections* of A over U.

Certain observations are in order. Since subcategories of X are certainly not in general filtering the functor, $\Sigma(U, ?)$, is right exact on $Car_R(X)$. Similarly $\Gamma(U, ?)$ is left exact. Furthermore $\Gamma(U, A) = \mathbf{Hom}_{R,U}(R_U, A)$. On the other hand, $\Sigma(U, A)$ cannot be represented as a homomorphism functor in any obvious way but its definition as a direct limit allows us to conclude that:

$$\mathbf{Hom}_R(\Sigma(X,A),M) = \mathbf{Hom}_{R,X}(A,M_X)$$

1.8 Definition. For any R-carapace on X, A, let

$$H_n(X,A) = L_n\Sigma(X,A)$$

and let

$$H^n(X,A) = R^n\Gamma(X,A)$$

the left and right derived functors of $\Sigma(X,-)$ and $\Gamma(X,-)$ respectively. These groups shall be referred to as the *exoskeletal homology and cohomology groups* of X with coefficients in A.

1.9 Example. The Koszul Resolution of the Constant Carapace. Choose an ordering on the vertices of X. For each $r \geq 0$, let $X(r)$ denote the set of simplices of dimension r. Let $K_q(X,R) = \coprod_{\sigma \in X(q)} R \uparrow_\sigma$. If A is any R-carapace on X, then $\bigwedge_R^q A$ is understood to be the carapace whose segment along σ is $\bigwedge_R^q(A(\sigma))$. Then, it is not at all difficult to see that $\bigwedge_R^{q+1} K_0(X,R) = K_q(X,R)$ Furthermore, when $\sigma \subseteq \tau$ there is always a natural map from $M \uparrow_\tau$ to $M \uparrow_\sigma$ obtained by applying 4,1 to $M \uparrow_\tau$ and noting that since $M \uparrow_\sigma (\tau) = M = M \uparrow_\tau (\tau)$ there is a map in $\mathbf{Hom}_{R,X}(M \uparrow_\tau, M \uparrow_\sigma$) corresponding to the identity. Make use of the ordering on the vertices of X to define an alternating sum of the maps corresponding to the faces of a simplex. It is easy to see that this gives a complex of carapaces:

$$\ldots \to K_q(X,R) \to K_{q-1}(X,R) \to \ldots \to K_0(X,R) \to R_X \to (0)$$

Then check that the sequence of segments on σ is just the standard Koszul resolution of the unit ideal which begins with a free module of rank $dim(\sigma)+1$ and the map which sends each of its generators to one. Consequently, this construction gives a resolution of the constant carapace by projectives. On the other hand it is evident that $\Sigma(X, K_q(X,R))$ is just the R-module of simplicial q-chains on X with coefficients in R and that the boundaries are the standard simplicial boundaries. In this way, one verifies that the exoskeletal homology and cohomology with coefficients in a constant carapace is just the simplicial homology and cohomology. This phenomenon has been observed and exploited by Casselman and Wigner in their work on admissible representations and the cohomology of arithmetic groups [CW].

2 Operations on Carapaces

In the category of R-carapaces on X there is a self-evident notion of tensor product:

2.1 Definition. Let A and B be two R-carapaces on X.Let:

$$(A \otimes_R B)(\sigma) = A(\sigma) \otimes_R B(\sigma)$$

Then $A \otimes_R B$ is a carapace which we will refer to as *the tensor product* of A and B.

The properties of the tensor product are for the most part clear. Most of them are stated in the following.

2.2 Proposition. *Let X be a simplicial complex. Then the tensor product of R-carapaces on X is an associative, symmetric, bi-additive functor right exact in both variables. Furthermore:*

(1) For any R-carapace, A, $R_X \otimes_R A \simeq A$

(2) If P is a projective R-carapace then tensoring with P on either side is an exact functor.

(3) For any two R-carapaces, A and B, there is a natural map:

$$t_{A,B} : \Sigma(X, A \otimes_R B) \to \Sigma(X, A) \otimes_R \Sigma(X, B)$$

Moreover, $t_{A,B}$ is a natural transformation in A and B and it is functorial in X as well.

Proof. The first statement is self-evident; the second follows trivially from 1.6 but the third might require some comment. To construct $t_{A,B}$ let $e_{A,\sigma}$: $A(\sigma) \to \Sigma(X, A)$ and $e_{B,\sigma} : B(\sigma) \to \Sigma(X, B)$ be the expansions. Then $e_{A,\sigma} \otimes_R e_{B,\sigma}$ maps $A \otimes_R B(\sigma)$ into $\Sigma(X, A) \otimes_R \Sigma(X, B)$ compatibly with respect to expansion. Since $\Sigma(X, A \otimes_R B)$ is a colimit, this defines $t_{A,B}$ uniquely and ensures that it is functorial as asserted. □

The construction of a tensor product is thus quite straightforward but the construction of an internal homomorphism functor with the requisite adjointness properties presents certain technical difficulties. For any $\sigma \in X$ let $X(\sigma)$ denote the subcategory of X whose objects are the simplices τ such that $\tau \supseteq \sigma$. The morphisms of $X(\sigma)$ are inclusions of simplices. Then $X(\sigma)$ is not a subcomplex of X. If $\sigma \subseteq \tau$ then $X(\tau) \subseteq X(\sigma)$. There is a natural restriction map from the group $\mathbf{Hom}_{R,X(\sigma)}(A|_{X(\sigma)}, B|_{X(\sigma)})$ to $\mathbf{Hom}_{R,X(\tau)}(A|_{X(\tau)}, B|_{X(\tau)})$. We will write $e^{\tau}_{\mathcal{H},\sigma}$ for this restriction map.

2.3 Definition. Let A and B be two R-carapaces on X. *The carapace of local homomorphisms* from A to B will be written $\mathcal{H}om_{R,X}(A,B)$. Its value on σ is

$$\mathcal{H}om_{R,X}(A,B)(\sigma) = \mathbf{Hom}_{R,X(\sigma)}(A|_{X(\sigma)}, B|_{X(\sigma)})$$

Its expansions are the maps $e^{\tau}_{\mathcal{H},\sigma}$

For want of a direct reference, we include some discussion of the basic properties of $\mathcal{H}om$.

2.4 Theorem. *The local homomorphism functor,* $\mathcal{H}om_{R,X}(A,B)$, *is additive, covariant in* B *and contravariant in* A *and left exact in both variables. Moreover*

(1) $\mathcal{H}om_{R,X}(R_X, A) \simeq A$, functorially in A

(2) $\Gamma(X, \mathcal{H}om_{R,X}(A,B)) = \mathbf{Hom}_{R,X}(A,B)$

(3) There is a canonical isomorphism functorial in A, B *and* C,

$$\phi : \mathbf{Hom}_{R,X}(A, \mathcal{H}om_{R,X}(B,C)) \to \mathbf{Hom}_{R,X}(A \otimes_R B, C)$$

(4) There is a functorial isomorphism:

$$\mathbf{Hom}_{R,X}(R_X, \mathcal{H}om_{R,X}(A,B)) \simeq \mathbf{Hom}_{R,X}(A,B)$$

Proof. Of the three preliminary statements, only left exactness requires comment. What must be shown is the left exactness of segments of $\mathcal{H}om_{R,X}(A,B)$ as A and B vary over short exact sequences. But $\mathcal{H}om_{R,X}(A,B)(\sigma) = \mathbf{Hom}_{R,X(\sigma)}(A|_{X(\sigma)}, B_{X(\sigma)})$. Restriction to $X(\sigma)$ is exact and $\mathbf{Hom}_{R,X(\sigma)}$ is left exact in both of its arguments. The requisite left exactness follows immediately.

To establish (1), we must establish the isomorphism on segments. But

$$\mathcal{H}om_{R,X}(R_X, A)(\sigma) = \mathbf{Hom}_{R,X(\sigma)}(R_X, A|_{X(\sigma)}),$$

but this last expression is equal to:

$$\varprojlim_{\tau \in X(\sigma)} A(\tau).$$

However the category, $X(\sigma)$ has an initial element and so the projective limit is just $A(\sigma)$.

For Statement (2), write:

$$\Gamma(X, \mathcal{H}om_{R,X}(A,B)) = \varprojlim_{\sigma\in X} \mathbf{Hom}_{R,X(\sigma)}(A|_{X(\sigma)}, B|_{X(\sigma)})$$

There is a natural map from $\Gamma(X, \mathcal{H}om_{R,X}(A,B))$ to this projective limit. Just send f to the element in the limit whose component at σ is the restriction (the pull-back actually) of f to the sub-category, $X(\sigma)$. It is a triviality to verify that this map is an isomorphism.

Rather than giving a fully detailed proof of (3), we will give complete definitions of ϕ and a map ψ which is inverse to it. The necessary verifications, though numerous and quite technical, contain no surprises and so we leave them to the reader. First suppose $f \in \mathbf{Hom}_{R,X}(A, \mathcal{H}om_{R,X}(B,C))$. Then for $a \in A(\sigma)$, $f_\sigma(a)$ is a family, $f_\sigma(a) = \{f(a)_\tau\}_{\tau\supseteq\sigma}$ where $f(a)_\tau \in Hom_R(B(\tau), C(\tau))$. The following equations express the facts that f is a carapace morphism and that $f(a)$ is a carapace morphism from $B|_{X(\sigma)}$ to $C|_{X(\sigma)}$:

$$f_\tau(e^\tau_{A,\sigma}(a))_\gamma = f_\sigma(a)_\gamma \tag{5}$$

$$e^\gamma_{C,\tau} \circ f_\tau(a)_\tau = f_\tau(a)_\gamma \circ e^\gamma_{B,\tau} \tag{6}$$

Then we may define $\phi : \mathbf{Hom}_{R,X}(A, \mathcal{H}om_{R,X}(B,C)) \to \mathbf{Hom}_{R,X}(A\otimes_R B, C)$ and ψ opposite to it by the equations:

$$\phi(f)_\sigma(a\otimes b) = [f_\sigma(a)_\sigma](b) \qquad a\in A(\sigma), b\in B(\sigma) \tag{7}$$

$$[\psi(f)_\sigma(a)]_\tau(b) = f_\tau([e^\tau_{A,\sigma}(a)]\otimes b) \qquad a\in A(\sigma), b\in B(\tau) \tag{8}$$

These are the two maps, inverse to one another, which establish (3).

The last statement is obtained by applying the third to the left hand side and observing that $R_X \otimes_R A \simeq A$. □

3 Derived Functors

In this section we introduce the most elementary derived functors on $Car_R(X)$. These include the derived functors of both the module valued and the carapace valued tensor and homomorphism functors and the relation between the two. First recall that by 1.4, whenever P is a projective R-module and σ is a simplex in X then $P\uparrow_\sigma$ is a projective R-carapace. Consequently a coproduct of carapaces of the form $F\uparrow_\sigma$, where F is a free R-module, is a projective R-carapace. We will refer to such carapaces as elementary projectives. Furthermore notice that if M is any R-module then $\Sigma(X, M\uparrow_\tau) = M$. Hence if Q is an elementary projective, $\Sigma(X,Q)$ is a direct sum of free modules and hence free.

3.1 Lemma. *Let X be any simplicial complex.*

(1) Every R-carapace on X is a surjective image of an elementary projective.

(2) If P is a projective R-carapace on X, then $\Sigma(X,P)$ is R-projective.

(3) A tensor product of elementary projectives is an elementary projective.

(4) A tensor product of projectives is projective.

Proof. Let A be an R-carapace on X. For each $\sigma \in X$ let F_σ be a free R-module and let $q_\sigma : F_\sigma \to A(\sigma)$ be a surjective morphism.Then, just as in 1.5, $\coprod_{\sigma\in X} F_\sigma \uparrow_\sigma$ is an elementary projective and $\coprod_{\sigma\in X} q_\sigma$ maps it onto A. Thus 1) is established.

To prove 2), choose an elementary projective, F, and a surjective map, $q : F \to P$. Since P is projective, it is a direct summand of F and so $\Sigma(X,P)$ is a direct summand of $\Sigma(X,F)$ which is free. This establishes 2.

The fourth statement follows from the third because, by 1, every projective is a direct summand of an elementary projective. Hence we must prove 3). But this reduces to proving that if σ and τ are two simplices, then $R\uparrow_\sigma \otimes_R R\uparrow_\tau$ is elementary projective. But $R\uparrow_\sigma \otimes_R R\uparrow_\tau (\gamma) \neq 0$ if and only if $\gamma \supseteq \sigma$ and $\gamma \supseteq \tau$. Thus $R\uparrow_\sigma \otimes_R R\uparrow_\tau \neq (0)$ if and only if $\sigma \cup \tau = \alpha$ is a simplex and then $R\uparrow_\sigma \otimes_R R\uparrow_\tau = R\uparrow_\alpha$. This establishes the result. □

There are at least four very obvious homological bifunctors on $Car_R(X)$

3.2 Definition. Let A and B be R-carapaces on X. Write $\mathbf{Ext}^r_{R,X}$ for the r'th right derived functor of the left exact module valued bifunctor, $\mathbf{Hom}_{R,X}(A,B)$. Write $\mathcal{E}xt^q_{R,X}$ for the carapace valued q'th right derived functor of the carapace valued local homomorhism functor. Write $\mathcal{T}or^{R,X}_q(A,B)$ for the q'th carapace valued left derived functor of the carapace valued tensor product, $A \otimes_R B$ and write $\mathbf{Tor}^{R,X}_q(A,B)$ for the q'th left derived functor of the right exact bifunctor $\Sigma(X, A\otimes_R B)$.

3.3 Lemma. *If P is a projective R-carapace on X, then for any R-carapace, A, $P\otimes_R A$ is Σ-acyclic and $\mathcal{H}om_{R,X}(P,A)$ is Γ-acyclic.*

Proof. Let $\{Q_j\}, j \geq 0$ be a projective resolution of A. Then $\{P \otimes_R Q_j\}$ is a projective resolution of $P\otimes_R A$. Apply the functor Σ to obtain $\mathbf{Tor}^{R,X}_j(A,P) = H_j(A\otimes_R P)$. But $\mathbf{Tor}^{R,X}_j(A,P) = (0)$ because P is projective. This establishes the Σ-acyclicity of $A\otimes_R P$. For the other acyclicity, let

$\{K_j\}$ be a projective resolution of R_X. Then $\mathbf{Hom}_{R,X}(K_j, \mathcal{H}om_{R,X}(P,A)) = \mathbf{Hom}_{R,X}(K_j \otimes_R P, A)$. But $K_j \otimes_R P$ is a projective resolution of P. Thus $H^j(X, \mathcal{H}om_{R,X}(P,A)) = \mathbf{Ext}^j_{R,X}(P,A)$ which is (0) because P is projective. □

The elementary properties of these four functors are described in the following.

3.4 Proposition. *Let A and B be R-carapaces on X. Then*

(1) $\mathcal{T}or_q^{R,X}(A,B)(\sigma) = \mathcal{T}or_q^{R}(A(\sigma), B(\sigma))$

(2) $\mathbf{Tor}_q^{R,X}(R_X, A) = H_q(X, A)$

(3) $\mathbf{Ext}^q_{R,X}(R_X, A) = H^q(X, A)$

(4) *There is a spectral sequence with $E_2^{p,q}$ term:*

$$E_2^{p,q} = H^p(X, \mathcal{E}xt^q_{R,X}(A,B))$$

and whose abutment is:

$$\mathbf{Ext}^{p+q}_{X,R}(A,B)$$

(5) *There is a spectral sequence with $E^2_{p,q}$ term:*

$$E^2_{p,q} = H_p(X, \mathcal{T}or_q^{R,X}(A,B))$$

and with abutment:

$$\mathbf{Tor}^{R,X}_{p+q}(A,B)$$

Proof. To prove 1), let $\{P_j\}$ be a projective resolution of A. Then for each σ, $\{P_j(\sigma)\}$ is a projective resolution of $A(\sigma)$. Moreover the segment of $P_j \otimes_R B$ along σ is $P_j(\sigma) \otimes_R B(\sigma)$. But the segment along σ is an exact functor on $Car_R(X)$ and so the σ-segment of the homology of the complex, $P_j \otimes_R B$ is the homology of the complex $P_j(\sigma) \otimes_R B(\sigma)$. This is just the result desired.

Statements 2) and 3) are both essentially trivial. Just note that $\mathbf{Tor}_q^{R,X}(R_X, A)$ (respectively $\mathbf{Ext}^q_{R,X}(R_X, A)$) is a connected sequence of homological functors acyclic on projectives (respectively injectives) and that $\mathbf{Tor}_0^{R,X}(R_X, A) = \Sigma(X, A)$ (respectively $\mathbf{Ext}^0_{R,X}(R_X, A) = \Gamma(X, A)$). The two statements follow.

The local global spectral sequences in 4) and 5) are just composition of two functor sequences as in [Gr]. Let $F_A(B) = A \otimes_R B$ and let $G_A(B) = \mathcal{H}om_{R,X}(B, A)$. By Lemma 3.3, F_A carries projectives to Σ-acyclics and G_A carries projectives to Γ-acyclics. The left derived functors of F_A are the functors, $\mathcal{T}or_q^{R,X}(A, -)$ while the right derived functors of G_A are the functors, $\mathcal{H}om^q_{R,X}(A, -)$. The construction of the spectral sequences is standard. □

4 Homological Dimension

In this section we will determine the homological dimension of $Car_R(X)$ for a finite dimensional simplicial complex, X. Write $X^{(r)}$ for the set of simplices in X of dimension r and write X_n for $\bigcup_{r\geq n} X^{(r)}$. If M is an R-module, write $pd(M)$ for the projective dimension of M and write $hd(R)$ for the homological dimension of R.

1 Definition. Let A be an R-carapace on X.

(1) Write $s(A) \geq n$ if $A(\sigma) = (0)$ whenever $dim(\sigma) < n$ and let $s(A) = inf\{n : s(A) \geq n\}$. Then $s(A)$ is called the *support dimension* of A.

(2) We say that A is *locally bounded* if $pd(A(\sigma)) \leq q$ for some fixed $q \geq 0$. In that case let $ld(A) = sup\{pd(A(\sigma)) : \sigma \in X\}$. When it exists, $ld(A)$ is called the *local projective dimension* of A.

2 Proposition. *Suppose that A is an R-carapace on X of support dimension at least n and local projective dimension r. Then there is an exact sequence:*

(4.3) $$0 \longrightarrow A^1 \longrightarrow P_r \longrightarrow \ldots \longrightarrow P_0 \longrightarrow A \longrightarrow 0$$

so that:

(1) $ld(A^1) = 0$

(2) P_i is projective.

(3) $s(A^1) \geq n+1$

(4) $s(P_i) \geq n$

Proof. For each $\sigma \in X_n$ choose a projective module, Q^σ and a surjective morphism, $\phi^\sigma : Q^\sigma \to A(\sigma) \to 0$. For each σ, let $\phi_\sigma : Q^\sigma \uparrow_\sigma \to A$ be the morphism of carapaces induced by ϕ^σ. Let $Q_0 = \coprod_{\sigma\in X_n}(Q^\sigma)\uparrow_\sigma$ and let $d_0 = \coprod_{\sigma\in X_n}\phi_\sigma$. Since $s(A) \geq n$, d_0 is surjective. Let $N_0 = ker(d_0)$. Then, clearly $s(N_0) \geq n$ but $ld(N_0) \leq r-1$. Thus we may repeat the process with A replaced by N_0 and continue inductively until we obtain an exact sequence:

$$0 \longrightarrow N_{r-1} \longrightarrow Q_{r-1} \longrightarrow \ldots \longrightarrow Q_0 \longrightarrow A \longrightarrow 0.$$

In this sequence, the Q_i are projective, the support dimensions of the Q_i and of N_{r-1} are at least n and $ld(N_{r-1}) = 0$. That is $N_{r-1}(\sigma)$ is projective for each σ. Now let $Q_r = \coprod_{\sigma \in X_n} (N_{r-1}(\sigma)) \uparrow_\sigma$. Clearly, Q_r maps onto N_{r-1} and the map is an isomorphism on segments over simplices of dimension n. Let d_r be the composition of the map onto N_{r-1} with the inclusion into Q_{r-1} and let $A^1 = ker(d_r)$. Clearly, A^1 and the Q_i answer the requirements of the proposition. □

4.4 Theorem. *Let X be a simplicial complex of dimension d and let A be a locally bounded R-carapace on X of local projective dimension r. Then $pd(A) \leq d + r$.*

Proof. Apply Proposition 4.2 with $n = 0$. The result is the exact sequence:

$$0 \longrightarrow A^1 \longrightarrow P_r \longrightarrow \ldots \longrightarrow P_0 \longrightarrow A \longrightarrow 0$$

Then apply 4.2 to A^1 observing that $ld(A^1) = 0$ and $s(A^1) \geq 1$. The result is a short exact sequence, $0 \longrightarrow A^2 \longrightarrow Q_1 \longrightarrow A^1 \longrightarrow 0$ where Q_1 is projective, $ld(A^2) = 0$ and $s(A^2) \geq 2$. We may continue until we reach $s(A^d) \geq d$. But for any B, if the segments of B are projective and $s(B) \geq dim(X)$ then B is projective. We may thus assemble these short sequences and the sequence of P_i to obtain a sequence:

$$0 \longrightarrow A^d \longrightarrow Q_{d-1} \longrightarrow \ldots \longrightarrow Q_1 \longrightarrow P_r \longrightarrow \ldots \longrightarrow P_0 \longrightarrow A \longrightarrow 0$$

This exact sequence is the projective resolution establishing the result. □

4.5 Corollary. *If $dim(X) = d$ and if M is an R-module of projective dimension r, then $pd(M_X) \leq r + d$. If M is projective then $pd(M_X) \leq d$.*

4.6 Corollary. *If R is of homological dimension r and $dim(X) = d$ then the homological dimension of $Car_R(X)$ is at most $d + r$.*

Neither of these corollaries requires so much as one word of proof.

5 Carapaces and Morphisms of Complexes

Recall that a morphism of complexes from X to Y is just a covariant functor from the category of simplices in X to the category of simplices in Y; it is simplicial if it carries vertices to vertices. If S is a subset of the vertex set of X then it admits a simplicial complex structure by taking as its set of simplices the set of simplices in X all of whose vertices lie in S. We will write $\tilde{S}$ for this complex. When we speak of a subcategory of X we will always, unless otherwise indicated, mean a full subcategory of the simplex category of X. If U and V are subcategories of X we will write $U \subset V$ to indicate that U is a full subcategory of V. In this case there is always a functorial map from $\Sigma(U, A)$ to $\Sigma(V, A)$ for any carapace, A.

Suppose that f is a morphism of complexes from X to Y and that Z is a simplicial sub-complex of Y. Let $f^{-1}(Z)$ denote the simplicial sub-complex of X which has as its set of simplices the set $\{\sigma \in X : f(\sigma) \in Z\}$. Clearly, f^{-1} is a functor from the subcomplexes in Y to those in X.

5.1 Definition. Let X and Y be simplicial complexes, let A be an R-carapace on X and let B be one on Y. Let $f : X \to Y$ be a morphism of complexes.

(1) Let $(f^*(B))(\sigma) = B(f(\sigma))$ and let $e^{\tau}_{f^*(B),\sigma} = e^{f(\tau)}_{B,f(\sigma)}$. Then $f^*(B)$ is an R-carapace on X and f^* is a covariant functor from $Car_R(Y)$ to $Car_R(X)$. The carapace $f^*(B)$ is called the *inverse image* of B under f.

(2) Let $((f_*(A))(\sigma) = \Sigma(f^{-1}(\tilde{\sigma}), A)$ and let $e^{\tau}_{f_*(A),\sigma}$ be the natural map of segments induced by the inclusion of categories, $f^{-1}(\tilde{\sigma}) \subset f^{-1}(\tilde{\tau})$ when $\sigma \subset \tau$. Then f_* is a covariant functor from carapaces on X to carapaces on Y. The carapace, $f_*(A)$ is called the *direct image* of A by f.

Both f^* and f_* are additive. In addition they satisfy the adjointness properties expected.

5.2 Theorem. *Let X and Y be simplicial complexes, let f be a morphism of complexes from X to Y, let A be an R-carapace on X and let B be one on Y.*

(1) f^* is exact.

(2) f_* is right exact.

(3) f_* is left adjoint to f^*. That is,

$$\mathbf{Hom}_{R,X}(A, f^*(B)) \simeq \mathbf{Hom}_{R,Y}(f_*(A), B)$$

functorially in A and B.

Proof. The first statement is a triviality. The second statement is nothing more than the right exactness of co-limits. Thus only the last statement requires attention.

To prove 3), we will give morphisms,

$$\psi : \mathbf{Hom}_{R,X}(A, f^*(B)) \to \mathbf{Hom}_{R,Y}(f_*(A), B)$$

and ϕ inverse to it. Begin with ψ. If $\alpha \in \mathbf{Hom}_{R,X}(A, f^*(B))$, write $\alpha = \{\alpha_\sigma\}_{\sigma \in X}$. Then α_σ maps $A(\sigma)$ to $B(f(\sigma))$ for each σ compatibly with respect to σ. Then $\sigma \in f^{-1}(\tilde{\rho})$ if and only if $f(\sigma) \subseteq \rho$. Thus the set of maps, $e^{\rho}_{B,f(\sigma)} \circ \alpha_\sigma$ is a direct system of maps giving a morphism from $[f_*(A)](\rho) = \Sigma(f^{-1}(\tilde{\rho}), A)$ to $B(\rho)$. For each ρ call this map β_ρ. Then since β_ρ is functorial in ρ, the family $\{\beta_\rho\}_{\rho \in Y}$ is a morphism, β, from $f_*(A)$ to B. Let $\psi(\alpha) = \beta$.

Now we wish to define ϕ. If $\beta \in \mathbf{Hom}_{R,Y}(f_*(A), B)$ then β is a family $\{\beta_\rho\}_{\rho \in Y}$ where β_ρ maps $\Sigma(f^{-1}(\tilde{\rho}), A)$ to $B(\rho)$. For any σ in X, let $\rho = f(\sigma)$ and let a_σ be the natural map from $A(\sigma)$ to $\Sigma(f^{-1}(\tilde{\rho}), A)$. Then $\beta_\rho \circ a_\sigma$ is a map from $A(\sigma)$ to $B(f(\sigma))$ for each σ. Let $\alpha_\sigma = \beta_{f(\sigma)} \circ a_\sigma$ for each σ and let $\phi(\beta)$ be the map $\alpha = \{\alpha_\sigma\}_{\sigma \in X}$. We leave the task of verifying that ψ and ϕ are maps of the requisite type and that they are inverse to one another to the reader. □

It is entirely expected that f^* has a left adjoint. It is a bit surprising, though not at all subtle, that it also has a right adjoint. Let $f : X \to Y$ be a morphism of complexes. For $\rho \in Y$ let $f^{\dagger}(\rho)$ denote the sub-category of X consisting of all $\sigma \in X$ such that $f(\sigma) \supseteq \rho$.

5.3 Definition. Let X and Y be simplicial complexes, let $f : X \to Y$ be a morphism of complexes and let A be an R-carapace on X. Define an R-carapace on Y by the equation:

$$f_{\dagger}(A)(\rho) = \Gamma(f^{\dagger}(\rho), A)$$

This is clearly an R-module valued functor on the simplex category of Y and so it is an R-carapace on Y. We will call it the *right direct image* of A under f.

4 Proposition. *Let $f : X \to Y$ be a morphism of complexes, let A be an R-carapace on X and let B be one on Y. Then $f_{\dagger}$ is left exact and right adjoint to f^*. That is,*

$$\mathbf{Hom}_{R,X}(f^*(B), A) \simeq \mathbf{Hom}_{R,Y}(B, f_{\dagger}(A))$$

functorially in A and B.

Proof. Left exactness follows from the left exactness of Γ and so we only need to establish the adjointness. We give the two morphisms. Let

$$\eta : \mathbf{Hom}_{R,Y}(B, f_\dagger(A)) \longrightarrow \mathbf{Hom}_{R,X}(f^*(B), A)$$

be one of the two morphisms and let ζ be its inverse.

Choose δ in $\mathbf{Hom}_{R,Y}(B, f_\dagger(A))$. For each $\rho \in Y$, δ takes each element, $b \in B(\rho)$ to a compatible family, $\{[\delta_\rho(b)]_\sigma\}_{f(\sigma)\supseteq\rho}$ where $[\delta_\rho(b)]_\sigma \in A(\sigma)$. For each σ we must give a map $\eta(\delta)_\sigma : B(f(\sigma)) \to A(\sigma)$. Let

$$[\eta(\delta)_\sigma](b) = [\delta_{f(\sigma)}(b)]_\sigma$$

This defines η.

To define ζ, choose $b \in B(\rho)$ and suppose that $\beta \in \mathbf{Hom}_{R,X}(f^*(B), A)$. If $f(\sigma) \supseteq \rho$ let $a_\sigma = \beta_\sigma(e^{f(\sigma)}_{B,\rho}(b))$. Then let

$$\zeta(\beta)_\rho(b) = \{a_\sigma\}_{f(\sigma)\supseteq\rho}$$

We leave the verifications involved to the reader.

5.5 Corollary. *Let X, Y and f be as above. Then:*

(1) f_* carries projectives on X to projectives on Y.

(2) For any R-carapace A on X, $\Sigma(Y, f_*(A)) = \Sigma(X, A)$

(3) $f_\dagger$ carries injectives on X to injectives on Y.

(4) For any A on X, $\Gamma(Y, f_\dagger(A)) = \Gamma(X, A)$

Proof. For the first statement, let P be a projective on X and let $M \to N \to 0$ be a surjective map in $Car_R(X)$. Consider the map, $\mathbf{Hom}_{R,Y}(f_*(P), M) \to \mathbf{Hom}_{R,Y}(f_*(P), N)$. By the adjointness statement in 5.2, 3), this is the same as the map $\mathbf{Hom}_{R,X}(P, f^*(M)) \to \mathbf{Hom}_{R,X}(P, f^*(N))$. But now f^* is exact and P is projective on X and so this map is surjective. This takes care of 1).

In general, if M and N are R-modules and there is an isomorphism $Hom_R(M, Q) \simeq Hom_R(N, Q)$ functorial in Q, then $M \simeq N$. Apply this to 2) using the definition of the functor $\Sigma(X, ?)$ and 5.2, 3) to obtain:

$$Hom_R(\Sigma(Y, f_*(A)), M) = \mathbf{Hom}_{R,Y}(f_*(A), M_Y)$$

$$= \mathbf{Hom}_{R,X}(A, M_X) = Hom_R(\Sigma(X, A), M)$$

Statement 2) follows.

The proof of 3) is precisely dual to the proof of 1). To establish 4), apply 5.4 and 2.4, 2). Write:

$$\Gamma(Y, f_{\dagger}(A)) = \mathbf{Hom}_{R,Y}(R_Y, f_{\dagger}(A)) = \mathbf{Hom}_{R,X}(R_X, A) = \Gamma(X, A)$$

Thus 4) is also proven. □

Corollary 5.5 establishes exactly what is necessary for two composition of functor spectral sequences. Many are possible but we content ourselves with the two most obvious.

5.6 Proposition. *Let X and Y be simplicial complexes, let A be an R-carapace on X and let $f : X \to Y$ be a morphism of complexes.*

(1) There is a spectral sequence with $E^2_{p,q}$ term:

$$E^2_{p,q} = H_p(Y, L_q fA)$$

and abutment:

$$H_r(X, A)$$

(2) There is a spectral sequence with $E_2^{p,q}$ term:

$$E_2^{p,q} = H^p(Y, R^q f_{\dagger} A)$$

and abutment:

$$H^r(X, A)$$

These spectral sequences are sufficiently standard that no proof is required. The proofs in [Gr], for example, apply.

6 Certain Special Carapaces

This section will be devoted to the study of certain acyclic carapaces. We will need certain conventions. If X is a simplicial complex, a complement in X is a full subcategory of its simplex category such that the complement of its collection of simplices is a simplicial complex. The reader may verify that C is a complement in X if, whenever $\sigma \in C$ and $\tau \supseteq \sigma$ then $\tau \in C$. Alternatively C is a complement in X if and only if whenever $\sigma \in C$, then $X(\sigma) \subseteq C$. These two conditions apply to arbitrary subcollections of the simplex set of X and we will use the term complement in this sense. Clearly arbitrary unions and intersections of complements are complements.

6.1 Definition. Let X be a simplicial complex.

(1) An R-carapace, B, is called *brittle* if for every sub-complex of X, Z, the natural map, $\Sigma(Z,B) \to \Sigma(X,B)$ is injective.

(1) An R-carapace, F, is called *flabby* if for every complement in X, C, the natural map, $\Gamma(X,F) \to \Gamma(C,F)$ is surjective.

Our development follows standard treatments of flabbyness for sheaves. On occasion something more is called for in the brittleness arguments. Flabbyness will be an entirely familiar concept, but brittleness might be a bit strange. We will begin with some descriptive comments. First notice that if $dim(X) > 0$ then R_X is not brittle. Suppose that σ is a positive dimensional simplex in X and that x and y are distinct vertices in it. Let $Z = \{x,y\}$. That is, Z is the disconnected two point complex. Then clearly, $\Sigma(Z, R_X) = R \oplus R$ and, since $\sigma \in X$ and $Z \subseteq \sigma$, the map, $\Sigma(Z,R_X) \to \Sigma(X,R_X)$ is not injective since it factors through $R_X(\sigma) = R$.

If $\sigma \in X$ and B is brittle then by definition, $B(\sigma) \subseteq \Sigma(X,B)$. But brittleness also forces the relation, $B(\sigma) \cap B(\tau) = B(\sigma \cap \tau)$ where the intersection is taken in $\Sigma(X,B)$. To see this just note that, because $\Sigma(Z,A)$ is nothing but the inductive limit over Z, there is an exact sequence,

$$0 \longrightarrow B(\sigma \cap \tau) \longrightarrow B(\sigma)\coprod B(\tau) \longrightarrow \Sigma(\sigma \cup \tau, B) \longrightarrow 0$$

and, by brittleness, an inclusion $\Sigma(\sigma \cup \tau, B) \subseteq \Sigma(X,B)$.

Before proceeding a convention is necessary. If σ is a simplex in X then write $\hat{\sigma}$ for the complex whose vertex set is σ but whose simplex set is the set of all proper subsets of σ. That is σ is not a simplex in $\hat{\sigma}$ which is a simplicial sphere. Then $\hat{\sigma} \subset \tilde{\sigma}$.

We will also require the following. Let $f : M \to N$ be a morphism of R-modules. Then f is injective if and only if, for each injective R-module, J, the induced map $Hom_R(N,J) \to Hom_R(M,J)$ is surjective.

Finally suppose that Z is a simplicial sub-complex of X. Let C be the set of simplices of X which are not simplices of Z. For any R-module, M, define R-carapaces M^*_Z and M^C_* by the equations:

$$(6.2) \qquad \begin{array}{rcll} M^*_Z(\sigma) & = & M & \text{if } \sigma \in Z \\ M^*_Z(\sigma) & = & (0) & \text{if } \sigma \notin Z \end{array}$$

Then M^C_* is defined by exactly the same equations, replacing M^*_Z by M^C_* and Z by C. As Z is a complex M^*_Z is naturally a quotient of M_X while M^C_*

is naturally a subobject. In fact, the following is exact:

$$0 \longrightarrow M_*^C \longrightarrow M_X \longrightarrow M_Z^* \longrightarrow 0$$

In addition, the following hold

$$\begin{array}{lrcl} (6.3) & Hom_{R,X}(A, M_Z^*) & = & Hom_R(\Sigma(Z,A), M) \\ & \mathbf{Hom}_{R,X}(M_*^C, A) & = & Hom_R(M, \Gamma(C,A)) \end{array}$$

6.4 Lemma. *Let X be a simplicial complex, let $Z \subseteq X$ be a subcomplex of X and let C be a complement in X. Then if A is brittle on X, $A|_Z$ is brittle on Z. If A is flabby on X, then $A|_C$ is flabby on C.*

Proof. If A is brittle and Z' is a subcomplex of Z then the composition, $\Sigma(Z', A) \to \Sigma(Z, A) \to \Sigma(X, A)$ is the map, $\Sigma(Z', A) \to \Sigma(X, A)$. If a composition is injective, each map in it is injective. This proves the first statement. The proof of the second statement is precisely dual to it and so we leave it to the reader. □

6.5 Theorem. *Let X be a simplicial complex and let*

$$0 \longrightarrow A' \longrightarrow A \longrightarrow A'' \longrightarrow 0$$

be exact.

(1) If A'' is brittle, then

$$0 \longrightarrow \Sigma(X, A') \longrightarrow \Sigma(X, A) \longrightarrow \Sigma(X, A'') \longrightarrow 0$$

is exact.

(2) If A' is flabby, then

$$0 \longrightarrow \Gamma(X, A') \longrightarrow \Gamma(X, A) \longrightarrow \Gamma(X, A'') \longrightarrow 0$$

is exact.

Proof. To prove (1) we need only show that $\Sigma(X, A') \rightarrow \Sigma(X, A)$ is injective. By the observation above, it would suffice to show that $Hom_{\mathbf{R}}(\Sigma(X, A), J) \rightarrow Hom_{\mathbf{R}}(\Sigma(X, A'), J)$ is surjective for any injective, J. But $Hom_{\mathbf{R}}(\Sigma(X, A), J) = \mathbf{Hom}_{R,X}(A, J_X)$ and the same for A'. Thus, to establish (1), it suffices to prove that every carapace morphism, $f : A' \rightarrow J_X$ extends to a morphism, $\tilde{f} : A \rightarrow J_X$.

Let $j : A' \rightarrow A$ be the injection and let $\pi : A \rightarrow A''$ be the surjection. Let $f : A' \rightarrow J_X$ be a morphism of carapaces. Let $\mathcal{F}$ be the family of pairs, (Z, f_Z) where Z is a subcomplex and $f_Z : A|_Z \rightarrow J_Z$ is a morphism such that $f_Z \circ j = f|_Z$. Order these by inclusion on Z and extension on f_Z. This orders $\mathcal{F}$ inductively and so Zorn's Lemma yields a maximal element, (W, f_W). If $W \neq X$ there is some $\sigma \in X$ such that $\sigma \notin W$. If $\sigma \cap W = \emptyset$ we may trivially extend f_W to $W \cup \{t\}$ where t is any vertex in σ. This contradicts maximality. Thus we may assume that $\sigma \cap W \neq \emptyset$. Let $\tilde{\sigma} \cap W = Y$. Consider $f_\sigma : A'(\sigma) \rightarrow J$. By the injectivity of J, we may choose $f^1_\sigma : A(\sigma) \rightarrow J$ such that $f^1_\sigma \circ j_\sigma = f_\sigma$. If $\gamma \subseteq \sigma$ let $f^1_\gamma = f^1_\sigma \circ e^\sigma_{A,\gamma}$. Since j is morphism, the following commutes:

$$\begin{array}{ccc} A'(\sigma) & \xrightarrow{j_\sigma} & A(\sigma) \\ e^\sigma_{A',\gamma} \uparrow & & \uparrow e^\sigma_{A,\gamma} A \\ A'(\gamma) & \underset{j_\gamma}{\longrightarrow} & A(\gamma) \end{array}$$

Hence $f^1_\gamma \circ j_\gamma = f^1_\sigma \circ e^\sigma_{A,\gamma} \circ j_\gamma = f^1_\sigma \circ j_\sigma \circ e^\sigma_{A',\gamma} = f_\gamma$. That is, $f^1 \circ j = f$ on σ. But on $\tilde{\sigma} \cap W = Y$, $f_W \circ j = f$. Thus on Y, $(f_W - f^1) \circ j = 0$. It follows that $f_W - f^1$ induces a map from $A''|_Y$ to J_Y. But $\Sigma(Y, A'') \rightarrow \Sigma(X, A'')$ is injective because A'' is brittle. Hence $\mathbf{Hom}_{R,X}(A'', J_X) \rightarrow \mathbf{Hom}_{R,Y}(A''|Y, J_Y)$ is surjective. Thus, there is an $f_2 \in \mathbf{Hom}_{R,X}(A'', J_X)$ such that $\pi \circ f_2|_Y = (f_W - f^1)|_Y$. Consequently, $(f^1 + \pi \circ f_2)|_{\tilde{\sigma} \cap W} = f_W|_{\tilde{\sigma} \cap W}$. Hence f_W can be extended to $W \cup \tilde{\sigma}$ contradicting the maximality of (W, f_W). That is $W = X$ and so 1) is established.

To prove 2) we must prove that $\Gamma(X, A) \rightarrow \Gamma(X, A'')$ is surjective. An element $a \in \Gamma(Z, A)$ is a function on Z such that $a(\sigma) \in A(\sigma)$ and $e^\tau_{A,\sigma}(a(\sigma)) = a(\tau)$. Suppose $a'' \in \Gamma(X, A'')$ is given. Order the pairs (C, a_C), where C is a complement and $a_C \in \Gamma(C, A)$, $\pi(a_C) = a''|_C$, by inclusion and extension. This being an inductive order, there is a maximal element, (U, a_U). If $U \neq X$, there is a simplex, τ not in U. Choose $\tilde{a}_\tau \in A(\tau)$ such that $\pi_\tau(\tilde{a}_\tau) = a''(\tau)$. Define $\tilde{a}_1$ on $X(\tau)$ by $\tilde{a}_1(\sigma) = e^\sigma_{A,\tau}(\tilde{a}_\tau)$. If $X(\tau) \cap U = \emptyset$ then $\tilde{a}_1$ extends a_U contradicting maximality of (U, a_U), and so we may assume that $X(\tau) \cap U \neq \emptyset$. This intersection is a complement. Consider the difference $\tilde{a}_1 - a_U$ on this intersection. Now $\pi(\tilde{a}_1 - a_U) = 0$ on $X(\tau) \cap U$ whence $(\tilde{a}_1 - a_U)|X(\tau) \cap U \in \Gamma(X(\tau) \cap U, A')$ Since A' is flabby there is an element $a' \in \Gamma(X, A')$ such that $a'|X(\tau) \cap U = (a_1 - a_U)|X(\tau) \cap U$. Clearly

$a_1 - (a'|X(\tau))$ extends a_U contradicting maximality. Thus $U = X$ and we have established 2). □

6.6 Corollary. *Let*

$$0 \longrightarrow A' \longrightarrow A \longrightarrow A'' \longrightarrow 0$$

be an exact sequence of R-carapaces on X.

(1) If A and A″ are brittle, then A′ is also.

(2) If A and A′ are flabby, then A″ is also.

Proof. We prove 1). Suppose Z is a subcomplex of X. Then, by 6.4, $A|_Z$ and $A''|_Z$ are both brittle and hence, $0 \to \Sigma(Z, A') \to \Sigma(Z, A)$ is exact. Thus the following diagram, which has exact rows and columns, commutes:

$$\begin{array}{ccccc} & & & & 0 \\ & & & & \downarrow \\ 0 & \longrightarrow & \Sigma(Z, A') & \longrightarrow & \Sigma(Z, A) \\ & & \downarrow & & \downarrow \\ 0 & \longrightarrow & \Sigma(X, A') & \longrightarrow & \Sigma(X, A) \end{array}$$

It is immediate that $\Sigma(Z, A') \to \Sigma(X, A')$ is monic. As for Statement 2), noting that Z must be replaced by a complement, the proof is both well known and strictly dual to the proof of 1). □

6.7 Proposition. *Let X and Y be simplicial complexes let $f : X \to Y$ be a morhism of complexes and let A be an R-carapace on X. Then*

(1) If A is brittle, then $f_*(A)$ is brittle.

(2) If A is flabby then $f_\dagger(A)$ is flabby.

Proof. To prove 1), let $U \subseteq Y$ be a subcomplex. Then, $f^{-1}(U)$ is a subcomplex of X and so, if A is brittle, then $\Sigma(f^{-1}(U), A) \to \Sigma(X, A)$ is injective. But $\Sigma(f^{-1}(U), A) = \Sigma(U, f_*A)and\Sigma(X, A) = \Sigma(Y, f_*A)$ by definition. That proves the first statement. The proof of 2) is completely parallel except that it uses 5.5, 4) in place of the corresponding properties of f_* □

6.8 Proposition. *Let X be a simplicial complex.*

(1) Projective carapaces are brittle; injective carapaces are flabby.

(2) A coproduct of brittle carapaces is brittle; a product of flabby carapaces is flabby.

(3) For any simplex, $\sigma \in X$ and any R-module, M, $M \uparrow_\sigma$ is brittle and $M \downarrow^\sigma$ is flabby.

Proof. Let Z be any subcomplex of X, let C be its complement and let M be any R-module. Then $0 \longrightarrow M_*^C \longrightarrow M_X \longrightarrow M_Z^* \longrightarrow 0$ is exact. Thus, if P is projective, $\mathbf{Hom}_{R,X}(P, M_X) \rightarrow \mathbf{Hom}_{R,X}(P, M_Z^*)$ is surjective. But this is the map, $Hom_R(\Sigma(P,X), M) \rightarrow Hom_R(\Sigma(P,Z), M)$. But this map will be surjective for every M if and only if the map $\Sigma(Z,P) \rightarrow \Sigma(X,P)$ is injective (in fact, it must be split).

If I is injective, we need only consider the case, $M = R$. Then $\mathbf{Hom}_{R,X}(R_X, I) \rightarrow \mathbf{Hom}_{R,X}(R_*^C, I)$ is surjective. This is the sequence, $\Gamma(X,I) \rightarrow \Gamma(C,I)$ and hence (2) is established.

To prove 2), let $\{A_i\}_{i\in I}$ be a family of R-carapaces on X. Since inductive limits of arbitrary co-products are co-products and projective limits of products are products, we may write:

$$(6.9)\qquad \begin{array}{rcl} \Sigma(Z, \coprod_{i\in I} A_i) & = & \coprod_{i\in I} \Sigma(Z, A_i) \\ \Gamma(C, \prod_{i\in I} A_i) & = & \prod_{i\in I} \Gamma(C, A_i) \end{array}$$

Since a co-product of monomorphisms is monic and a product of surjections is surjective, 2) follows at once.

Statement 3) is quite clear. □

6.10 Proposition. *If A is a brittle R-carapace on X, then $H_i(X, A) = 0$ for all $i > 0$. If F is flabby, then $H^i(X, F) = 0$ for all $i > 0$.*

Proof. First choose a projective, P and a surjective map so that there is an exact sequence:

$$0 \longrightarrow A_0 \longrightarrow P \longrightarrow A \longrightarrow 0$$

The acyclicity of P and Theorem 6.5 together imply that for any brittle A, $H_1(X, A)$ $= 0$. Then choose a projective resolution of A. Break this into a series of short exact sequences, use Corollary 6.6 and apply induction. The same technique, applied dually, gives the second statement. □

We conclude with local criteria for which there are no immediate applications but which are somewhat interesting.

6.11 Proposition. *Let A be an R-carapace on X.*

(1) If for each simplex $\sigma \in X$ the map, $\Sigma(\hat{\sigma}, A) \to \Sigma(\tilde{\sigma}, A)$, is injective then A is brittle.

(2) If for each simplex $\sigma \in X$ the restriction $A|X(\sigma)$ is flabby, then A is flabby.

Proof. Proofs of these statements use Zorn's lemma as it was used in Theorem 6.5. First we prove 1). Let Z be an arbitrary subcomplex of X. We must show that $\Sigma(Z, A) \to \Sigma(X, A)$ is injective. As in the proof of theorem 5, this comes to proving that for any injective module, J, any morphism, $f : A|Z \to J_Z$, admits an extension, $f : A \to J_X$. Applying Zorn one finds a maximal subcomplex on which f admits an extension and so, replacing Z by this maximal subcomplex, we may assume that f does not extend to any subcomplex containing Z. If the vertex x is not in Z then f clearly extends to the disconnected union and so we may assume that every vertex is in Z. Choose a simplex, σ of minimal dimension among the simplices not in Z. Then $\hat{\sigma} \subseteq Z$. Making use of the condition in (1), we obtain a diagram:

$$\begin{array}{ccccc} 0 & \longrightarrow & \Sigma(\hat{\sigma}, A) & \longrightarrow & \Sigma(\tilde{\sigma}, A) \\ & & \downarrow & & \\ & & \Sigma(Z, A) & & \\ & & \downarrow f_Z & & \\ & & J & & \end{array}$$

Hence there is a map, $f_1 : \Sigma(\tilde{\sigma}, A) \to J$ extending f_Z and so one may extend f to $Z \cup \sigma$ contradicting maximality. It follows that it must be that $Z = X$.

The proof of 2), by duality, is entirely straightforward and so we omit it. □

7 Canonical Resolutions

In this section we give canonical chain and co-chain complexes which can be used to compute the exoskeletal homology and cohomology groups. They arise from canonical resolutions and they are sufficiently canonical that they will be seen to be equivariant when there is a group action involved.

Let A be an R-carapace on X. Then by 1.4, 1) and 2), the identity map on $A(\sigma)$ induces a map, $\pi_\sigma : (A(\sigma))\uparrow_\sigma \rightarrow A$ and a map $j_\sigma : A \rightarrow (A(\sigma))\downarrow^\sigma$.

7.1 Definition. Let X be a simplicial complex and let A be an R-carapace on X.

(1) Let

$$\mathcal{T}_0(A) = \coprod_{\sigma\in X} (A(\sigma))\uparrow_\sigma \qquad \text{and let} \qquad \pi_A = \coprod_{\sigma\in X} \pi_\sigma$$

(2) Let

$$\mathcal{S}^0(A) = \prod_{\sigma\in X} (A(\sigma))\downarrow^\sigma \qquad \textit{and let} \qquad j_A = \prod_{\sigma\in X} j_\sigma.$$

(3) Let $\mathcal{K}_0(A) = Ker(\pi_A)$.

(4) Let $\mathcal{C}^0(A) = Coker(j_A)$.

This definition has certain immediate consequences.

7.2 Lemma. *Let X be a simplicial complex and let A be an R-carapace on X.*

(1) The four functors, $\mathcal{T}_0, \mathcal{K}_0, \mathcal{S}^0$, and *$\mathcal{C}^0$ are exact additive functors.*

(2) Both π_A and j_A are natural transformations in the argument A. Further π_A is always surjective and j_A is always monic.

(3) For all A, $\mathcal{T}_0(A)$ is brittle and $\mathcal{S}^0(A)$ is flabby.

(4) If A is brittle, then $\mathcal{K}_0(A)$ is brittle; if A is flabby, $\mathcal{C}^0(A)$ is flabby.

Proof. That $\mathcal{T}_0$ and $\mathcal{S}^0$ are exact and additive is a trivial observation. Since $\mathcal{T}_0(A)$ is a coproduct of carapaces of the form $M\uparrow_\sigma$, Proposition 6.8, 2) and 3) guarantee that it is brittle. The flabbyness of $\mathcal{S}^0(A)$ follows similarly from the fact that it is a product of carapaces of the form $M\downarrow^\sigma$. That $\mathcal{K}_0$ and $\mathcal{C}^0$ are exact is little more than the snake lemma. Statement 2) is a triviality and so only 4) remains to be proven. This follows from 3) and Corollary 6.6. □

Definition 7.1 and Lemma 7.2 are just what is necessary to construct standard resolutions.

7.3 Definition. Let A be an R-carapace on X. Let $\mathcal{K}_n(A) = \mathcal{K}_0(\mathcal{K}_{n-1}(A))$ and let $\mathcal{C}^n(A) = \mathcal{C}^0(\mathcal{C}^{n-1}(A))$. That is $\mathcal{K}_n$ is the $(n+1)'st$ iterate of $\mathcal{K}_0$ and the same, mutatis mutandis, is true for $\mathcal{C}^n$. Let $\mathcal{T}_{n+1}(A) = \mathcal{T}_0(\mathcal{K}_n(A))$ and let $\mathcal{S}^{n+1}(A) = \mathcal{S}^0(\mathcal{C}^n(A))$ for $n \geq 0$. Define maps, $\delta_n : \mathcal{T}_{n+1}(A) \to \mathcal{T}_n(A)$ and $\delta^n : \mathcal{S}^n(A) \to \mathcal{S}^{n+1}(A)$ as follows. The map, δ_n is the composition of the natural surjection, $\mathcal{T}_{n+1}(A) \to \mathcal{K}_n(A)$, with the inclusion, $\mathcal{K}_n(A) \hookrightarrow \mathcal{T}_n$. Similarly δ^n is the composition of the surjection, $\mathcal{S}^n(A) \to \mathcal{C}^n(A)$, and the inclusion, $\mathcal{C}^n(A) \hookrightarrow \mathcal{S}^{n+1}(A)$. Then $\{\mathcal{T}_n(A), \delta_n\}$ is called the *canonical brittle resolution* and $\{\mathcal{S}^n(A), \delta^n\}$ is called the *canonical flabby resolution* of A.

Some remarks are in order. First of all, since each of the functors, $\mathcal{T}_n$ and $\mathcal{S}^n$, are compositions of exact functors, they are themselves exact functors. Further, by Lemma 7.2, for any A, each of the carapaces $\mathcal{T}_n(A)$ is brittle while the $\mathcal{S}^n(A)$ are flabby. Thus, letting $C_n(X, A) = \Sigma(X, \mathcal{T}_n(A))$ and $C^n(X, A) = \Gamma(X, \mathcal{S}^n(A))$, whenever $0 \longrightarrow A' \longrightarrow A \longrightarrow A'' \longrightarrow 0$ is exact,

$$0 \longrightarrow C_n(X.A') \longrightarrow C_n(X, A) \longrightarrow C_n(X, A'') \longrightarrow 0$$

and

$$0 \longrightarrow C^n(X, A') \longrightarrow C^n(X, A) \longrightarrow C^n(X, A'') \longrightarrow 0$$

are exact. Abusing language, use δ_n and δ^n for the maps of segments and sections respectively as well as maps of carapaces, the homology groups of the complexes, $\{C_n(X, A), \delta_n\}$ and $\{C^n(X, A), \delta^n\}$ are connected sequences of homological functors.

7.4 Definition. Let A be an R-carapace on X. The complex, $\{C_n(X, A), \delta_n\}$ will be called the *complex of Alexander chains* on X with coefficients in A; $\{C^n(X, A), \delta^n\}$ will be called the Alexander co-chains. The homology of the complex of Alexander chains will be called the *Alexander homology* and it will be written, $H^a_n(X, A)$. The homology of the Alexander co-chain complex will be called the *Alexander cohomology* and it will be written $H^n_a(X, A)$.

7.5 Proposition. *The Alexander homology and cohomology of the simplicial complex, X, with coefficients in A are isomorphic, respectively, to the exoskeletal homology and cohomology of X with coefficients in A, functorially in A.*

Proof. By Proposition 6.10, the exoskeletal homology groups vanish on brittle carapaces while the cohomology groups vanish on flabby carapaces. Hence the Alexander groups are the homology groups of the segments (respectively sections) over an acyclic resolution. The proposition follows. □

The following is an interesting footnote.

7.6 Proposition. *If A is projective, the canonical brittle resolution of A consists of projective carapaces. If A is injective, each term in the canonical flabby resolution is injective.*

Proof. It suffices to prove that if A is projective then $\mathcal{T}_0(A)$ and $\mathcal{K}_0(A)$ are projective and the corresponding statement for an injective A and $\mathcal{S}^0$ and $\mathcal{C}^0$. Suppose that P is projective and that I is injective. Then by Proposition 1.6, $P(\sigma)$ is projective and $I(\sigma)$ is injective for each $\sigma \in X$. But then, by 1.4, $(P(\sigma)) \uparrow_\sigma$ is projective and $(I(\sigma)) \downarrow^\sigma$ is injective. By the definition of $\mathcal{T}_0$ and $\mathcal{S}^0$ and because coproducts of projectives are projective and products of injectives are injective, $\mathcal{T}_0(P)$ is projective and $\mathcal{S}^0(I)$ is injective. But then

$$0 \longrightarrow \mathcal{K}_0(P) \longrightarrow \mathcal{T}_0(P) \xrightarrow{\pi_P} P \longrightarrow 0$$

and

$$0 \longrightarrow I \xrightarrow{j_I} \mathcal{S}^0(I) \longrightarrow \mathcal{C}^0(I) \longrightarrow 0$$

are exact. The last two terms of the first sequence are projective while the first two terms of the second sequence are injective. Hence $\mathcal{K}_0(P)$ is projective and $\mathcal{C}^0(I)$ is injective. An iterative application of these facts establishes the result. □

8 G-carapaces and their Homology

In this section we consider a simplicial complex, X, with a G-action for some group, G. Then there is a corresponding notion of G-carapace and several ways of constructing G-representations on the homology and cohomology of a G-carapace. One of our main purposes in this section is to show that all of these representations are the same. The method is standard "relative homological algebra".

If X is a simplicial complex and G is a group, a simplicial action of G on X is an action of G on the vertex set of X which carries simplices to simplices. If σ is a simplex in X, write G_σ for the setwise stabilizer of σ and $\hat{G}_\sigma$ for the pointwise stabilizer of σ. We will usually write t_g for the translation map, $t_g(x) = gx$. Then, if A is an R-carapace on X, the inverse image of A under t_g is the carapace, $[t_g^*(A)](\sigma) = A(g\sigma)$. When space does not permit otherwise, write g^*A for $t_g^*(A)$. Recall that the expansions on $t_g^*(A)$ are the maps, $e^{\tau}_{g^*A,\sigma} = e^{g\tau}_{A,g\sigma}$. Notice that $t_g^*(t_h^*A) = t_{hg}^*(A)$.

8.1 Definition. A *G-carapace* on X is an R-carapace, A together with a family of isomorphisms, $\Phi = \{\Phi_g\}_{g\in G}$, $\Phi_g : A \to t_g^* A$, such that for any pair, $g, h \in G$, the diagram:

$$(8.2)\qquad \begin{array}{ccc} A & \xrightarrow{\Phi_g} & t_g^* A \\ \Phi_{hg}\downarrow & & \downarrow t_g^*\Phi_h \\ t_{hg}^* A & & t_g^*(t_h^* A) \end{array}$$

commutes. That is, $t_g^*(\Phi_h) \circ \Phi_g = \Phi_{hg}$.

If (A, Φ) and (B, Ψ) are two G-carapaces a *G-morphism* $f : A \to B$ is just a morphism such that $\Psi_g \circ f = t_g^*(f) \circ \Phi_g$. Clearly G-carapaces are an Abelian category.

We now apologize for a digression which some would prefer to conceal in an "obviously". Write $\prod A$ for the functor $\prod_{\sigma\in X} A(\sigma)$. An element, a, in $\prod A$ is a function such that $a(\sigma) \in A(\sigma)$. If α is an automorphism of X, there is an isomorphism, $I_A^\alpha : \prod \alpha^* A \to \prod A$. It is defined by $[I_A^\alpha(a)](\sigma) = a(\alpha^{-1}(\sigma))$. Now the copoduct, $\coprod_{\sigma\in X} A(\sigma)$, the module of relations ,N, such that $\coprod_{\sigma\in X} A(\sigma)/N = \Sigma(X, A)$ and $\Gamma(X, A)$ are all submodules of $\prod A$ preserved by I_A^α and hence I_A^α induces two other maps, both of which we will denote I_A^α, from $\Sigma(X, \alpha^* A)$ to $\Sigma(X, A)$ and from $\Gamma(X, \alpha^* A)$ to $\Gamma(X, A)$. These maps are functorial in the same ways and so we will describe their properties for $\prod A$ and consider them established for all three functors. Let A and B be two R-carapaces and let α and β be automorphisms of X.Let $\phi : A \to B$ be a morphism. The following equations, whose proof we leave to the reader, are the properties of interest. We emphasize that we shall use these equations for Σ and Γ rather than $\prod$.

$$(8.3)\qquad \begin{array}{rcl} I_A^\alpha \circ I_{\alpha^* A}^\beta & = & I_A^{\alpha\circ\beta} \\ (\prod \phi) \circ I_A^\alpha & = & I_B^\alpha \circ \prod \alpha^* \phi \end{array}$$

In general, if $f : A \to B$ is a morphism, write f^Σ and f^Γ for the induced morphisms on the segment and the sections respectively. Let (A, Φ) be a G-carapace on X. Then there are natural representations of G on $\Sigma(X, A)$ and $\Gamma(X, A)$ respectively denoted Φ^Σ and Φ^Γ defined by the equations:

$$(8.4) \qquad \begin{aligned} (\Phi^{\Sigma})_g &= I^g_A \circ (\Phi_g)^{\Sigma} \\ (\Phi^{\Gamma})_g &= I^g_A \circ (\Phi_g)^{\Gamma} \end{aligned}$$

To see that these are representations, we just apply 8.3. Then, $(\Phi^{\Sigma})_{gh} = I^{gh}_A \circ (\Phi_{gh})^{\Sigma} = I^g_A \circ I^h_{g^*A} \circ ((h^*\Phi_g)^{\Sigma}) \circ \Phi^{\Sigma}_h = I^g_A \circ (I^h_{g^*A} \circ (h^*\Phi_g)^{\Sigma}) \circ \Phi^{\Sigma}_h = I^g_A \circ (\Phi^{\Sigma}_g \circ I^h_A) \circ \Phi^{\Sigma}_h = (\Phi^{\Sigma})_g \circ (\Phi^{\Sigma})_h$ The computation for Φ^{Γ} is virtually identical.

It is also clear that this argument gives canonically determined representations of G on the left derived functors of $\Sigma(X,?)$ and the right derived functors of $\Gamma(X,?)$. We give the argument for $\Sigma(X,?)$. Let α be an automorphism of X. Then α^* is an automorphism of $Car_R(X)$ and so it carries projectives to projectives and injectives to injectives. It is moreover an exact functor. Let $\ldots \to P_r \to P_{r-1} \to P_0 \to A \to 0$ be a projective resolution of A. Then $\ldots \to \alpha^*P_r \to \alpha^*P_{r-1} \to \ldots \to \alpha^*P_0 \to \alpha^*A \to 0$ is a projective resolution of A and, applying $\Sigma(X,-)$, deleting α^*A and taking homology yields the left derived functors of $\Sigma(X, \alpha^*(-))$. Making use of the functoriality expressed by the second equation of 8.3), we obtain a commutative diagram.

$$\begin{array}{ccccccccc} \cdots & \longrightarrow & \Sigma(X,\alpha^*P_r) & \longrightarrow & \Sigma(X,\alpha^*P_{r-1}) & \longrightarrow & \cdots & \longrightarrow & \Sigma(X,\alpha^*P_0) \\ & & I^{\alpha}_{P_r}\downarrow & & I^{\alpha}_{P_{r-1}}\downarrow & & & & I^{\alpha}_{P_0} \\ \cdots & \longrightarrow & \Sigma(X,P_r) & \longrightarrow & \Sigma(X,P_{r-1}) & \longrightarrow & \cdots & \longrightarrow & \Sigma(X,P_0) \end{array}$$

Passing to the homology of these complexes, we obtain unique, canonically defined morphisms, $L_rI^{\alpha}_A : H_r(X,\alpha^*A) \to H_r(X,A)$. Clearly, the dual construction will yield canonical morphisms, $R^qI^{\alpha}_A : H^q(X,\alpha^*A) \to H^q(X,A)$.

Suppose that G acts on X and that A is a G-carapace with G-structure, Φ. It is now clear that there is a canonical representation of G on the exoskeletal homology and cohomology groups of X in A. Write $L_r\Phi_g : H_r(X,A) \to H_r(X,g^*A)$ and $R^q\Phi_g : H^q(X,A) \to H^q(X,g^*A)$ for the map induced by Φ_g on the homology and the cohomology respectively. Let $\Phi^q_g = R^qI^g_A \circ R^q\Phi_g$ and $\Phi^g_r = L_rI^g_A \circ L_r\Phi_g$. Then Φ^g_r and Φ^q_g are easily seen to give the unique representations on the homology and cohomology groups making them into homological functors with values in the category of G-modules.

What remains is the question of natural G-structures on carapace valued functors applied to G-carapaces. Let A be a carapace with G-structure, Φ, and let B be one with G-structure, Ψ. First consider the tensor product,

$A \otimes_R B$. Inverse image preserves tensor product. That is, $t_g^*(A \otimes_R B) \simeq t_g^*(A) \otimes_R t_g^*(B)$. Consequently, the family of isomorphisms, $\{\Phi_g \otimes_R \Psi_g : g \in G\}$, is a G-structure on $A \otimes_R B$. We denote it $\Phi \otimes_R \Psi$.

Before considering $\mathcal{H}om_{X,R}(A, B)$ observe that the commutativity expressed by diagram 8.2) can also be described by the equation:

$$\Phi_{h,g\sigma} \circ \Phi_{g,\sigma} = \Phi_{hg,\sigma} \tag{8.5}$$

Consider the carapace of local homomorphisms. By definition, $\mathcal{H}om_{X,R}(A, B)(\sigma) = \mathbf{Hom}_{X(\sigma),R}(A|_{X(\sigma)}, B|_{X(\sigma)})$. We define a map, $\Theta_{g,\sigma} : \mathbf{Hom}_{X(\sigma),R}(A|_{X(\sigma)}, B|_{X(\sigma)}) \to \mathbf{Hom}_{X(g\sigma),R}(A|_{X(g\sigma)}, B|_{X(g\sigma)})$. The translation ,$t_g$ maps $X(\sigma)$ to $X(g\sigma)$. Hence Φ_g maps $A|_{X(\sigma)}$ to $t_g^*(A|_{X(g\sigma)})$ and similarly for B. Consequently, for any $f \in \mathbf{Hom}_{X(\sigma),R}(A|_{X(\sigma)}, B|_{X(\sigma)})$, $\Psi_{g,\sigma} \circ f \circ (\Phi_{g,\sigma})^{-1}$ maps $t_g^*(A|_{X(g\sigma)})$ to $t_g^*(B|_{X(g\sigma)})$. Hence, $t_{g^{-1}}^*(\Psi_{g,\sigma} \circ f \circ (\Phi_{g,\sigma})^{-1}) \in \mathbf{Hom}_{X(g\sigma),R}(A|_{X(g\sigma)}, B|_{X(g\sigma)})$.But, this last group is just $\mathcal{H}om_{X,R}(A, B)(g\sigma)$. Hence, define the G-structure on $\mathcal{H}om_{X,R}$ by the equation:

$$\Theta_{g,\sigma}(f) = t_{g^{-1}}^*(\Psi_g|_{X(\sigma)} \circ f \circ (\Phi_g|_{X(\sigma)})^{-1}) \tag{8.6}$$

This equation is to be understood in the following sense. The map $\Phi_g|_{X(\sigma)}$ maps the restriction of A to $X(\sigma)$ to the corresponding restriction of t_g^*A while $\Psi_g|_{X(\sigma)}$ does the same for B. Hence the composition in parentheses takes $t_g^*(A)|_{X(\sigma)}$ to $t_g^*(B)|_{X(\sigma)}$. Thus the inverse image of this map under $t_{g^{-1}}$ yields an element of $\mathcal{H}om_{X,R}(A, B)(g\sigma) = \mathbf{Hom}_{X(g\sigma),R}(A|_{X(g\sigma)}, B|_{X(g\sigma)})$ which is what is needed.

We check that Θ is a G-structure by establishing 8.5 for it by direct computation. The computation is:

$$\begin{aligned}
\Theta_{hg,\sigma}(f) &= t_{(hg)^{-1}}^*(\Psi_{hg} \circ f \circ (\Phi_{hg})^{-1}) \\
&= t_{h^{-1}}^*(t_{g^{-1}}^*(t_g^*(\Psi_h) \circ \Psi_g \circ f \circ \Phi_g^{-1} \circ t_g^*(\Phi_h^{-1}))) \\
&= t_{h^{-1}}^*(\Psi_h \circ t_{g^{-1}}^*(\Psi_g \circ f \circ \Phi_g^{-1}) \circ \Phi_h^{-1}) \\
&= \Theta_{h,g\sigma}(\Theta_{g,\sigma}(f))
\end{aligned}$$

Thus Θ is a natural G-structure on $\mathcal{H}om_{X,R}(A, B)$. But $\Gamma(X, \mathcal{H}om_{X,R}(A, B)) = \mathbf{Hom}_{X,R}(A, B)$. Hence 8.4 determines a representation of G on $\mathbf{Hom}_{X,R}(A, B)$. This is what we will call the natural representation of G on $\mathbf{Hom}_{X,R}(A, B)$. The explicit description of this action is:

$$g \cdot f = t_{g^{-1}}^*(\Psi_g \circ f \circ (\Phi_g)^{-1}) \qquad g \in G \quad f \in \mathbf{Hom}_{X,R}(A, B) \tag{8.7}$$

Establish this as follows. If T is and G-carapace with G-structure Θ, then if $\tau \in \Gamma(X,T)$ the action of G on $\Gamma(X,T)$ defined by 8.4 is described by the equation:

$$(g\cdot\tau)_\sigma = \Theta_{g,g^{-1}\sigma}(\tau_{g^{-1}\sigma}) \tag{8.8}$$

When $T = \mathcal{H}om_{X,R}(A,B)$, this becomes $(g\cdot f)_\sigma = \Theta_{g,g^{-1}\sigma}(f_{g^{-1}\sigma})$. Under the identification of $\mathbf{Hom}_{X,R}(A,B)$ with $\Gamma(X,\mathcal{H}om_{X,R}(A,B))$, the σ component of the map, f, is $f|_{X(\sigma)}$. Using this, apply 8.6 to compute the right hand side of 8.8 for $T = \mathcal{H}om_{X,R}(A,B)$. We obtain:

$$(g\cdot f)_\sigma = t^*_{g^{-1}}(\Psi_g|_{X(g^{-1}\sigma)} \circ (f|_{X(g^{-1}\sigma)}) \circ (\Phi_g|_{X(g^{-1}\sigma)})^{-1}).$$

By the definition of inverse image, this is $(t^*_{g^{-1}}(\Psi_g \circ f \circ (\Phi_g)^{-1}))|_{X(\sigma)}$ and this is just the right hand side of 8.7 restricted to $X(\sigma)$. This proves the truth of 8.7.

8.9 Proposition. *Let A, B and C be G-carapaces with G-structures, Φ, Ψ and Υ respectively. Then, the natural adjointness isomorphism*

$$\phi : \mathbf{Hom}_{X,R}(A,\mathcal{H}om_{X,R}(B,C)) \to \mathbf{Hom}_{X,R}(A\otimes_R B, C)$$

is a G-morphism.

Proof. Fix $\sigma \in X$, $a\in A(\sigma)$, $b \in B(\sigma)$ and $f \in \mathbf{Hom}_{X,R}(A,\mathcal{H}om_{X,R}(B,C))$ and recall the definition of ϕ. It is $\phi(f)_\sigma(a\otimes b) = [f_\sigma(a)]_\sigma(b)$ and in interpreting this formula one must remember that $f_\sigma(a) \in \mathbf{Hom}_{X(\sigma),R}(B|_{X(\sigma)}, C|X(\sigma))$. We will show that $\phi(g\cdot f) = g\cdot\phi(f)$ and we will prove this by evaluating both sides of this equation on $a\otimes b \in A(\sigma)\otimes B(\sigma)$. Starting with the left hand side:

$$\begin{aligned}
[\phi(g\cdot f)_\sigma](a\otimes b) &= [(t^*_{g^{-1}}(\Theta_g\circ f\circ(\Phi_g)^{-1}))_\sigma(a)]_\sigma(b) \qquad \text{by 8.7)}\\
&= [(\Theta_{g,g^{-1}\sigma}\circ f_{g^{-1}\sigma}\circ(\Phi_{g,g^{-1}\sigma})^{-1})(a)]_\sigma(b)\\
&= \Upsilon_{g,g^{-1}\sigma}([f_{g^{-1}\sigma}(\Phi^{-1}_{g,g^{-1}\sigma}(a))]_{g^{-1}\sigma}](\Psi^{-1}_{g,g^{-1}\sigma}(b)))\\
&= \Upsilon_{g,g^{-1}\sigma}[\phi(f)(\Phi^{-1}_{g,g^{-1}\sigma}(a)\otimes\Psi^{-1}_{g,g^{-1}\sigma}(b))] \text{ by the def. of } \phi\\
&= [g\cdot\phi(f)](a\otimes b) \qquad \text{by 8.7)}
\end{aligned}$$

That is $\phi(g\cdot f)_\sigma(a\otimes b) = [g\cdot\phi(f)]_\sigma(a\otimes b)$ as asserted. This proves the result. □

We conclude this section by showing that the canonical brittle and flabby resolutions of a G-carapace, A, are naturally equivariant. A cosideration of the definitions of these resolutions shows that it suffices to show that there are canonical G-structures on $\mathcal{T}_0(A)$ and $\mathcal{S}^0(A)$ so that the maps, $\mathcal{T}_0(A)\to A$ and $A\to\mathcal{S}^0(A)$ are G-equivariant.

8.10 Lemma. *Let $\alpha : X \to X$ be an automorphism. Let M be an R-module, let A be an R-carapace and let $\Phi : A \to \alpha^* A$ be an isomorphism.*

(1) $\alpha^*(M \uparrow_\sigma) = M \uparrow_{\alpha^{-1}\sigma}$

(2) $\alpha^*(M \downarrow^\sigma) = M \downarrow^{\alpha^{-1}\sigma}$

(3) There is a natural equality $\mathcal{T}_0(\alpha^* A) = \alpha^{*f}(\mathcal{T}_0(A))$.

(4) There is a natural equality $\mathcal{S}^0(\alpha^* A) = \alpha^* \mathcal{S}^0(A)$.

Proof. The first two statements are trivially true. As for the third and fourth statements, the proofs are nearly identical and so we prove only the first. Write:

$$\begin{aligned} \mathcal{T}_0(\alpha^* A) &= \coprod_{\sigma \in X} \alpha^*(A)(\sigma) \uparrow_\sigma \\ &= \coprod_{\sigma \in X} (A(\alpha\sigma)) \uparrow_\sigma \\ &= \coprod_{\sigma \in X} A(\sigma) \uparrow_{\alpha^{-1}\sigma} \\ &= \coprod_{\sigma \in X} \alpha^*(A(\sigma) \uparrow_\sigma) \\ &= \alpha^* \mathcal{T}_0(A) \end{aligned}$$

Thus $\alpha^*(\mathcal{T}_0(A) = \mathcal{T}_0(\alpha^* A)$. That is, the two apparently different constructions applied to A, $\alpha^*\mathcal{T}_0(A)$ and $\mathcal{T}_0(\alpha^*(A))$ result in identically the same object. □

8.11 Proposition. *Let A be a G-carapace on X with G-structure Φ. For each $g \in G$, let $\Phi^{\mathcal{T}}_g = \mathcal{T}_0(\Phi_g)$ and let $\Phi^{\mathcal{S}}_g = \mathcal{S}^0(\Phi_g)$. Then $\Phi^{\mathcal{T}}$ and $\Phi^{\mathcal{S}}$ are G-structures. Furthermore the surjection, $\pi_A : \mathcal{T}_0(A) \to A$ and the injection, $j_A : A \to \mathcal{S}^0(A)$ are G-equivariant.*

Proof. First, we show that $\Phi^{\mathcal{T}}$ and $\Phi^{\mathcal{S}}$ are G-structures. For $\Phi^{\mathcal{T}}$ the calculation is:

$$\begin{aligned} \Phi^{\mathcal{T}}_{gh} &= \mathcal{T}_0(t^*_h(\Phi_g) \circ \Phi_h) \\ &= \mathcal{T}_0(t^*_h(\Phi_g)) \circ \mathcal{T}_0(\Phi_h) \\ &= t^*_h(\mathcal{T}_0(\Phi_g)) \circ \mathcal{T}_0(\Phi_h) \end{aligned}$$

To prove that π_A and j_A are equivariant just note that π is a natural transformation from $\mathcal{T}_0$ to the identity functor while j is one from the identity functor to $\mathcal{S}^0$. Then note that $\Phi^{\mathcal{T}}_g$ and $\Phi^{\mathcal{S}}_g$ are just the values of $\mathcal{T}_0$ and $\mathcal{S}^0$ on the morphism, Φ_g. □

8.12 Corollary. *Let A be a G-carapace on X. Then there are canonical G-structures on the canonical brittle and flabby resolutions of A so that the natural morphisms are G-equivariant.*

Proof. This is nothing more than an iterative application of 8.11). The details are left to the reader. □

9 Induced and Co-Induced Carapaces

Let X be a simplicial complex. Recall that $X(r)$ denotes the set of simplices of dimension r. We will use X_n to denote the collection of simplices of dimension at least n. That is, $X_n = \bigcup_{r \geq n} X(r)$.

Let G be a group acting simplicially on X. Recall that for any simplex, $\sigma \in X$, $G_\sigma = \{g : g \in G, g\sigma = \sigma\}$ and $\widehat{G}_\sigma = \{g : g \in G_\sigma gx = x \quad \forall x \in \sigma\}$. We will call the action of G on X *separated* if whenever $\tau \subseteq \sigma$ and $g\tau \subseteq \sigma$ for some $g \in G$, then $g\tau = \tau$.

If G acts on X, let $Y(0) = X(0)/G$ be the orbit space and let $\pi : X(0) \to Y(0)$ be the quotient map. Construct a simplicial complex, Y, with vertex set , $Y(0)$, by taking as simplices in Y all finite subsets, $\tau \subseteq Y$ such that $\tau = \pi(\sigma)$ for some simplex σ in X. If the action of G on X is separated, then for each simplex σ in X, $G_\sigma \subseteq \widehat{G}_\sigma$ and the dimension of $\pi(\sigma)$ is equal to the dimension of σ.

If G is acting on X so that the action is separated and if Y is the quotient with quotient map $\pi : X \to Y$, then a section to π is a simplicial map, $s : Y \to X$, such that $\pi \circ s = id_Y$. A separated action admitting a section, $s : Y \to X$, will be called an *excellent* action. If the action of G on X is excellent with section, $s : Y \to X$, we will identify Y with its image, $s(Y)$, in X and we shall refer to π as the retraction onto Y. We will describe this situation by saying that (X, G) is an excellent pair with retraction $\pi : X \to Y$. Notice that the action of G on X is separated if and only if whenever $X(\sigma) \cap X(g\sigma) \neq \emptyset$ then $g\sigma = \sigma$.

If $\mathcal{C}$ is a category and X is a simplicial complex, then a $\mathcal{C}$-valued sheaf on X is just a contravariant functor from X to $\mathcal{C}$. If $\sigma \subseteq \tau$ and $\mathbf{S}$ is a sheaf on X write $r^\sigma_{\mathbf{S},\tau} : \mathbf{S}(\tau) \to \mathbf{S}(\sigma)$ for the corresponding map and call it the restriction. If G operates simplicially on X, then the assignment, $\widehat{G}(\sigma) = \widehat{G}_\sigma$ is a sheaf of groups on X. If the action is separated, then $G_\sigma = \widehat{G}_\sigma$, and so this also is a sheaf of groups. In any case we will refer to $\widehat{G}$ as the stabilizer of the action.

If $f : X \to Y$ is a morphism of complexes and $\mathbf{S}$ is a sheaf on Y then $f^*\mathbf{S}$, defined by the equation $f^*\mathbf{S}(\sigma) = \mathbf{S}(f(\sigma))$ with the corresponding restrictions is called the inverse image of $\mathbf{S}$. When f is the inclusion of a subcomplex, we call $f^*\mathbf{S}$ the restriction of $\mathbf{S}$ to X and we may on occasion write it, $\mathbf{S}_X$.

Suppose now that G is a group and that H is a subgroup. We wish to fix notation for induced and co-induced modules. Write $R[G]$ for the group algebra of G over R and write $R[G]^\ell$ for the free rank one $R[G]$-module isomorphic to $R[G]$ as an R-module but with $R[G]$ structure defined by the equation, $g \cdot x = xg^{-1}$ for $x \in R[G]$ and $g \in G$. We simply write gx for the product in $R[G]$. Let M be an H-module. Then the G-module induced by M is $R[G] \otimes_{R[H]} M$; the G-module co-induced by M is $Hom_{R[H]}(R[G]^\ell, M)$. The G-structure on the induced module is just that obtained from left multiplication on $R[G]$. The structure on the co-induced module is just the structure described by $(gf)(x) = f(g^{-1}x)$. Write $I_{G/H}M$ for the induced module and $C_{G/H}M$ for the co-induced module.

Choose a complete set of coset representatives $Q \subset G$ for the space of left cosets, G/H. Then $I_{G/H}M = \coprod_{g \in Q} g \otimes M$. Write gM for $g \otimes M$. The R-module, gM depends only on the coset gH and not on the particular representative, g.

Suppose now that $H = G_\sigma$ for some simplex, σ. If $\gamma \in G\sigma$ write M^γ to denote xM for any x such that $x\sigma = \gamma$. Then M^γ depends only on γ for $x\sigma = y\sigma = \gamma$ if and only if $x \in yG_\sigma = yH$. If $g\gamma = \tau$ then $gM^\gamma = M^\tau$. Dually, $R[G]^\ell = \coprod_{x \in Q} R[H] \cdot x = \coprod_{x \in Q} xR[H]$. Hence $C_{x \in Q}(M) = \prod_{x \in Q} Hom_{R[H]}(xR[H], M)$. If $\gamma \in G\sigma$ let $M_\gamma = Hom_{R[H]}(xR[H], M)$ for any x such that $x\sigma = \gamma$. This is well defined. Moreover, if $y\gamma = \lambda$ then left translation by y carries M_γ to M_λ.

9.1 Definition. Suppose G acts on X, that σ is a simplex in X and that M is a G_σ representation over R. Let:

$$\begin{aligned} T_\sigma(M) &= \textstyle\coprod_{\tau \in G\sigma} M^\tau \uparrow_\tau \\ S^\sigma(M) &= \textstyle\prod_{\tau \in G\sigma} M_\tau \downarrow^\tau \end{aligned} \tag{9.2}$$

Then $T_\sigma(M)$ is called the *carapace induced by M* and $S^\sigma(M)$ is called the *carapace coinduced by M*.

9.3 Proposition. *Let G act on X and let M be a representation of G_σ over R. Then*

(1) $T_\sigma M$ and $S^\sigma M$ both admit canonical G-structures.

(2) Let A be any G-carapace on X. Then

$$\mathbf{Hom}_{X,R}(T_\sigma M, A)^G = Hom_{G_\sigma}(M, A(\sigma))$$

and

$$\mathbf{Hom}_{X,R}(A, S^\sigma M)^G = Hom_{G_\sigma}(A(\sigma), M)$$

(3) $T_\sigma(M)$ is brittle; $S^\sigma(M)$ is flabby.

Proof. First observe that 3) is just a consequence of Definition 9.1 and Proposition 6.8. We pass to the construction of G-structures on $T_\sigma(M)$ and $S^\sigma(M)$.

First we evaluate these carapaces on a typical simplex, τ. Recall that $\tilde{\tau}$ denotes the full simplicial complex underlying τ. Then:

$$T_\sigma(M)(\tau) = \coprod_{\gamma \in \tilde{\tau} \cap G\sigma} M^\gamma$$

$$S^\sigma(M)(\tau) = \prod_{\gamma \in X(\tau) \cap G\sigma} M_\gamma \tag{9.4}$$

.

If $g \in G$, then g carries distinct simplices in $\tilde{\tau} \cap G\sigma$ to distinct simplices in $\widetilde{g\tau} \cap G\sigma$. Hence left multipication by g carries separate summands in $T_\sigma(M)(\tau)$ to the corresponding summands in $T_\sigma(M)(g\tau)$. Let $\Phi_{g,\tau}$ be the sum of left multiplication by g on the separate components of $T_\sigma(M)(\tau)$. Similarly define a map, $\Psi_{g,\tau}$, from $S^\sigma(M)(\tau)$ to $S^\sigma(M)(g\tau)$ by taking it to be left translation by g on each of the factors. Then, using Equation 8.5, one verifies that Φ and Ψ are G-structures.

Only 2) remains to be proved. We compute directly.

$$\begin{aligned}
\mathbf{Hom}_{X,R}(T_\sigma(M), A) &= \mathbf{Hom}_{X,R}(\textstyle\coprod_{\tau \in G\sigma} M^\tau, A) \\
&= \textstyle\prod_{\tau \in G\sigma} \mathbf{Hom}_{X,R}(M^\tau \uparrow_\tau, A) \\
&= \textstyle\prod_{\tau \in G\sigma} Hom_R(M^\tau, A(\tau)) \qquad \text{by4, (1)of§1}
\end{aligned}$$

Let $\Upsilon = \{\Upsilon_g\}_{g \in G}$ be the G-structure on A and let $\{f_\tau\}_{\tau \in G\sigma}$ be an element of $\prod_{\tau \in G\sigma} Hom_R(M^\tau, A(\tau))$. The element f_τ may be thought of as the τ segment of a morphism, $f : T_\sigma(M) \to A$. Moreover these segments may be chosen freely because $T_\sigma(M)$ is the direct sum of the carapaces, $M^\gamma \uparrow_\gamma$ as γ ranges over $G\sigma$. Then, by Equation 8.7, $g \cdot \{f_\tau\}_{\tau \in G\sigma}$ is the element of the same product whose τ-component is $\Upsilon_{g,g^{-1}\tau} \circ f_{g^{-1}\tau} \circ g^{-1}$. This element of the product is G-stable if and only if

$$\Upsilon_{g,\sigma} \circ f_\sigma \circ g^{-1} = f_{g\sigma}. \tag{9.5}$$

That is, if the family, $\{f_\tau\}$ is G-invariant, each component is uniquely determined by f_σ. Conversely, given $f_\sigma \in Hom_{G_\sigma}(M, A(\sigma))$ one may use Equation 9.5 to define $f_{g\sigma}$ for eac g.All that must be verified is that this is independent of the element, g, chosen that carries σ to $g\sigma$. Any other such element is of the form gh for some $h \in G_\sigma$. But then replacing g by gh in 9.5) yields the same result because f_σ is,by hypothesis, a G_σ-morphism. The proof for $S^\sigma(M)$ is too similar to bear repetition. □

9.6 Proposition. *Let G act on X and let M be a representation of G_σ over R. Then:*

$$\begin{aligned}\Sigma(X, T_\sigma(M)) &= I_{G/G_\sigma}(M)\\ \Gamma(X, S^\sigma(M)) &= C_{G/G_\sigma}(M)\end{aligned}$$

Proof. Begin by observing that for any R-module, N, $\Sigma(X, N\uparrow_\sigma) = N$ and $\Gamma(X, N\downarrow^\sigma) = N$. Let U be some complete set of representatives of the cosets, gG_σ. Then,

$$\begin{aligned}\Sigma(X, T_\sigma(M)) &= \Sigma(X, \textstyle\coprod_{\gamma\in G\sigma} M^\gamma\uparrow_\gamma) \qquad \text{by Definition 9.1}\\ &= \textstyle\coprod_{\gamma\in G\sigma}\Sigma(X, M^\gamma\uparrow_\gamma)\\ &= \textstyle\coprod_{x\in U} M^{x\sigma}\\ &= \textstyle\coprod_{x\in U} x\otimes M = I_{G/G_\sigma}(M)\end{aligned}$$

The corresponding computation for $S^\sigma(M)$ is:

$$\begin{aligned}\Gamma(X, S^\sigma(M)) &= \Gamma(X, \textstyle\prod_{\gamma\in G\sigma} M_\gamma\downarrow^\gamma)\\ &= \textstyle\prod_{x\in U}\Gamma(X, M_{x\sigma}\downarrow^{x\sigma})\\ &= \textstyle\prod_{x\in U} M_{x\sigma}\\ &= \textstyle\prod_{x\in U} Hom_{R[G_\sigma]}(xR[G_\sigma], M)\\ &= Hom_{R[G_\sigma]}(\textstyle\coprod_{x\in U} xR[G_\sigma], M)\\ &= Hom_{R[G_\sigma]}(R[G]^\ell, M) = C_{G/G_\sigma}(M).\end{aligned}$$

□

9.7 Proposition. *Let G act excellently on X with retraction $\pi : X \to Y \subseteq X$. Then*

$$\begin{aligned}T_\sigma(A(\sigma)) &= \textstyle\coprod_{\gamma\in G\sigma} A(\gamma)\uparrow_\gamma\\ S^\sigma(A(\sigma)) &= \textstyle\prod_{\gamma\in G\sigma} A(\gamma)\downarrow^\gamma.\end{aligned}$$

Proof. Let Φ be the G-structure on A. Then $\coprod_{\gamma\in G\sigma} A(\gamma)$ admits a natural representation of G. If $g \in G$ let $\phi_g = \prod_{\gamma\in G\sigma}\Phi_{g,\gamma}$. It is understood that application of ϕ_g must be followed by reindexing of components. Furthermore, the natural injection, $j_\sigma : A(\sigma) \to \coprod_{\gamma\in G\sigma} A(\gamma)$ is G_σ equivariant. Thus, j_σ extends to a G-morphism, $j_A : I_{G/G_\sigma} \to \coprod_{\gamma\in G\sigma} A(\gamma)$. Then $j_A(g\otimes A(\sigma)) = \phi_g(j_A(1\otimes A(\sigma)) = \Phi_{g,\sigma}(A(\sigma)) = A(g\sigma)$. Since $I_{G/G_\sigma}(A(\sigma))$ is a coproduct of the R-submodules, $g\otimes A(\sigma)$, it is clear that j_A is an isomorphism. Since $A(\sigma)^\gamma = g\otimes A(\sigma)$ one sees that j_A restricts to isomorphisms

$j_\gamma : A(\sigma)^\gamma \to A(\gamma)$ which comprise an equivariant family in the sense that $\phi_g \circ j_\gamma = j_{g\gamma} \circ h_g$ where h_g is left homothety by g.

Now consider $A(\sigma)_\gamma$ By definition it is $Hom_{R[G_\sigma]}(gR[G_\sigma], A(\sigma))$ with G_σ-action, $xf(u) = f(x^{-1} \cdot u) = f(ux)$ and where g is any element carrying σ to γ. For some choice of g, let $\beta_\gamma(f) = \Phi_{g,\gamma}(f(g))$. Choose $x \in G_\sigma$. Then $\Phi_{gx,\gamma} f(gx) = \Phi_{gx,\gamma}(x^{-1}f)(g) = \Phi_{g,\sigma} \circ \Phi_{x,\sigma}(x^{-1}f)(g) = \Phi_{g,\sigma} f(g)$. Consquently, β_γ is independent of the choice of g and the maps $\{\beta_\gamma\}$ are an equivariant family as above.

To prove the two statements of the proposition, note that, by definition, $\mathcal{T}_\sigma(A(\sigma))$
$= \coprod_{\gamma \in G\sigma} A(\sigma)_\gamma \uparrow_\gamma$ and $S^\sigma(A(\sigma)) = \prod_{\gamma \in G\sigma} A(\sigma)^\gamma \downarrow^\gamma$. The isomorphisms in question, then, are $\coprod_{\gamma \in G\sigma} j_\gamma \uparrow_\gamma$ and $\prod_{\gamma \in G\sigma} \beta_\gamma \downarrow^\gamma$. □

9.8 Proposition. *Let (X,G) be excellent with quotient, $Y \subseteq X$, and retraction, π. Let A be any G-carapace of R-modules on X. Then:*

(1) $\mathcal{T}_0(A) = \coprod_{\sigma \in Y} \mathcal{T}_\sigma(A(\sigma))$

(2) $\mathcal{S}^0(A) = \prod_{\sigma \in Y} S^\sigma(A(\sigma))$

(3) $\Sigma(X, \mathcal{T}_0(A)) = \coprod_{\sigma \in Y} I_{G/G_\sigma}(A(\sigma))$. *That is $\Sigma(X, \mathcal{T}_0(A))$ is a coproduct of induced modules.*

(4) $\Gamma(X, \mathcal{S}^0(A)) = \prod_{\sigma \in Y} C_{G/G_\sigma}(A(\sigma))$. *That is $\Gamma(X, \mathcal{S}^0(A))$ is a product of coinduced modules.*

Proof. Statements 3) and 4) follow from 1) and 2) by a direct application of Proposition 9.6 and so we need only prove 1) and 2).

To prove 1) and 2), first notice that since G acts excellently, we may write the collection of simplices in X as the disjoint union $\bigcup_{\sigma \in Y} G\sigma$. Then, just apply Proposition 9.7. One obtains:

$$\begin{aligned} \mathcal{T}_0(A) &= \textstyle\coprod_{\tau \in X} A(\tau) \\ &= \textstyle\coprod_{\sigma \in Y} \coprod_{\tau \in G\sigma} A(\tau) \\ &= \textstyle\coprod_{\sigma \in Y} \mathcal{T}_\sigma(A(\sigma)) \qquad \text{by Proposition 9.7} \end{aligned}$$

For $\mathcal{S}^0(A)$, the proof is:

$$\begin{aligned} \mathcal{S}^0(A) &= \textstyle\prod_{\sigma \in Y} \prod_{\tau \in G\sigma} A(\tau) \\ &= \textstyle\prod_{\sigma \in Y} S^\sigma(A(\sigma)) \end{aligned}$$

We leave the G-structures to the reader. □

Before proceeding note that if $Y \subseteq X$, if $\sigma \in Y$ and if M is an R-module, we may perform the constructions $M \uparrow_\sigma$ and $M \downarrow^\sigma$ both in X and in Y

and that the results might differ. Rather than lingering upon unnecessary distinctions or unnecessarily complicating notation, we caution the reader to maintain a certain vigilance in reading the next proof.

9.9 Lemma. *Let G act excellently on X with retraction $\pi : X \to Y$ and section, s. Then for any $\sigma \in Y$, and any G_σ-module, M, $s^*(T_\sigma(M)) = M \uparrow^\sigma$.*

Proof. By definition, $T_\sigma(M) = \coprod_{\gamma \in G\sigma} M^\gamma \uparrow_\gamma$. Evaluating,

$$T_\sigma(M)(\lambda) = \coprod_{\gamma \in \tilde{\lambda} \cap G\sigma} M^\gamma.$$

Suppose that two simplices, $g\sigma$ and $h\sigma$ both lie in $\tilde{\lambda}$. The action is excellent and so separated. It follows that $g\sigma = h\sigma$. If $\lambda \in Y$ then $G\sigma \cap \tilde{\lambda}$ is just one simplex and if $\sigma \in Y$ that simplex must be σ. That is, for any $\lambda \in Y$, $T_\sigma(M)(\lambda) = M \uparrow_\sigma (\lambda)$ which establishes the result. □

Suppose X is a simplicial complex and that M is a sheaf of groups on X. Let V be a carapace of R-modules on X. Suppose that for each $\sigma \in X$, we are given a representation, $\rho_\sigma : M(\sigma) \to Aut_R(V)$. Then for each pair, $\sigma \subseteq \tau$ notice that $V(\sigma)$ is naturally an $M(\tau)$ module simply by pulling back by the restriction, $r^\sigma_{M,\tau}$.

9.10 Definition. Let X be a simplicial complex and let M be a sheaf of groups on X. Then a *carapace of representations* of M on X over R is a carapace of R-modules, V, together with a family of representations, $\rho_\sigma : M(\sigma) \to Aut_R(V)$ such that for any pair, $\sigma \subseteq \tau$, the expansion, $e^\tau_{V,\sigma}$, is an $M(\tau)$-morphism. If A and B are two carapaces of representations of M on X over R, a *morphism* of carapaces of representations is a morphism of carapaces which is an $M(\sigma)$-morphism for each simplex, σ.

Notice that carapaces of representations of M are clearly an Abeliam category. Generally, when there is no danger of confusion, we will just say M-carapace to denote a carapace of representations of M on X over R.

Let G operate excellently on X with retraction, $\pi : X \to Y$. Write $s : Y \to X$ for the inclusion. Since excellent actions are separated, $G_\sigma = \widehat{G}_\sigma$, is a sheaf of groups on X which we will write G_*. If A is a G-carapace of R-modules on X, then A is certainly a carapace of representations over the sheaf of groups, G_*. Finally notice that if $f : X \to Y$ is a morhism of complexes and if V is a carapace of representations over the sheaf of groups, M, on Y then it is purely formal to check that $f^*(M)$ is a carapace of representations over the sheaf of groups, f^*M.

9.11 Definition. Suppose that G operates excellently on X with retraction, $\pi : X \to Y$, and section, $s : Y \to X$. Let A be a G-carapace of R-modules on X. Then, the *prototype* of X on Y is $s^*(A)$. It is a carapace of representations over s^*G_*, the restriction of the sheaf of stabilizers.

It is patently obvious that s^* is an exact functor from the category of G-carapaces to the category of carapaces of representations of s^*G_* on Y over R. More can be said. We will write G_* for the restriction of the stabilizer sheaf to Y if there is no danger of confusion.

9.12 Theorem. *Let G act excellently on X with retraction $\pi : X \to Y$ and section, s. Then s^* is an isomorphism of categories from the category of G-carapaces of R-modules on X to the category of carapaces of representations of s^*G_* on Y over R.*

Proof. In this discussion the functors $\mathcal{T}_i$ and $\mathcal{K}_i$ on X as well as the corresponding functors associated to Y occur. Consequently we will use $\mathcal{T}_i$ and $\mathcal{K}_i$ exclusively for the functors associated to X. The corresponding functors on Y will be written $\overline{\mathcal{T}}_i$ and $\overline{\mathcal{K}}_i$.

Write $Car_{\mathbf{R}}(G, X)$ for the category of G carapaces of R-modules on X and $Car_{\mathbf{R}}(G_*, Y)$ for the category of carapaces of representations of G_* in R-modules on Y. Then s^* is an exact functor from $Car_{\mathbf{R}}(G, X)$ to $Car_{\mathbf{R}}(G_*, Y)$. Define a functor from $Car_{\mathbf{R}}(G_*, Y)$ to $Car_{\mathbf{R}}(G, X)$ by the equation:

9.13) $$\mathcal{T}^0_Y(A) = \coprod_{\sigma \in Y} \mathcal{T}_\sigma(A(\sigma))$$

Then $s^*(\mathcal{T}^0_Y(A)) = \coprod_{\sigma \in Y} s^*(\mathcal{T}_\sigma(A(\sigma))) = \coprod_{\sigma \in Y} A(\sigma) \uparrow_\sigma$ by Lemma 9.9. But then by the definition of the functor $\overline{\mathcal{T}}_0$, this says that $s^*(\mathcal{T}^0_Y(A)) = \overline{\mathcal{T}}_0(A)$. Let $\mathcal{T}^1_Y(A) = \mathcal{T}^0_Y(\overline{\mathcal{K}}_0(A))$. For each $\sigma \in Y$, there is a natural inclusion, $\overline{\mathcal{K}}_0(A)(\sigma) \to \mathcal{T}^0_Y(A)(\sigma)$. By Proposition 9.3, 2), this inclusion gives a unique map $q_\sigma : \mathcal{T}_\sigma(\overline{\mathcal{K}}_0(A)(\sigma)) \to \mathcal{T}^0_Y(A)$. Let $q_A = \coprod_{\sigma \in Y} q_\sigma$. Then q_A maps $\mathcal{T}^1_Y(A)$ to $\mathcal{T}^0_Y(A)$ and its restriction to Y is just the natural map from $\overline{\mathcal{T}}_1(A)$ to $\overline{\mathcal{T}}_0(A)$. Define a functor on $Car_{\mathbf{R}}(G_*, Y)$ to $Car_{\mathbf{R}}(G, X)$ by the equation:

9.14) $$\mathcal{I}_Y(A) = Coker(q_A)$$

We will show that $\mathcal{I}_Y$ is inverse to s^*. First suppose that A is in $Car_{\mathbf{R}}(G^*, Y)$. Then the sequence,

$$\mathcal{T}^1_Y(A) > q_A >> \mathcal{T}^0_Y(A) \to \mathcal{I}_Y(A) \to 0$$

is exact. Apply the exact functor, s^*. The result is a commutative diagram:

$$\begin{array}{ccccccc} s^*T^1_Y(A) & \xrightarrow{q_A} & s^*T^0_Y & \longrightarrow & s^*(\mathcal{I}_Y(A)) & \rightarrow & 0 \\ \| & & \| & & \downarrow & & \\ \overline{T}_1(A) & \underset{\delta_1}{\longrightarrow} & \overline{T}_0(A) & \underset{\pi_A}{\longrightarrow} & A & \longrightarrow & 0 \end{array}$$

By exactness of the rows and commutativity, $s^*\mathcal{I}_Y(A)$ is isomorphic to A.

Now consider B in $Car_{\mathbf{R}}(G, X)$. First notict that proposition 3, (2) gives canonical maps, $T_\sigma(B(\sigma)) \to B$. Sum to obtain a canonical map, $d_0T^0_Y(s^*B) \to B \to 0$. This map clearly vanishes on the image of $T^1_Y(s^*B)$. Hence it induces a mapping, $\xi_B : \mathcal{I}_Y(s^*B) \to B$. Apply the exact functor, s^*, and use what we have just proven. Then $s^*(\mathcal{I}_Y(s^*B)) = s^*(B)$ and ξ_B induces the identity on segments in Y. Now finally note that a G-morphism of G-carapaces which is an isomorphism on Y is an isomorphism. That is $\mathcal{I}_Y(s^*(B)) \simeq B$ functorially in B. It follows that $\mathcal{I}_Y$ is inverse to s^*. □

In what follows, Theorem 9.12 will be a very essential and fundamental tool for analyzing $Car_{\mathbf{R}}(G, X)$.

10 Recollections and Fundamentals; Buildings

For the most part we follow the notation and conventions of [BT-I] and [BT-II]. Our purpose here is a brief review which will establish notation and emphasize one or two differences. Throughout K is a field complete with respect to the discrete rank one valuation, $\omega : K^* \to \mathbb{Z}$. Then $\mathcal{O}$ will be the center of ω, $\overline{k}$ will be the residue field of $\mathcal{O}$ and $\xi : \mathcal{O} \to \overline{k}$ is the natural map.

Let $\mathbf{G}$ be a Chevalley group scheme defined over $\mathbb{Z}$. Assume it to be split, simply connected, connected and of simple type. Let $\mathbf{T}$ be a maximal torus, let $\mathbf{N}$ be its Cartan subgroup and let $\mathbf{B}$ be a Borel subgroup containing $\mathbf{T}$ all given as group subschemes of $\mathbf{G}$ defined over $\mathbb{Z}$.

Each of these group schemes being a functor, applying any one of them to ξ gives a morphism which, in all cases, we will also call ξ from $\mathbf{G}(\mathcal{O})$ to $\mathbf{G}(\overline{k})$, $\mathbf{B}(\mathcal{O})$ to $\mathbf{B}(\overline{k})$, etc.. Let $G = \mathbf{G}(K)$, let $G_0 = \mathbf{G}(\mathcal{O})$ and let $\overline{G} = \mathbf{G}(\overline{k})$. Let $\mathcal{B} = \mathbf{B}(K)$, let $\overline{B} = \mathbf{B}(\overline{k})$ but let $B = \{x \in G_0 : \xi(x) \in \overline{B}\}$. Let $N = \mathbf{N}(K)$, let $T = \mathbf{T}(K)$ and let $H = N \cap B$.

Let $\mathbf{X}$ denote the finite free $\mathbb{Z}$-module of characters of $\mathbf{T}$, let Φ denote the roots of $\mathbf{G}$ with respect to $\mathbf{T}$; let Φ_+ denote those positive with respect to $\mathbf{B}$; let Δ be a basis of simple roots in Φ_+ and let $\tilde{\alpha}$ denote the largest root. Let $\Gamma = Hom_{\mathbb{Z}}(\mathbf{X}, Z)$ be the group of one parameter subgroups of $\mathbf{T}$ and let $\check{\Phi} \subseteq \Gamma$ be the set of co-roots.

Let $\{\mathbf{U}_\alpha : \alpha \in \Phi\}$ be a set of root subgroups let $U_\alpha = \mathbf{U}_\alpha(K)$ and let $x_\alpha : \mathbf{U}_\alpha \to G_{a,Z}$ be the natural isomorphism. Let $U_{\alpha,n} = \{u : \omega(x_\alpha(u)) \geq n\}$ and observe that $U_{\alpha,0} = \mathbf{U}(\mathcal{O})$.

Let $N_0 = N \cap G_0$. Then Γ can be identified with the group scheme morphisms, $\gamma : G_{m,Z} \to \mathbf{T}$. Write $< \gamma, \chi >$ for the value, $\gamma(\chi)$ when $\gamma \in \Gamma$ and $\chi \in \mathbf{X}$. We may always think of γ as a map, $\gamma : K^* \to T$ and so we may write $\chi(\gamma(y)) = y^{<\gamma,\chi>}$. Following this convention, we may define the action of N on Γ by the equation, $({}^n\gamma)(x) = n(\gamma(x))n^{-1}$. For $t \in T$ write t^χ for $\chi(t)$. Then N acts on $\mathbf{X}$ by the equation $\chi^n(t) = \chi(ntn^{-1})$. Then N_0 normalizes T and so acts on it by conjugation. Consider the semi-direct product, $T \triangleleft N_0$. For any $(t,n) \in T \triangleleft N_0$, define a mapping, $\tau_{(t,n)} : \Gamma \to \Gamma$ by the equation,

$$< \tau_{(t,n)}(\gamma), \chi >= -\omega(t^\chi) + < {}^n\gamma, \chi >$$

Then $\tau_{(t,n)}$ is an affine transformation and $(t,n) \mapsto \tau_{(t,n)}$ is an action of $T \triangleleft N_0$ on Γ. If $n \in T \cap N_0$ it is clear that $\tau_{(n,n^{-1})} = id_\Gamma$ and so this action reduces to an affine action of $T \cdot N_0 = N$ on Γ. If $n \in N, \gamma \in \Gamma$ write $n = tn_0$ and let ${}^n\gamma = \tau_{(t,n_0)}(\gamma)$. It is straightforward to verify that if ${}^n\gamma = \gamma$ for all $\gamma \in \Gamma$ then $n \in H$.

Let $\mathbf{A} = \Gamma \otimes_Z R$ and extend the action of N to $\mathbf{A}$ by linearity. Let $\mathbf{X}_R = \mathbf{X} \otimes_Z R$ and choose a form on $\mathbf{X}_R$ invariant under the vector Weyl group. For $\lambda, \chi \in \mathbf{X}_R$ write the form, $(\lambda|\chi)$. With this form, identify $\mathbf{A}$ with $\mathbf{X}_R$ and write $\check{\alpha} = \frac{2\alpha}{(\alpha|\alpha)}$ for each $\alpha \in \Phi$.

Since H acts trivially under τ, and since N/H is naturally isomorphic to the affine Weyl group τ induces an action of the affine Weyl group on $\mathbf{A}$ and it is a triviality that this is the canonical action.

For any pair, $(\alpha, r) \in \Phi \times \mathbb{Z}$, let $\alpha^* = \{x \in \mathbf{A} : \alpha(x) + r \geq 0\}$. These are closed half spaces and they are in bijective correspondence with $\Phi \times \mathbb{Z}$. If the half space, α^* corresponds to the pair (α, r), write U_{α^*} for the group, $U_{\alpha,r}$ defined above. Write $\partial\alpha^*$ for the boundary of α^* and write s_{α^*} for reflection through $\partial\alpha^*$. The closed half spaces, α^*, are called the affine roots of G in $\mathbf{A}$ and we write Σ for the set of all affine roots of G in $\mathbf{A}$.

Define an equivalence on $\mathbf{A}$ by saying that $x \sim y$ if and only if $x \in \alpha^*$ if and only if $y \in \alpha^*$ for all $\alpha \in \Sigma$. These equivalence classes are the facets of $\mathbf{A}$ and they are the interiors of simplices (because $\mathbf{G}$ is of simple type). Their closures give a simplicial decomposition of $\mathbf{A}$. The maximal dimensional facets are called chambers. They are the connected components of $\mathbf{A} \setminus \bigcup_{\alpha^* \in \Sigma} \partial\alpha^*$.

Let $\tilde{\Delta}$ denote the set of affine roots $(\alpha, 0)$, $\alpha \in \Delta$ and $(-\beta, 1)$ where β is the unique largest root relative to the dual Weyl chamber. Let S denote the set of reflections through the hyperplanes, $\partial\alpha^*$ for all $\alpha^* \in \tilde{\Delta}$. These reflections afford a Coxeter presentation of the affine Weyl group, N/H, so that (G, B, N, S) is a Tits system in G. Then N/H acts transitively on the chambers of **A**.

Let $\overline{C}_0 = \bigcap_{\alpha^*\in\tilde{\Delta}} \alpha^*$ and let C_0 be its interior. Let F be any facet contained in $\overline{C}_0$ and let $S_F = \{s \in S : s(F) = F\}$. In particular $S_{C_0} = \emptyset$. For any F, let P_F be the subgroup of G generated by B and some arbitrarily chosen set of representatives of S_F in N. Then P_F is independent of the particular set of representatives of S_F chosen and it is called the parahoric subgroup of G associated to F. By definition, the parahoric subgroups of G relative to the Tits system, (G, B, N, S), are the conjugates of the subgroups, P_F. We shall call the facets, $F \subseteq \overline{C}_0$ the types of G with respect to the Tits system, (G, B, N, S). If P is a parahoric subgroup of G, then P is conjugate to a unique group of the form P_F for some F. Then we call F the type of P and we write $F = \tau(P)$. Notice that our terminology differs slightly from [BT I] in which the subsets, S_F are called the types.

We may now describe $\mathcal{I} = \mathcal{I}(G, B, N, S)$, the building of G with respect to the Tits system, (G, B, N, S). As a point set, $\mathcal{I}(G, B, N, S)$ is the set of pairs, (P, x) where P is parahoric and $x \in \tau(P)$. Now G acts on this set by the equation, $g(P, x) = (gPg^{-1}, x)$.

Let x be any point of **A**. Then for some $n \in N$, $nx \in \overline{C}_0$. Then the point nx is uniquely determined. Let F be the smallest open facet containing nx. Let $a(x) = (n^{-1}P_F n, nx)$. The group, $n^{-1}P_F n$ is uniquely determined just as nx is and so a maps **A** into $\mathcal{I}(G, B, N, S)$. (This map is called j in [BT I].) We may now make $\mathcal{I}(G, B, N, S)$ into a geometric simplicial complex. Its simplices are just G translates of closed facets in $a(\mathbf{A})$ and its vertices are translates of the special points. The G-translates of $a(\mathbf{A})$ are called the apartments of $\mathcal{I}(G, B, N, S)$.

There are several structures on $\mathcal{I} = \mathcal{I}(G, B, N, S)$. First there is what is called an affine structure in [BT I]. If x and y are any two points in $\mathcal{I}$ they are contained in one apartment. Thus for any $\lambda \in [0, 1]$, there is a point, $\lambda x + (1-\lambda)y$ determined by the apartment. This point is, however, independent of the apartment chosen and the operation which assigns to each pair, (x, y) together with a real $\lambda \in [0, 1]$, the point $\lambda x + (1-\lambda)y$ is the affine structure. There is a G invariant metric whose restriction to any apartment is the metric associated to the W invariant form (?|?). For any pair (x, y), this metric is written $d(x, y)$. Finally there is a "bornology" on subsets of G. A set is called bounded if it is contained in a finite union of double cosets, BwB for $w \in N$. Now let us record some statements particularly useful to us. They are for the most part simple rearrangements of statements in [BT I] and [BT II].

10.1 Lemma. *The action of G on $\mathcal{I}(G,B,N,S)$ is excellent with section, $\overline{C}_0$.*

Proof. First, $\overline{C}_0$ is a fundamental domain for the G action and G is transitive on maximal dimensional simplices. This implies that the action is separated. The closed chamber, $\overline{C}_0$, is isomorphic to the quotient of $\mathcal{I}$ by the G action and the inclusion of $\overline{C}_0$ into $\mathcal{I}$ is clearly a section. □

10.2 Lemma. *G acts transitively on the pairs $(F, \mathbf{C})$ where F is a facet and $\mathbf{C}$ is an apartment containing F.*

Proof. This is just 2.26, p. 36 in [BT I].

10.3 Proposition. *The action of B on $\mathcal{I}(G,B,N,S)$ is excellent with section,*
$a : \mathbf{A} \to \mathcal{I}$.

Proof. The action of G on $\mathcal{I}(G,B,N,S) = \mathcal{I}$ is separated and so, a fortiori, the action of B is as well. Let C_0 be the chamber associated to B. Recall the definition of the retraction of $\mathcal{I}$ on $\mathbf{A}$ with center, C_0 ([BT I] 2.3.5, p. 38). As we remarked, given any two facets, there is an apartment containing them. Thus for any facet, $F \subseteq \mathcal{I}$, there is an apartment $\mathbf{A}_1$ containing F and C_0. By 10.2, there is an element g in G so that $g(C_0, \mathbf{A}_1) = (C_0, \mathbf{A})$ whence $gF \subseteq \mathbf{A}$. Let $\rho_{C_0,\mathbf{A}}(F) = gF$. We show that $\rho^{-1}_{C_0,\mathbf{A}}(F) = B \cdot F$.

Suppose that $\rho_{C_0,\mathbf{A}}(F') = F$. Then, by definition, there is an apartment, $\mathbf{A}'$, containing C_0 and F' and an elment, $g \in G$, so that $g\mathbf{A}' = \mathbf{A}, gF' = F$ and $gC_0 = C_0$. Since $gC_0 = C_0$, $g \in B$ and $gF' = F$. Hence $\rho^{-1}_{C_0,\mathbf{A}}(F) \subseteq B{\cdot}F$. The opposite inclusion is clear.

To complete the proof that $\mathcal{I}/B = \mathbf{A}$, we use, nearly unmodified, the proof of 2.3.2, p. 37 of [BT I]. Clearly, for any facet, F, the orbit, $B \cdot F$ meets $\mathbf{A}$. What remains to be shown is that $B \cdot F$ contains exactly one facet in $\mathbf{A}$.

Suppose that F and F' are in $\mathbf{A}$ and that $F = bF'$, $b \in B$. Then, $F = nF_0$, $F' = n'F_0$ for some $F_0 \subseteq \overline{C}_0$. The stabilizer of F (respectively, F') is $nP_{F_0}n^{-1}$ ($n'P_{F_0}(n')^{-1}$ respectively). Then, $bn'P_{F_0}(n')^{-1}b^{-1} = nP_{F_0}n^{-1}$. Since parahoric subgroups are self normalizing, $n^{-1}bn' \in P_{F_0}$ and hence, $b^{-1}n \in n'P_{F_0}$. Let $X = S_{F_0}$ and let W_X be the subgroup of W generated by X. Then $P_{F_0} = BW_XB$. By [NB I], IV, 2.5, Proposition 2, $n'P_{F_0} \subseteq Bn'W_XB$. If $\nu : N \to W$ is the natural surjection, this implies that $\nu(n) \in \nu(n')W_X$ and so $nP_{F_0}n^{-1} = n'P_{F_0}(n')^{-1}$. But F is the fixed point set of $nP_{F_0}n^{-1}$ and

F' is the fixed point set of $n' P_{F_0}(n')^{-1}$. Consequently, since the two groups are equal, $F = F'$. Thus for any facet, $B \cdot F$ contains exactly one facet in $\mathbf{A}$. As this is true, in particular, for vertices, $\mathbf{A}$ is the quotient of $\mathcal{I}$ by the action of B and so the proposition is proven. □

11 Monumental Complexes

In this section the term monumental is thought of as meaning resembling or having the scale of a building as it is used in the history of art. We use it to describe certain G actions which have all the properties of the natural actions on buildings which are of interest to us. If X is a finite dimensional simplicial complex and if G is a group acting on it then a subcomplex, $Y \subseteq X$ will be called homogeneous if,whenever Y contains two simplices, τ and γ such that $\tau = g\gamma$ for some $g \in G$, there is an element $s \in G$ such that $sY = Y$ and $\tau = s\gamma$.

11.1 Definition. Let G be a group and let X be a finite dimensional simplicial complex on which G is acting with a separared action. Then the action of G on X will be called *monumental* if and only if the following conditions hold:

(1) Every simplex in X is contained in a maximal dimensional simplex.

(2) G acts transitively on the maximal dimensional simplices.

(3) For any simplex, σ in X the stabilizer, G_σ is self normalizing in G.

(4) If σ is a maximal dimensional simplex, the stabilizer, G_σ, acts excellently and it admits a homogeneous section, $Y \subseteq X$.

Suppose that G is a group acting monumentally on X. Then a maximal dimensional simplex will be called a closed chamber; its interior will be called a chamber. Let $\overline{C}$ be a closed chamber in X. Then separation of the action implies that for any simplex σ in X, the orbit, $G\sigma$ contains at most one simplex which is a face of $\overline{C}$ while homogeneity implies that there is always at least one. Consequently the action of G is excellent with quotient $\overline{C}$ and section equal to the inclusion of $\overline{C}$ in X. Let $B = B(\overline{C})$ denote the G stabilizer of $\overline{C}$ and let Y be a homogeneous subcomplex of X mapping isomorphically onto the quotient, X/B. Then any translate of Y, gY will be called an apartment of type Y. In general we will think of the type, Y, as chosen and fixed once and for all and so we will often speak merely of apartments.

Let F be a field. Let H be a commutative Hopf algebra over F with comultiplication, $\mu_H : H \to H \otimes_F H$, augmentation, $\epsilon_H : H \to F$, and antipode, $s_H : H \to H$. Then H will be called *proalgebraic* if it is reduced and a direct limit of sub-Hopf algebras finitely generated over F. Then $Spec(H)$ is a proalgebraic group scheme over F.

If H is proalgebraic over F let $\mathcal{G}_H$ denote the group of F-valued points of H. That is $\mathcal{G}_H$ is the group of F-homomorphisms from H to F. Let $\mathcal{F}(\mathcal{G}_H, F)$ denote the ring of F valued functions on $\mathcal{G}_H$ and let γ_H be the natural map from H to $\mathcal{F}(\mathcal{G}_H, F)$, namely $\gamma_H(a)(\phi) = \phi(a)$. We will say that H separates k-points if γ_H is injective.

11.2 Definition. Let G be a group. A *monumental G-complex* is a simplicial complex, X, on which G is acting monumentally together with a G-carapace of commutative F-algebras with unit, $(\mathcal{A}, \{\Phi_g\}_{g \in G})$, and G-morphisms, $\mu_{\mathcal{A}} : \mathcal{A} \to \mathcal{A} \otimes_F \mathcal{A}$, $s_{\mathcal{A}} : \mathcal{A} \to \mathcal{A}$ and $\epsilon_{\mathcal{A}} \to F_X$ so that the following conditions are satisfied:

(1) Each of the morphisms, $\mu_{\mathcal{A}}$, $s_{\mathcal{A}}$ and $\epsilon_{\mathcal{A}}$ is a morphism of G-carapaces of commutative F-algebras and for any simplex, σ, $\mathcal{A}(\sigma), \mu_{\mathcal{A},\sigma}, s_{\mathcal{A},\sigma}, \epsilon_{\mathcal{A},\sigma}$ is a profinite Hopf algebra over F.

(2) For each pair of simplices, $\sigma \subseteq \tau$, the expansion, $e^{\tau}_{\mathcal{A},\sigma}$ is a surjective morphism of proalgebraic Hopf algebras.

(3) The G structure, $\{\Phi_g\}_{g \in G}$ acts by isomorphisms of carapaces of proalgebraic Hopf algebras. That is $\Phi_{g,\sigma}$ is a Hopf algebra isomorphism for each g and σ.

(4) For each σ let $\mathcal{G}(\sigma) = \mathcal{G}_{\mathcal{A}(\sigma)}$. Then $\mathcal{G}$ is naturally a sheaf of groups with the G structure induced by Φ. Then $\mathcal{A}(\sigma)$ is reduced for each σ and there is an isomorphism of G-sheaves, $\alpha : G_* \to \mathcal{G}$.

The first three conditions of Definition 11.2 are self explanatory but the third requires some amplification. First of all, if M is a simplicial sheaf on X and ψ is an automorphism of X, $\psi^* M(\sigma) = M(\psi(\sigma))$. The G-structure on G_* is that arising from conjugation. That is, define $c_{g,\sigma} : G_\sigma \to G_{g\sigma}$ by the equation:

$$c_{g,\sigma}(x) = gxg^{-1} \tag{11.3}$$

Now we explain the G-structure on $\mathcal{G}$ induced by Φ. The functor $\mathcal{G}$ is contravariant and so $\mathcal{G}(\Phi_{g,\sigma})$ maps $\mathcal{G}(g\sigma)$ to $\mathcal{G}(\sigma)$. Thus define a G-structure, $\{\Gamma_g\}_{g \in G}$ by the equation:

$$\Gamma_{g,\sigma} = \mathcal{G}(\Phi_{g^{-1},g\sigma}) \tag{11.4}$$

Recalling that elements of $\mathcal{G}(\sigma)$ are the F-homomorphisms from $\mathcal{A}(\sigma)$ to F, this map can be more explicitly written, $\Gamma_{g,\sigma}(x) = x \circ \Phi_{g^{-1},g\sigma}$.

It is customary to write $a(x)$ for $x(a)$ when a is in a ring and x is a F-point of that ring. We may use α to identify G_σ with $\mathcal{G}(\sigma)$, writing, for $a \in \mathcal{A}(\sigma)$ and $g \in G_\sigma$, $a(g)$ to denote $[\alpha_\sigma(g)](a)$. With these conventions, Condition 3) is nothing more than the equation:

$$[\Phi_{g,\sigma}(a)](x) = a(g^{-1}xg) \tag{11.5}$$

Henceforth if X is a monumental G-complex with carapace of F-Hopf algebras $\mathcal{A}$ we will just say that $(G, X, \mathcal{A})$ is a monumental G-complex over F. Further we will use α to identify G_* with $\mathcal{G}$ and we will always view $\mathcal{A}(\sigma)$ as a ring of functions on G_σ. The definitions of this section contain all the properties of buildings which will be used later on. The next section explains how the affine building of the group of K-valued points of some semi-simple group for some valued field, K, satisfies all of these conditions for an appropriate choice of the carapace $\mathcal{A}$.

12 The Main Examples

In this section we will discuss three monumental complexes. The first two are quite straightforward but the third requires a short dicussion of some classical results of M. Greenberg. Such symbols as $K, \omega, \overline{k}, \mathcal{O}, \mathbf{G}, \mathbf{B}, \mathcal{B}, \overline{B}$ etc. mean just what they did in the previous section.

The Admissible Complex

Let K be a locally compact, non-Archimedean field and let F be any (discrete) field. Let $\mathcal{I}(G, B, N, S)$ be the building associated to G. For each $\sigma \in \mathcal{I}$, G_σ, the parahoric subgroup associated to the facet, σ, is a profinite group. For each σ let $\mathcal{A}^0_F(\sigma)$ denote the ring of locally constant F-valued functions on G_σ. For any set T, T^F will denote the set of all F-valued functions on T. Then

$$\mathcal{A}^0_F(\sigma) = \varinjlim_M (G_\sigma/M)^F$$

where M varies over the open normal subgroups of G_σ. Since each of the algebras, $(G_\sigma/M)^F$ is in fact a finite dimensional Hopf algebra with augmentation and antipode and since the inclusions $(G_\sigma/M_1)^F \subseteq (G_\sigma/M_2)^F$ when $M_2 \subseteq M_1$, is a Hopf morphism, $\mathcal{A}^0_F(\sigma)$ is a Hopf algebra with antipode and augmentation. It is clearly profinite and it is also clear that the expansions are surjective Hopf morphisms. Let $\mathcal{G}_F$ be the sheaf of F-points of the carapace $\mathcal{A}_F^0$ as in §11.2, 4). Then $\mathcal{G}_F(\sigma)$ is the set of algebra homomorphisms,

$$\begin{aligned} Hom_F^{al}(\varinjlim_M (G_\sigma/M)^F &= \varinjlim_M Hom_F^{al}((G_\sigma/M)^F, F) \\ &= \varinjlim_M G_\sigma/M \\ &= G_\sigma \end{aligned}$$

That is, there is a canonical isomorphism of sheaves of groups, $\alpha : G_* \to \mathcal{G}_F$. Finally define the G structure $\{\Phi_g\}_{g\in G}$ by the equation, $[\Phi_{g,\sigma}(a)](x) = a(g^{-1}xg)$. Since $\mathcal{I}$ is the building of G, 1), 2) and 3) of §11 show that G acts monumentally on it. We have just observed that $\mathcal{A}_F^0$ and $\mathcal{I}$ satisfy Conditions 1) through 4) of Definition 10.2. Hence $(\mathcal{I}, \mathcal{A}_F^0)$ is a monumental G complex. We shall call it the admissible complex of G over F. We note that the group. G, may be replaced by a central extension, $\widetilde{G}$.

The Spherical Complex

For this example we depart somewhat from usual terminology. Let F be an algebraically closed field and let $G_F = \mathbf{G}(F)$ be the group of F points of the Chevalley scheme, $\mathbf{G}$ which we assume to be of simple type. Construct a complex as follows. The vertices of $\mathcal{S}$ are the proper reduced maximal parabolic subgroup schemes of G_F which we regard as the base extension of $\mathbf{G}$ to F. The set $P_1, \ldots P_n$ is a simplex in $\mathcal{S}$ if and only if the intersection $P_1 \cap \ldots \cap P_n$ is parabolic. Let G_F act on $\mathcal{S}$ by conjugation. For any $\sigma \in \mathcal{S}$, we write G_σ for the stabilizer. Then G_σ is just the intersection of the groups corresponding to the vertices of σ. A chamber is the set of maximal parabolics containing a Borel subgroup; an apartment is the set of maximal parabolics containing a maximal torus. We must first establish that the action of G on $\mathcal{S}$ is monumental. The first three conditions of Definition 11.1 are quite well known. The proof of the fourth condition is again a modification of the proof of 2.3.2 of [BT I]. Let B be a Borel subgroup containing the maximal torus, T, and let N be the normalizer of T. Let P be any parabolic subgroup of G. By [H], 6.1, the intersection of any two parabolic subgroups contains a maximal torus. Thus there is a maximal torus, S, in $P \cap B$. Hence there is some element, $b \in B$, so that $bSb^{-1} = T$. Hence bPb^{-1} contains T. Let $\mathbf{A}(T)$ denote the apartment corresponding to T. We have shown that for any facet, τ, the B orbit, $B\tau$ meets $\mathbf{A}(T)$.

Now suppose that P and Q are two B-conjugate parabolics both of which contain T. Then there are elements m and n in N and b in B and a parabolic P_0 containing B so that $P = nP_0n^{-1}$, $Q = mP_0m^{-1}$ and $bPb^{-1} = Q$. Thus, $bnP_0n^{-1}b^{-1} = mP_0m^{-1}$ and so, since P_0 is self normalizing, $bn \in mP_0$. By

[Bo], IV, 2.5.2, $bn \in BmW_0B$ where W_0 is the Weyl group of P_0. Hence $nP_0n^{-1} = mP_0m^{-1}$. That is any two B conjugate parabolics containing T are necessarily equal. It follows that any B orbit $B_\mathcal{T}$ meets $\mathbf{A}(T)$ in exactly one facet. Condition 4) of the definition is hence established.

For any $\sigma \in \mathcal{S}$ let $\mathcal{A}(\sigma)$ be the coordinate ring of the parabolic subgroup, G_σ. It is an elementary exercise in the theory of algebraic groups to see that $(G_\mathcal{F}, \mathcal{S}, \mathcal{A})$ is a monumental $G_\mathcal{F}$ complex.

The Affine Complex

To describe this complex, we must recall some classical results of M. Greenberg. Let $\overline{k}$ be a perfect field and let R be a ring scheme over $\overline{k}$. Let X be any $\overline{k}$ scheme. Define a ringed space, $\widetilde{R}(X)$ as follows. Its topological space is the underlying topological space of X and we denote it $b(X)$. If U is open in $b(X)$, let $\mathcal{G}_{R,X}(U) = R(Spec(\mathcal{O}_X(U))$. As this functor is representible, it is a sheaf of rings on $b(X)$. Call R a scheme of local rings if $\mathcal{G}_{R,X}$ is a sheaf of local rings for each scheme, X and assume this to be the case.

Let $V = R(\overline{k})$. Then for each X, $\mathcal{G}_{R,X}$ is a sheaf of V algebras. Let $\tilde{R}(X) = (b(X), \mathcal{G}_{R,X})$. Then $\tilde{R}$ is a covariant functor from $\overline{k}$ schemes to V local ringed spaces. The category of V local ringed spaces contains the category of $Spec(V)$ schemes. Though we call the following the first theorem of Greenberg, it is not given as one theorem in [MG I] but it largely summarizes the content of §4 of that work, especially Propositions 1 to 4 of §4 and the extensions thereof in §6.

12.1 First Theorem of Greenberg. *Suppose the ring scheme, R, is a projective limit of schemes each isomorphic to affine n space over $\overline{k}$ for varying n. Then there is a right adjoint to the functor $\mathcal{G}_{R,X}$ from the category of $\overline{k}$ schemes to the category of V-schemes. That is, there is a functor $\mathbf{F}$ so that for any $\overline{k}$ scheme, X, and any $Spec(V)$ scheme, Y, the following holds:*

$$Hom_V(\mathcal{G}_R(X), Y) = Hom_{\overline{k}}(X, \mathbf{F}(Y))$$

Moreover $\mathbf{F}$ satisfies:

(1) If Y is of finite type over $Spec(V)$, then $\mathbf{F}(Y)$ is a projective limit of schemes of finite type over $\overline{k}$.

(2) If Y is affine then so is $\mathbf{F}(Y)$.

(3) If Y is a group scheme over $Spec(V)$, then $\mathbf{F}(Y)$ is a group scheme over $\overline{k}$ in such a way that the adjointness isomorphism of (2) is a group morphism functorially in X.

This brings us to what we will call the second theorem of Greenberg. In this case we are assembling parts of §6, Proposition 1 of [MG I] and Proposition 2 and the structure theorem of [MG II], §2. Assume that $R = \varinjlim_n R_n$ where R_n is a ring scheme over $\overline{k}$ which is $\overline{k}$ isomorphic to $\mathbf{A}^n_{\overline{k}}$, affine $n+1$ space over $\overline{k}$. Let I_n be the scheme of ideals in R corresponding to the kernel of the projection $R \to R_n$. Let I^r_n be the scheme of ideals in R_r such that $0 \to I^r_n \to R_r \to R_n \to 0$ is an exact sequence of group schemes for the additive structure. Assume that I^r_n is affine $n-r$ space over $\overline{k}$ and that R_r is a locally trivial fiber space over R_n with fibre I^R_n and that it is in fact a vector bundle over R_n. Let $\mathcal{G}_n = \mathcal{G}_{R_n}$ and let $\mathbf{F_n}$ be its right adjoint. Let $\mathbf{U}_n(Y)$ be the kernel of $\mathbf{F}(Y) \to \mathbf{F}_n(Y)$ and let $\mathbf{U}^r_n(Y)$ be the kernel of $\mathbf{F}_r(Y) \to \mathbf{F}_n(Y)$, $(r > n)$.

12.2 Second Theorem of Greenberg. *Let Y be a smooth group scheme with connected fibers over $Spec(V)$. If $r \geq n \geq 0$, $\mathbf{U}^r_n$ is a finite dimensional unipotent group scheme. Moreover, $\mathbf{F}_r(Y)$ is $Spec(V)$ isomorphic to the total space of a vector bundle over $\mathbf{F}_0(Y)$.*

We will refer to $\mathbf{U}_n(Y)$ (respectively, $\mathbf{U}^r_n(Y)$) as the congruence subscheme of $\mathbf{F}(Y)$ (respectively $\mathbf{F}_r(Y)$) of level n, and we shall call $\mathbf{F}(Y)$ the realization of Y over $\overline{k}$.

Now we return to the notation and conventions of §10. Assume that $\overline{k}$ is perfect. Let $X = \mathcal{I}(G, B, N, S)$. By Proposition 10.3, the G action is monumental. We will show that X is a monumental complex over G. One of the more astonishing results in [BT II] is that for any $\sigma \in X$, there is a group scheme over $\mathcal{O}$, M_σ, so that the generic fiber of M_σ is the base extension to K of $\mathbf{G}$ and such that $M_\sigma(\mathcal{O}) = G_\sigma$. Since $\mathbf{G}$ is split and simply connected and ω is discrete, M_σ can be assumed to be smooth with connected fibers. These group schemes are determined up to isomorphism if one requires that they admit Bruhat decompositions of a particular type. (see section 4.6 of [BT II]) We may assume that the schemes M_σ are carried to each other by the conjugation action on the generic fibre. Finally we will assume that $\mathcal{O}$ is the ring of $\overline{k}$ points of a ring scheme which is a projective limit of affine spaces over $\overline{k}$. This is the case when $\mathcal{O}$ is the ring of Witt vectors of $\overline{k}$ or the ring of formal power series in one variable over $\overline{k}$. Then for any $\sigma \in X$, G_σ is the set of $\overline{k}$ points of $\mathbf{F}M_\sigma$, the Greenberg realization of M_σ over $\overline{k}$.

Now we may describe the affine complex. Choose $\mathcal{O}$ as above to be the $\overline{k}$ points of a ring scheme isomorphic to an inverse limit of affine spaces. Choose $\mathbf{G}, \omega, B$ etc. as in §11 and let $X = \mathcal{I}(G, B, N, S)$. For each $\sigma \in X$, let $\mathcal{A}(\sigma)$ be the $\overline{k}$ coordinate ring of $\mathbf{F}M_\sigma$. The verification that $(G, X, \mathcal{A})$ is a monumental complex is now an entirely routine affair. This monumental complex over G is what we shall call the affine complex of G over $\overline{k}$

13 Locally Rational Carapaces

In this section and for the remainder of this discussion $(G, X, \mathcal{A})$ will always denote a monumental G-complex over the field, k. Recall that the action of the G-structure on $\mathcal{A}$ can be described by the equation $\Phi_{g,\sigma}(f) = f \circ c_{g^{-1}}$ where c_y denotes conjugation by y. Recall also that if M is any proalgebraic group with coordinate ring, A, over k, and if V is a rational representation of M with structure map $\beta : V \to V \otimes_k A$, then if M acts on A by conjugation, $\gamma_g(f) = f \circ c_{g^{-1}}$, then β is equivariant as a map from the representation, V, to the representation $V \otimes_k A$, where the actions on V and $\mathcal{A}$ are determined by β and γ respectively.

13.1 Definition. Let $(G, X, \mathcal{A})$ be a monmental complex and let Φ be the G-structure on $\mathcal{A}$. A *locally rational carapace of representations* on $(G, X, \mathcal{A})$ is a G-carapace of k-vector spaces on X, V, with G structure, Ψ and a morphism of G-carapaces, $\beta : V \to V \otimes_k \mathcal{A}$ so that for each σ, the map, $\beta_\sigma : V(\sigma) \to V(\sigma) \otimes_k \mathcal{A}(\sigma)$ is a comodule structure map in such a way that for each g, $\Psi_{g,\sigma}$ is a morphism of comodules.

A morphism of locally rational carapaces is a G-morphism of G-carapaces, $f : V \to U$ which is a comodule morphism on each segment. It is clear that the locally rational carapaces on $(G, X, \mathcal{A})$ are an abelian category.

The reader is cautioned to note that that the G-structures on $\mathcal{A}$ and V exist quite apart from the local comodule structure map, β. In particular $\mathcal{A}$ admits several local comodule structures corresponding to left translation, right translation and conjugation. These are non-isomorphic locally rational structures on the same G-carapace. The comultiplication, $\mu : \mathcal{A} \to \mathcal{A} \otimes_k \mathcal{A}$ may be regarded as the local structure corresponding locally to right translation. We will write $\mathcal{A}^r$ for $\mathcal{A}$ with this local comodule structure.

13.2 Proposition. *Let $(G, X, \mathcal{A})$ be a monumental complex and let V, β be a locally rational carapace of representations on $(G, X, \mathcal{A})$. Let μ, ϵ, s denote the structure morphisms on $\mathcal{A}$. Then*

(1) *The structures on $\mathcal{A}$ endow the segment, $\Sigma(X, \mathcal{A})$, with the structure of a co-algebra with co-unit and antipode.*

(2) *The structure morphism, β, makes the segment, $\Sigma(X, V)$ into a comodule over $\Sigma(X, \mathcal{A})$.*

(3) *Both the co-algebra structure data on $\Sigma(X, \mathcal{A})$ and the comodule structure map on $\Sigma(X, V)$ are G morphisms of representations.*

Proof. In this proof, we will make use of two functorial properties of the map, $t_{A,B}$ of Proposition 2.2, 3) which have not been established. Let A, A', B, B' and C be k-carapaces on X and let $\alpha : A \to A'$ and $\beta : B \to B'$ be morphisms. The two functorial properties, whose proofs we leave to the reader, are these:

$$t_{A',B'} \circ \Sigma(\alpha \otimes \beta) = (\Sigma\alpha \otimes \Sigma\beta) \circ t_{A,B}$$

$$(t_{A,B} \otimes id_{\Sigma C}) \circ t_{A\otimes B,C} = (id_{\Sigma A} \otimes t_{B,C}) \circ t_{A,B\otimes C} \tag{13.3}$$

Now define structure data on $\Sigma\mathcal{A}$ as follows. Let $\mu^{\Sigma} = t_{\mathcal{A},\mathcal{A}} \circ \Sigma\mu$, let $e^{\Sigma} = \Sigma e$ and let $s^{\Sigma} = \Sigma s$.

The proof, for example, of co-associativity is the following computation:

$$\begin{aligned} (\mu^{\Sigma} \otimes id_{\Sigma\mathcal{A}}) \circ \mu^{\Sigma} &= (t_{\mathcal{A},\mathcal{A}} \otimes id_{\Sigma\mathcal{A}}) \circ ((\Sigma\mu) \otimes id_{\Sigma\mathcal{A}}) \circ t_{\mathcal{A},\mathcal{A}} \circ \Sigma\mu \\ &= (t_{\mathcal{A},\mathcal{A}} \otimes id_{\Sigma\mathcal{A}}) \circ t_{\mathcal{A}\otimes\mathcal{A},\mathcal{A}} \circ \Sigma(\mu \otimes id_{\mathcal{A}}) \circ \Sigma\mu \\ &= (t_{\mathcal{A},\mathcal{A}} \otimes id_{\Sigma\mathcal{A}}) \circ t_{\mathcal{A}\otimes\mathcal{A},\mathcal{A}} \circ \Sigma(id_{\mathcal{A}} \otimes \mu) \circ \Sigma\mu \end{aligned}$$

by the co-associativity of μ

$$\begin{aligned} &= (id_{\Sigma\mathcal{A}} \otimes t_{\mathcal{A},\mathcal{A}}) \circ t_{\mathcal{A},\mathcal{A}\otimes\mathcal{A}} \circ \Sigma(id_{\mathcal{A}} \otimes \mu) \circ \Sigma\mu \quad \text{— by (3)} \\ &= (id_{\Sigma\mathcal{A}} \otimes t_{\mathcal{A},\mathcal{A}}) \circ (id_{\Sigma\mathcal{A}} \otimes \Sigma\mu) \circ t_{\mathcal{A},\mathcal{A}} \circ \Sigma\mu \\ &= [id_{\Sigma\mathcal{A}} \otimes (t_{\mathcal{A},\mathcal{A}} \circ \Sigma\mu)] \circ (t_{\mathcal{A},\mathcal{A}} \circ \Sigma\mu) \\ &= (id_{\Sigma\mathcal{A}} \otimes \mu^{\Sigma}) \circ \mu^{\Sigma} \end{aligned}$$

The proofs of the remaining axioms making $\Sigma\mathcal{A}$ into a coalgebra with co-unit and antipode are similar and are, for that reason, left to the reader.

Define the co-action on ΣV by the equation, $\beta^{\Sigma} = t_{V,\mathcal{A}} \circ \Sigma\beta$. The proof that this is co-associative is truly identical to the above computation with the first $\mathcal{A}$'s in the expressions there replaced by V's.

The last numbered assertion just follows from functoriality.

13.4 Definition. Let $(G, X, \mathcal{A})$ be a monumental complex. The *algebra of measures* on $(G, X, \mathcal{A})$, which we write $\mathcal{L}^1(G, X, \mathcal{A})$ or, more briefly as $\mathcal{L}^1(X)$ when no confusion will result, is the algebra, $(\Sigma\mathcal{A})^*$, the k-linear dual of $\Sigma\mathcal{A}$.

Much of what follows depends on a natural $\mathcal{L}^1(G, X, \mathcal{A})$-structure on the exoskeletal homology groups of a locally rational carapace. To construct this action we must examine the functor, $\mathcal{T}_0$ of §7, (1), the first term in the canonical brittle resolution. Recall that for any carapace A, $[\mathcal{T}_0(A)](\sigma) = \coprod_{\tau\subseteq\sigma} A(\tau)$ and that the expansions are the natural inclusions. Moreover, the boundary map, δ_0 naturally maps $\mathcal{T}_0(A)$ onto A.

13.5 Proposition. *Let $(G, X, \mathcal{A})$ be a monumental complex and let V be a locally rational carapace on X with structure map, β. Then*

(1) The carapace, $\mathcal{T}_0(V)$, admits a natural structure map, β_0, making it into a locally rational carapace.

(2) The structure map, β_0, is uniquely determined by the requirement that δ_0 be a morphism of locally rational carapaces.

(3) If $V(\sigma)$ is finite dimensional for each σ then the same is true of $\mathcal{T}_0(V)$.

Proof. Suppose that (V, β) is locally rational and that $\tau \subseteq \sigma$ are two simplices in X. Then, $(id_{V(\tau)} \otimes e^{\sigma}_{\mathcal{A},\tau}) \circ \beta_\tau = \beta_{\tau,\sigma}$ makes $V(\tau)$ into a $\mathcal{A}(\sigma)$ comodule in such a way that the diagram,

$$(*) \qquad \begin{array}{ccc} V(\tau) & \xrightarrow{\beta_{\tau,\sigma}} & V(\tau) \otimes \mathcal{A}(\sigma) \\ V,\tau^{\sigma} \downarrow & & \downarrow e^{\sigma}_{V,\tau} \\ V(\sigma) & \xrightarrow{\beta_\sigma} & V(\sigma) \otimes \mathcal{A}(\sigma) \end{array}$$

commutes. Thus $(\beta_0)_\sigma = \coprod_{\tau \subseteq \sigma} \beta_{\tau,\sigma}$ is a comodule structure map on $[\mathcal{T}_0(V)](\sigma)$ Then (*) may be applied twice, once to prove that $\mathcal{T}_0(V)$ is a carapace of co-modules over $\mathcal{A}$ and again to prove that δ_0 commutes with the two comodule structure morphisms. One verifies directly that β_0 is a G-morphism. The uniqueness statement is clear as is the finiteness statement. □

13.6 Proposition. *Let $(G, X, \mathcal{A})$ be a monumental complex. Then for each $r \geq 0$ the r'th exoskeletal cohomology, $H_r(X, -)$, is a covariant functor from the category of locally rational carapaces on X to the category of left $\mathcal{L}^1(G, X, \mathcal{A})$-modules.*

Proof. By Proposition 13.2, Σ is a functor from the category of locally rational carapaces to the category of $\Sigma(\mathcal{A})$ co-modules. The identity functor takes $\Sigma(\mathcal{A})$-co-modules to $\mathcal{L}^1(X) = (\Sigma(\mathcal{A}))^*$-modules. Thus Σ is a covariant functor to the category of left $\mathcal{L}^1(X)$ modules.

By Proposition 13.5 and the definition of the canonical brittle resolution (Definition 7.1), the canonical brittle resolution of V is a resolution by locally rational carapaces with boundary maps which are morphisms of locally rational carapaces. Thus the Alexander chains are a complex of $\mathcal{L}^1(X)$-modules. The result follows immediately. □

We shall be working with subalgebras of $\mathcal{L}^1(X)$. In consequence, a "working description" of it might be of use. First notice that $\mathcal{L}^1(X) = Hom_k(\Sigma\mathcal{A}, k) = \mathbf{Hom}_{X,k}(\mathcal{A}, k_X)$. Thus a typical element of $\mathcal{L}^1(X)$ is a family, $\{\partial_\sigma\}_{\sigma\in X}$ where $\partial_\sigma \in \mathcal{A}(\sigma)^*$, the linear dual of $\mathcal{A}(\sigma)$. The coherence condition on the family, $\{\partial_\sigma\}_{\sigma\in X}$ is the commutativity of:

$$\begin{array}{ccc} \mathcal{A}(\sigma) & \xrightarrow{\partial_\sigma} & k \\ e^{\tau}_{\mathcal{A},\sigma} \downarrow & & \| \\ \mathcal{A}(\tau) & \xrightarrow[\partial_\tau]{} & k \end{array} \tag{13.7}$$

for every pair, $\sigma \subseteq \tau$. This should be understood in the following sense. For any pro-algebraic group, the linear dual of its coordinate ring is an algebra under convolution. If one group contains another as a closed subgroup, then the dual of its coordinate ring contains the dual of the coordinate ring of the subgroup. Thus, 13.7) says that whenever $\sigma \subseteq \tau$ then $\partial_\sigma \in (\mathcal{A}(\tau))^*$. In particular $\partial_\sigma \in \bigcap_{\gamma\supseteq\sigma}$ as γ ranges over the chambers containing σ. In each of the main examples, this means that ∂_σ is in the linear dual of the coordinate ring of the radical of G_σ. This prompts us to define the X-radical of a stabilizer, G_σ as the intersection, $\mathcal{R}G(\sigma) = \bigcap_{\gamma\supseteq\sigma} G_\gamma$ where the intersection is taken over all chambers, γ, which contain σ.

Henceforth we will write $\mathcal{A}^*(\sigma)$ for the linear dual of $\mathcal{A}(\sigma)$. Recall that for any commutative Hopf algebra A, the dual algebra, A^* can be identified with the algebra of k-linear endomorphisms of A which commute with left translation or alternatively with those that commute with right translation. That is, if $\omega : A \to A$ is an endomorphism commuting with left translations, then $\phi = e_A \circ \omega \in A^*$ is an element of the dual such that $\phi * a = \omega(a)$ for all $a \in A$ and where the operation of ϕ is by right convolution. There is a similar statement with left and right interchanged. With this in mind, it is clear that $\mathcal{L}^1(X)$ is the algebra of k endomorphisms of $\mathcal{A}$ which are co-module morphisms for left translation (but not necessarily G-morphisms).

The G-action on $\mathcal{L}^1$ can now be described. If $g \in G_\sigma$ then, g can be thought of as a k-homomorphism from $\mathcal{A}(\sigma)$ to k and so as an element of $\mathcal{A}^*(\sigma)$ and G_σ can be thought of as a subgroup of the unit group of $\mathcal{A}^*(\sigma)$. Hence in this case g acts by true conjugation, $g \cdot \partial = g\partial g^{-1}$. More generally the action is $(g \cdot \partial)_\sigma = \partial_{g\sigma} \circ \Phi_{g,\sigma}$.

14 The Injective Co-generator

Let $(G, X, \mathcal{A})$ be a monumental complex. In this section we show that the category of locally rational carapaces admits an injective cogenerator. We give a particular injective cogenerator and we use it to construct a module category containing an image of the locally rational carapaces.

Let C denote a chamber in X and let A denote an apartment containing C.The unadorned symbols, $\mathcal{T}_i$ and $\mathcal{S}^i$ will denote the canonical brittle and flabby resolutions on X (see 8.3) while $\mathcal{T}_i^C$ and $\mathcal{S}_C^i$ will denote those on C and $\mathcal{T}_i^A$ and $\mathcal{S}_A^i$ those on A. Let $\iota : C \to X$ be the inclusion. By Theorem 9.12, ι^* is an isomorphism of categories from G-carapaces on X to G_*-carapaces on C. Recall that $\mathcal{I}_C$ denotes its inverse.

The locally rational carapaces on X are a subcategory of the G-carapaces on X and so ι^* carries them to a subcategory of the G_* carapaces on C which we will call the rational G_*-carapaces on C. The reader can work out their properties when necessary; we will frequently argue in that category rather than the category of locally rational carapaces on X. For a G-carapace on X, V, we will write V_C to denote its restriction to C.

If R is the coordinate ring of the proalgebraic group, H, it admits several structures as a rational representation. Let μ, e and s be the structural data for R. Write R^ℓ for the left translation module, R^r for the right translation module and R unadorned for the conjugating action, $\gamma_x(a)(g) = a(x^{-1}gx)$. Notice that s establishes an isomorphism between R^ℓ and R^r. Consequently except when the particular features of a calculation or proof demand otherwise we will always write R^r for this representation.

For a representation of H, M, write $M^\flat$ to mean the vector space, M refurbished with the trivial representation. If $\beta : M \to M \otimes_k R$ is the coaction on M, then the coassociativity of β is precisely equivalent to the statement that β is an H morphism from M to $M^\flat \otimes_k R^r$ It is also true that β is an H-morphism from M to $M \otimes_k R$ where the tensor product is with respect to the given structure on M and the conjugating representation on R.

If M is any vector space equipped with the trivial representation, then $M \otimes_k R^r$ is H-injective. (It is, in fact, co-free.) To prove it, let $f : N \to M \otimes_k R^r$ be an H-map, and let $j : N \to Q$ be an H-monomorphism. Let $\xi : Q \to Q \otimes_k R$ be the coaction. Just choose $\phi : Q \to M$ so that $\phi \circ j = (id_M \otimes e) \circ f$. Then it is straightforward to verify that $(\phi \otimes id_R) \circ \xi$ is a comodule map from Q to $M \otimes_k R$ whose composition with j is just f.

Suppose that P is a closed subgroup of H of finite codimension. If N is a representation of P then the induced algebraic representation is a representation of H, $I_{H/P}(N)$ together with a P morphism ,$\epsilon_N : I_{H/P}(N) \to$

N inducing the Frobenius reciprocity isomorphism, $Hom_H(W, I_{H/P}(N)) = Hom_P(W|_P, N)$. To construct $I_{H/P}(N)$ consider $N \otimes_k R$. The subgroup P acts on this product diagonally through the given representation on N and right translation on R. In addition, H acts by left translation on R and the trivial action on N. The actions of H and P on $N \otimes_k R$ commute and so $(N \otimes_k R)^P$ is an H subrepresentation of $N^\flat \otimes_k R^\ell$. Then, $I_{H/P}(N) = (N \otimes_k R)^P$, and the map, ϵ_N is the restriction of $id_N \otimes e$. Composing $id_N \otimes s$ with the inclusion of $I_{H/P}(N)$ in $N \otimes_k R$, we obtain a functorial map:

(14.1) $$\imath_N : I_{H/P}(N) \hookrightarrow N^\flat \otimes_k R^r$$

The following is crucial.

14.2 Lemma. *Let H be profinite with structural data as above. Let V be a rational representation of H. There is a vector space, U, and an exact sequence,*

$$0 \to V \to V \otimes_k R^r \to U \otimes_k R^r$$

Proof. Let $\alpha : V \to V^\flat \otimes_k R^r$ be the coaction viewed as an H-morphism. We will construct a morphism $\psi : V^\flat \otimes_k R^r \to (V \otimes_k R)^\flat \otimes_k R^r$ so that $ker(\psi) = im(\alpha)$. Let $\eta(v \otimes a) = v \otimes 1 \otimes a$ and let $(\lambda = id_V \otimes m \otimes id_R) \circ (\alpha \otimes id_R \otimes id_R) \circ id_V \otimes [(s \otimes id_R) \circ \mu]$. These are both H morphisms for right translation on the last factor and $ker(\eta - \lambda) = im(\alpha)$. To see why, notice that if $\tau(v \otimes r \otimes s) = v \otimes s \otimes r$, then $\tau \circ \lambda$ is a comodule structure on $V^\flat \otimes_k R^r$. With this interpretation, the kernel of $\tau \circ (\eta - \lambda)$ is just the space of invariants for this action. But then $V \otimes_k R$ is the module of sections for a homogeneous bundle on H and the image of α is the subspace of invariant sections. Applying τ again, we have shown that the kernel of $\eta - \lambda$ is the image.

Notice that as a G_*-carapace, $\mathcal{A}_C$ is equipped with the conjugating representation. If V is a rational G_*-carapace, then $V \otimes_k \mathcal{A}_C$ always denotes the tensor product with respect to this structure. The comodule structure, $\alpha : V \to V \otimes_k \mathcal{A}_C$ is a G_*-morphism with respect to this structure. It is not a G_*-morphism with repect to the right translation action.

Define the carapace, $\mathcal{A}_C^r$ by the equation, $\mathcal{A}_C^r(\sigma) = \mathcal{A}(\sigma)^r$. Let $\mathcal{A}^r = \mathcal{I}_C(\mathcal{A}_C^r$. Clearly, $\mathcal{A}$ and $\mathcal{A}^r$ are not isomorphic as G carapaces.

Let $\mathcal{S}_C^0(k_C) = \mathbf{I}_C$. Since k is a field, $\mathbf{I}_C$ is injective. Let $\mathbf{I}\mathcal{A}_C^r = \mathbf{I}_C \otimes_k \mathcal{A}_C^r$. Then $id_{\mathbf{I}_C} \otimes \mu$ makes $\mathbf{I}\mathcal{A}_C^r$ into a locally rational carapace of G_*-modules. Tensoring the natural inclusion, $k_C \hookrightarrow \mathbf{I}_C$, with $\mathcal{A}^r$ yields the natural inclusion:

$$\jmath : \mathcal{A}_C^r \to \mathbf{I}\mathcal{A}_C^r$$

In general we will call a carapace locally finite if each of its segments is a finite module.

14.3 Proposition. *Let $(G, X, \mathcal{A})$ be monumental with chamber, C. Let J be an injective k-carapace on C. Then $J \otimes_k \mathcal{A}_C^r$ is injective in the category of rational G_*-carapaces.*

Proof. The corresponding proof for a proalgebraic group globalizes. Let U and V be two rational G_* carapaces with co-actions, $\alpha : U \to U \otimes_k \mathcal{A}_C$ and $\beta : V \to V \otimes_k \mathcal{A}_C$ respectively. Let $\eta : U \to V$ be a G_*-monomorphism and let $f : U \to J \otimes_k \mathcal{A}_C^r$ be any G_* map. Let $e_{\mathcal{A}} : \mathcal{A} \to k_C$ be the counit. Then $(id_J \otimes e_{\mathcal{A}}) \circ f$ maps U to the k injective, J. Hence there is a map, $\phi : V \to J$ such that $(id_J \otimes e_{\mathcal{A}}) \circ f = \phi \circ \eta$. Then $(\phi \otimes id_{\mathcal{A}}) \circ \beta$ maps V to $J \otimes_k \mathcal{A}^r$ and a routine computation shows it to be a G_*-morphism extending f. □

14.4 Definition. Let $(G, X, \mathcal{A})$ be monumental with chamber, C. Let $\sigma \subseteq C$ be a face and let V be a rational G_σ-module. Let $\widetilde{V}^\sigma$ be the carapace on C with values:

$$\text{(14.5)} \qquad \begin{aligned} \widetilde{V}^\sigma(\tau) &= I_{G_\tau/G_\sigma}(V) \quad \text{if } \tau \subseteq \sigma \\ \widetilde{V}^\sigma(\tau) &= (0) \qquad\qquad \text{otherwise} \end{aligned}$$

The expansions are the canonical maps corresponding to transitivity of induction. The locally rational carapace on X, $\mathcal{I}_C(\widetilde{V}^\sigma)$ will be written, $V_\sigma^!$.

14.6 Lemma. *Let $(G, X, \mathcal{A})$ be monumental with chamber, C. Let V be a representation of G_σ for some face σ in C. Then if U is any locally rational G-carapace,*

$$Hom_{X,G}(U, V_\sigma^!) = Hom_{C,G}(\iota^* U, \widetilde{V}^\sigma) = Hom_{G_\sigma}(U(\sigma), V)$$

Proof. The first equality is just the isomorphism induced by the isomorphism of categories, ι^*. To prove the second, suppose that $f \in Hom_{C,G_*}(\iota^* U, \widetilde{V}^\sigma)$. We must show that f is uniquely determined by f_σ. If τ is not a face of σ then, $f_\tau = 0$. Write $e_{V,\tau}^\sigma$ for the expansion in $\widetilde{V}^\sigma$. If $\tau \subseteq \sigma$, then $e_{V,\tau}^\sigma \circ f_\tau = f_\sigma \circ e_{U,\tau}^\sigma$. But $e_{V,\tau}^\sigma = \epsilon_V$ and so f_τ is the G_τ-morphism uniquely determined by Frobenius reciprocity. The result follows. □

14.7 Lemma. *Let $(G, X, \mathcal{A})$ be monumental with chamber, C. Then, for each face, σ,*

$$\widetilde{\mathcal{A}(\sigma)^r}^{\sigma} = (k\downarrow^{\sigma}) \otimes_k \mathcal{A}_C^r$$

Proof. This is just the following observation. Let H be a group and let P be a closed subgroup with coordinate rings, R and S respectively. Then $I_{H/P}(S^r) = R^r$ and the identification is functorial. □

14.8 Theorem. *Let $(G, X, \mathcal{A})$ be monumental with chamber, C. Then:*

(1) If V is an injective rational G_σ-module then, $\widetilde{V}^\sigma$ (respectively, $V_\sigma^\sharp$) is an injective rational G_* carapace on C (respectively an injective locally rational G-carapace on X).

(2) The carapace, $\mathbf{I}_X\mathcal{A}_C^r = \mathcal{I}_C(\mathbf{I}\mathcal{A}_C)$ is injective in the category of locally rational carapaces on X.

(3) Let V be a locally rational carapace on X. Then there are k-vector spaces, M and N and an exact sequence of locally rational G-carapaces:

$$0 \to V \to M \otimes_k \mathbf{I}_X\mathcal{A}^r \to N \otimes_k \mathbf{I}_X\mathcal{A}^r$$

If V is locally finite, then M may be taken to be finite dimensional.

Proof. To prove 1), notice that it suffices to prove the statement for rational G_* carapaces on C. Notice that Lemma 14.6 implies that the functor, $\widetilde{V}^\sigma$, carrying rational G_σ modules, V, to G_*-carapaces is right adjoint to the exact functor which associates to a carapace its σ segment. The statement is an immediate consequence.

To prove 2), notice that $\mathbf{I}_X\mathcal{A}_C^r = \mathcal{I}_C(\mathbf{I}_C) \otimes_k \mathcal{A}^r$ and so by Proposition 14.3 it is injective.

Now we establish 3). Making use of the canonical equivalence, we may prove the corresponding statement for rational G_* carapaces on C. Thus let V be rational on C.

Let $\alpha : V \to V \otimes_k \mathcal{A}$ be the coaction on V. For each face in C, σ, we may use the correspondence of lemma 6 to construct the unique map, $\psi : V \to \widetilde{V(\sigma)}^{\sigma}$, such that $\psi_\sigma = id$. Now consider the canonical morphisms $\imath_{V(\sigma)}$ of (1). These morphisms induce a monomorphism of G_*-carapaces, $\imath^\sigma : \widetilde{V(\sigma)}^{\sigma} \to [V(\sigma)^\flat \downarrow^\sigma] \otimes_k \mathcal{A}^r$. Let $f^\sigma = \imath^\sigma \circ \psi$ and let $f = \prod_{\sigma \subseteq C} f^\sigma$. This is an injective map from V to

$$\prod_{\sigma \subseteq C} (V(\sigma)^\flat \downarrow^\sigma \otimes \mathcal{A}^r)$$

Let $M = \coprod_{\sigma \subseteq C} V(\sigma)^\flat$ and let $j_\sigma : V(\sigma)^\flat \to M$ be the natural inclusion. Let π_σ be the natural projection of

$$\prod_{\sigma \subseteq C} (V(\sigma)^\flat \downarrow^\sigma \otimes \mathcal{A}^r)$$

on $V(\sigma) \downarrow^\sigma \otimes \mathcal{A}^r$. Then, $\mathbf{I}_X \coprod_{\sigma \subseteq C} (\pi_\sigma \circ f_\sigma)$ is the required map. If V is locally finite, M is finite. To conclude the proof just note that this construction may be applied to the cokernel of the map we have just defined. □

14.9 Corollary. *The carapace $\mathbf{I}_X \mathcal{A}^r_C$ is an injective cogenerator for the category of locally rational G-carapaces on X.*

Proof. This is just Statement 3) of the theorem. □

14.10 Corollary. *Let $\mathcal{E}_{G,X} = End_{X,G}(\mathbf{I}_X \mathcal{A}^r)$. Then the functor $\mathcal{P}_{X,G}$ defined by $\mathcal{P}_{X,G}(V) = \mathbf{Hom}_{X,G}(V, \mathbf{I}_X \mathcal{A}^r)$ is a fully faithful contravariant embedding of the category of locally rational G-carapaces on X in the category of left $\mathcal{E}_{G,X}$ modules.*

Proof. This follows from Corollary 14.9 by a direct application of the Gabriel Mitchell embedding theorem. □

15 Locally Rational Carapaces

In this section we will construct a section from a certain subcategory of the category of $\mathcal{E}_{G,X}$ modules to the category of locally finite locally rational carapaces. We begin with a preliminary description of $\mathcal{E}_{G,X}$. Recall first that if R is the coordinate ring of a proalgebraic group then thelinear endomorphisms of R commuting with right convolutions are just convolutions on the left with elements of the linear dual of R. If ω is such an involution, then let $\overline{\omega}$ be the element of the dual defined by the equation, $\overline{\omega}(f) = [\omega(f)](e)$. Then left convolution with $\overline{\omega}$ is ω. Also notice that Frobenius reciprocity for the identity subgroup is the equation, $Hom_G(V, M \otimes R^r) = Hom_k(V, M)$.

Now observe that the carapace (on C), $\mathbf{I}\mathcal{A}^r_C$ may be written as the finite product, $\prod_{\sigma \subseteq C} \widetilde{\mathcal{A}(\sigma)^r}^\sigma$. Hence,

$$\mathbf{Hom}_{X,G}(\mathbf{I}\mathcal{A}^r_C, \mathbf{I}\mathcal{A}^r_C) = \prod_{\sigma,\tau \subseteq C} \mathbf{Hom}_X, G(\widetilde{\mathcal{A}(\sigma)^r}^\sigma, \widetilde{\mathcal{A}(\sigma)^r}^\tau) \tag{15.1}$$

By Lemma 14.6, $\mathbf{Hom}_X, G(\widetilde{\mathcal{A}(\sigma)^r}^{\sigma}, \widetilde{\mathcal{A}(\tau)^r}^{\tau}) = (0)$ whenever $\widetilde{\mathcal{A}(\sigma)^r}^{\sigma}(\tau) = (0)$. That is, the **Hom** is null unless $\sigma \supseteq \tau$. When this condition does hold:

$$\mathbf{Hom}_{X,G}(\widetilde{\mathcal{A}(\sigma)^r}^{\sigma}, \widetilde{\mathcal{A}(\tau)^r}^{\tau}) = Hom_{G(\tau)}(\mathcal{A}(\tau)^r, \mathcal{A}(\tau)^r)$$

because $\widetilde{\mathcal{A}(\sigma)^r}^{\sigma}(\tau) = \mathcal{A}(\tau)^r$. Thus the σ, τ component of the homomorphism is an element $\partial \in \mathcal{A}(\tau)^*$ operating by left convolution. Hence an element of $\mathcal{E}_{G,X}$ can be represented as a matrix, $(\partial_{\tau,\sigma})_{\tau\subseteq\sigma}$, where $\partial_{\tau,\sigma} \in (\mathcal{A}(\tau)^*$. Notice also that if $\tau \subseteq \sigma$ then $\mathcal{A}(\sigma)^* \subseteq \mathcal{A}(\tau)^*$. Consequently, if $(\partial_{\tau,\sigma})_{\tau\subseteq\sigma}$ and $(\overline{\partial}_{\tau,\sigma})_{\tau\subseteq\sigma}$ correspond to two elements of $\mathcal{E}_{X,G}$, for any τ, γ, σ such that $\tau \subseteq \gamma \subseteq \sigma$, the product, $\partial_{\tau,\gamma}\overline{\partial}_{\gamma,\sigma}$ is defined and is an element of $\mathcal{A}(\tau)^*$.

For a locally algebraic carapace, V,$\mathcal{P}_{X,G}(V)$ may also be calculated directly:

$$\mathcal{P}_{X,G}(V) = \mathbf{Hom}_{X,G}(V, \mathbf{I}\mathcal{A}_C) = \prod_{\sigma\subseteq C} Hom_{G(\sigma)}(V(\sigma), \mathcal{A}(\sigma))$$

By the remarks above, $Hom_{G(\sigma)}(V(\sigma), \mathcal{A}(\sigma)) = V(\sigma)^*$, the contragredient. Hence, $\mathcal{P}_{X,G}(V) = \prod_{\sigma\subseteq C} V(\sigma)^*$. Moreover, if $\partial_{\tau,\sigma}$ is an element of $\mathcal{E}_{X,G}$ then $\partial_{\tau,\sigma}$ operates on $(e^{\overline{\sigma}}_{V,\tau})^*(u)$ for $u \in V(\sigma)^*$ and $(e^{\sigma}_{V,\tau})^*$ the adjoint of the expansion in V. Having established this much, we leave the remainder of the proof of the following to the reader.

15.2 Proposition. *Let $(G, X, \mathcal{A})$ be monumental and let V be locally rational on X. Then:*

(1) The ring $\mathcal{E}_{X,G}$ is equal to the ring of matrices, $(\partial_{\tau,\sigma})_{\tau\subseteq\sigma}$. The product of two such matrices is given by $(\partial_{\tau,\sigma}) \cdot (\overline{\partial}_{\tau,\sigma}) = (\delta_{\tau,\sigma})$ where

$$\delta_{\tau,\sigma} = \sum_{\tau\subseteq\gamma\subseteq\sigma} \partial_{\tau,\gamma}\overline{\partial}_{\gamma,\sigma}$$

(2) The module $\mathcal{P}_{X,G}(V)$ is isomorphic as an additive group to the set of vectors $(v_\sigma)_{\sigma\subseteq C}$ where the component, v_σ is in the contragredient module, $V(\sigma)^*$. The action of $(\partial_{\tau,\sigma})$ on (v_σ) yields (u_σ) where

$$u_\sigma = \sum_{\sigma\subseteq\gamma\subseteq C} \partial_{\sigma,\gamma}(e^{\gamma}_{V,\sigma})^*(v_\gamma)$$

For the remainder of this section, we will write $\mathbf{I}$ to denote the canonical co-generator, $\mathbf{I}\mathcal{A}^r_C$. Then, $\mathcal{E}_{X,G} = End_{X,G}(\mathbf{I})$. First, restriction to the chamber, C, is an isomorphism to the category of G_*-carapaces. On the category of locally algebraic G_*-carapaces on C, the segment over σ is an exact functor to the category of algebraic representations of $G(\sigma)$. Consequently, for each σ, the segment, $\mathbf{I}(\sigma)$ evaluated as a carapace on C, is a left $\mathcal{E}_{X,G}$-module and the action commutes with the co-action, $\mathbf{I}(\sigma) \to \mathbf{I}(\sigma) \otimes \mathcal{A}(\sigma)$. Evaluating the restriction of $\mathbf{I}$ to C on the facet, σ, we obtain the module of vectors, $(a_\tau)_{\tau\supseteq\sigma} : a_\tau \in \mathcal{A}(\sigma)$. By the description of $\mathcal{E}_{X,G}$ above the action of the matrix, $(\partial_{\gamma,\lambda})_{\gamma\subseteq\lambda}$, on the vector, $(a_\tau)_{\tau\supset}$

$$(15.3)\qquad \begin{array}{rcl} (\partial_{\gamma,\lambda})\cdot(a_\tau)_{\tau\supseteq\sigma} & = & (b_\tau)_{\tau\supseteq\sigma} \\ \text{where } b_\tau & = & \sum_{C\supseteq\gamma\supseteq\tau} \partial_{\tau,\gamma}\cdot a_\gamma \end{array}$$

Furthermore, since elements of $\mathcal{E}_{X,G}$ act as morphisms of carapaces, the expansions, which are compositions of projection with restriction of functions, are left $\mathcal{E}_{X,G}$-morphisms. Notice that the result would have been quite different if we had evaluated before restricting to C. We write $\mathbf{I}_C(\sigma)$ to denote the $\mathcal{E}_{X,G}$-module, $\mathbf{I}|_C(\sigma)$. Since X and G are fixed throughout this section, we will write $\mathcal{E}$ for $\mathcal{E}_{X,G}$ as long as no ambiguity will result.

15.4 Lemma. *Let $\mathcal{M}$ be a finitely generated $\mathcal{E}$-module. Let $\widetilde{\mathcal{M}}$ be the G-carapace on X whose prototype on C has the σ-segment, $Hom_{\mathcal{E}}(\mathcal{M}, \mathbf{I}_C(\sigma))$ and the expansions, $Hom_{\mathcal{E}}(id_{\mathcal{M}}, e^{\lambda}_{\mathbf{I},\gamma})$. Then $\widetilde{\mathcal{M}}$ is a locally algebraic carapace on X. Furthermore $\widetilde{\mathcal{M}}$ is a G subcarapace of a finite direct sum of carapaces each isomorphic to $\mathbf{I}$.*

Proof. As usual we need only consider the restriction of $\widetilde{M}$ to C. Being the composition of a covariant functor with a carapace, it is certainly a carapace. We propose to show that it is locally algebraic.

Write $\alpha : \mathbf{I}_C(\sigma) \to \mathbf{I}_C(\sigma) \otimes \mathcal{A}(\sigma)$ be the coaction commuting with the $\mathcal{E}$ action. Finite generation over $\mathcal{E}$ means that there is an exact sequence, $\mathcal{E}^{\oplus r} \to \mathcal{M} \to (0)$. Moreover tensoring with $\mathcal{A}(\sigma)$ also results in an exact sequence.

Write α_1 for the map from $Hom_{\mathcal{E}}(\mathcal{M}, \mathbf{I}_C(\sigma))$ to $Hom_{\mathcal{E}}(\mathcal{M}, \mathbf{I}(\sigma) \otimes \mathcal{A}(\sigma))$ and write α_2 for the map induced when $\mathcal{M}$ is replaced by $\mathcal{E}^{\oplus r}$. Also notice that, for any $\mathcal{E}$-module, N, we may write $Hom_{\mathcal{E}}(\mathcal{E}^{\oplus r}, N \otimes_k U) = (N \otimes_k U)^{\oplus r} = Hom_{\mathcal{E}}(\mathcal{E}^{\oplus r}, N) \otimes_k U$ for any k-vector space, U. Putting this all togther, we obtain a commutative diagram:

$$(15.5)\qquad \begin{array}{ccccc} 0 & \longrightarrow & Hom_{\mathcal{E}}(\mathcal{M}, \mathbf{I}_C(\sigma)) & \longrightarrow & Hom_{\mathcal{E}}(\mathcal{E}^{\oplus r}, \mathbf{I}_C(\sigma)) \\ & & \alpha_1 \downarrow & & \alpha_2 \downarrow \\ 0 & \longrightarrow & Hom_{\mathcal{E}}(\mathcal{M}, \mathbf{I}_C(\sigma) \otimes \mathcal{A}(\sigma)) & \longrightarrow & Hom_{\mathcal{E}}(\mathcal{E}^{\oplus r}, \mathbf{I}_C(\sigma)) \otimes \mathcal{A}(\sigma) \end{array}$$

First observe that there is a natural embedding of $Hom_{\mathcal{E}}(\mathcal{M}, \mathbf{I}_C(\sigma)) \otimes \mathcal{A}(\sigma)$ in $Hom_{\mathcal{E}}(\mathcal{M}, \mathbf{I}_C(\sigma) \otimes \mathcal{A}(\sigma))$. Then all that must be proven is that the image of α_1 is contained in $Hom_{\mathcal{E}}(\mathcal{M}, \mathbf{I}_C(\sigma)) \otimes \mathcal{A}(\sigma)$ for then α_1 will a fortiori be a coaction while the horizontal arrow in the first row of 15.5 is the embedding establishing the last assertion of the lemma.

Let j denote the upper horizontal arrow in 15.5 and let j' denote the lower one. Let $\{a_\gamma : \gamma \in \Gamma\}$ be a basis for $\mathcal{A}(\sigma)$. If $u \in Hom_{\mathcal{E}}(\mathcal{M}, \mathbf{I}_C(\sigma) \otimes \mathcal{A}(\sigma))$, then we may write $j'(u) = \sum_\gamma \phi_\gamma \otimes a_\gamma$ where almost all the ϕ_γ are null. But then since j' is injective, $u = \sum_\gamma (\phi_\gamma)|_{\mathcal{M}} \otimes a_\gamma$. But this shows that each element of $Hom_{\mathcal{E}}(\mathcal{M}, \mathbf{I}_C(\sigma) \otimes \mathcal{A}(\sigma))$ is in $Hom_{\mathcal{E}}(\mathcal{M}, \mathbf{I}_C(\sigma)) \otimes \mathcal{A}(\sigma)$. That is the two spaces are equal and so the result is established. □

15.6 Definition. Let V be a locally algebraic G-carapace on X. We shall say that V is *finitely cogenerated* if for some integer, $r \geq 0$, there is an exact sequence, $0 \to V \to \mathbf{I}^{\oplus r}$

Since $\mathcal{P}_{X,G}(\mathbf{I}) = \mathcal{E}$, and since $\mathbf{I}$ is injective, it is clear that $\mathcal{P}_{X,G}$ carries finitely cogenerated carapaces to finitely generated $\mathcal{E}$-modules. Further, by lemma 4, the functor $\mathcal{M} \mapsto \widetilde{\mathcal{M}}$ carries finitely generated $\mathcal{E}$-modules to finitely cogenerated carapaces. Notice that finitely cogenerated carapaces are not an Abelian category. While subcarapaces of finitely cogenerated carapaces are finitely cogenerated it is not clear that quotients of finitely cogenerated carapaces are finitely cogenerated. This is of course dual to the question of whether submodules of finitely generated $\mathcal{E}$-modules are finitely generated. That is, it would imply left Noetherianness of $\mathcal{E}$.

15.7 Theorem. *Let $(G, X, \mathcal{A})$ be monumental. Then*

(1) The functor $\mathcal{P}_{X,G}$ carries the category of finitely cogenerated carapaces on X to the category of finitely generated $\mathcal{E}$-modules.

(2) The functor $\mathcal{M} \mapsto \widetilde{\mathcal{M}}$ carries the category of finitely generated $\mathcal{E}$-modules to the category of finitely cogenerated carapaces on X.

(3) If V is a locally finite carapace on X, there is a natural isomorphism from V to $\widetilde{\mathcal{P}_{X,G}(V)}$. Moreover these two functors are isomorphisms between the category of locally finite carapaces and the category of $\mathcal{E}$-modules which are finite dimensional over k.

Proof. The first statement in the theorem was proven in the paragraph above. The second statement follows immediately from Lemma 15.4. Only the last statement requires proof.

Since $\mathcal{P}_{X,G}(V) = \coprod_{\sigma \subseteq C} V(\sigma)^*$, it is clear that when V is locally finite, $\mathcal{P}_{X,G}(V)$ is finite dimensional. For the converse, let $e_{\lambda,\gamma}$ for $\lambda \subseteq \gamma \subseteq C$ denote the element of $\mathcal{E}$ corresponding to the matrix whose μ, ν entry is 0 when $(\mu,\nu) \neq (\lambda,\gamma)$ and whose λ,γ entry is the identity in $\mathcal{A}(\lambda)^*$. The elements $e_{\lambda,\lambda}$ are a complete set of orthogonal idempotents. Further for each λ such that $\lambda \not\supseteq \sigma$, $e_{\lambda,\lambda}\mathbf{I}_C(\sigma) = (0)$. If $\mathcal{M}$ is any left $\mathcal{E}$-module, write $\mathcal{M}_\sigma = e_{\sigma,\sigma}\mathcal{M}$ Then, $\mathcal{M} = \coprod_{\tau \subseteq C} \mathcal{M}_\tau$. By simple restriction there is a map from $Hom_{\mathcal{E}}(\mathcal{M}, \mathbf{I}_C(\sigma))$ to $Hom_{\mathcal{A}(\sigma)^*}(\mathcal{M}_\sigma, \mathbf{I}_C(\sigma)_\sigma) = Hom_{\mathcal{A}(\sigma)^*}(\mathcal{M}_\sigma, \mathcal{A}(\sigma))$. We propose to show that it is an isomorphism.

By the remarks above, if $f \in Hom_{\mathcal{E}}(\mathcal{M}, \mathbf{I}_C(\sigma))$, then $f(\mathcal{M}_\gamma) = (0)$ whenever $\gamma \not\supseteq \sigma$. Hence it suffices to determine f on $\mathcal{M}_\tau$ for $\tau \supseteq \sigma$. If $u = e_{\tau,\tau}u \in \mathcal{M}_\tau$ then, $f(e_{\sigma,\tau}u) = e_{\sigma,\tau}f(u)$. But then $f(u)$ is just an element of $e_{\tau,\tau}\mathbf{I}_C(\sigma) = \mathcal{A}(\sigma)$ and $e_{\sigma,\tau}$ acts on it as the identity and so the equation becomes $f(u) = f(e_{\sigma,\tau}u)$. But $e_{\sigma,\tau}u \in \mathcal{M}_\sigma$ and so f is determined by its restriction to $\mathcal{M}_\sigma$. Thus,

$$Hom_{\mathcal{E}}(\mathcal{M}, \mathbf{I}_C(\sigma)) = Hom_{\mathcal{A}(\sigma)^*}(\mathcal{M}, \mathcal{A}(\sigma)) \tag{15.8}$$

This has two consequences. First, for general $\mathcal{M}$, there is a natural inclusion $Hom_{\mathcal{A}(\sigma)^*}(\mathcal{M}_\sigma, \mathcal{A}(\sigma)) \hookrightarrow \mathcal{M}_\sigma^*$ given by $f \mapsto e \circ f$ where e just denotes evaluation at the identity. The image of this map is always algebraic and when $\mathcal{M}_\sigma$ is itself algebraic it is an isomorphism as remarked in the first paragraph of this section. Thus if $\mathcal{M}$ is finite dimensional, the same can be said of $\widetilde{\mathcal{M}}(\sigma)$. Moreover, when V is locally algebraic, $\mathcal{P}_{X,G}(V)_\sigma = V(\sigma)^*$. Thus if V is locally finite, $V(\sigma)^*$ is finite dimensional and so the last part of Statement 3) follows at once. □

References

[B1] A. Borel, *Admissible representations of a semi-simple group over a local field with vectors fixed under an Iwahori subgroup*, Invent. Math. **35** (1976) 233–259.

[B2] A. Borel, *Linear Algebraic Groups*, second enlarged edition, Springer-Verlag, New York (1991).

[BW] A. Borel and N. Wallach, *Continuous Cohomology, Discrete Subgroups and Representations of Reductive Groups*, Annals of Mathematics Studies, **94** Princeton University Press. Princeton, New Jersey (1980) 3–381.

[Bo] N. Bourbaki, *Éléments de mathématique*, Fasc. XXXIV, Groupes et Algèbres de Lie, IV, V and VI, Actualités Scientifiques et Industrielles, **1337** Hermann, Paris (1968).

[Br] K. S. Brown, *Buildings,* Springer-Verlag, New York (1989) 1–209.

[BTI] F. Bruhat and J. Tits, *Groupes Réductifs sur un Corps Local I. Données radicielles valuées* Publ. Math. IHES, **41** (1972) 5–252.

[BT II] F. Bruhat and J. Tits, *Groupes Réductifs sur un Corps Local II. Schémas en groupes. Existence d'une donée radicielle valuée*, Publ. Math. IHES, **60** (1984) 5–184.

[CE] H. Cartan and S. Eilenberg, *Homological Algebra*, Princeton University Press, Princeton, New Jersey (1956)

[C] P. Cartier, *Representations of p-adic groups,* A survey in *Automorphic forms, representations and L-functions I*, Proceedings of the Symposium in Pure Mathematics, **33** American Mathematical Society, Providence, Rhode Island (1979) 111–155.

[CW] W. Casselman and D. Wigner, *Continuous cohomology and a conjecture of Serre's*, Invent. Math. **25** (1974) 199–211.

[Go] R. Godement, *Théorie des faisceaux*, Act. Sci. Ind. **1252** Hermann, Paris (1958).

[MGI] M. J. Greenberg, *Schemata over local rings*, Ann. Math. **73** (1961) 624–648.

[MG II] M. J. Greenberg, *Schemata over local rings II*, Ann. Math. **78** (1963) 256–266.

[Gr] A. Grothendieck *Sur quelques points d'algèbre homologique*, Tohoku Math. Jour. **9** (1957) 119–221.

[H] W. J. Haboush, *Reductive groups are geometrically reductive*, Ann. Math. **102** (1975) 67–83.

[Hu] J. E. Humphreys. *Introduction to Lie Algebras and Representation Theory*, Springer-Verlag, New York (1991)

[NM] N. Iwahori and H. Matsumoto, *On some Bruhat decomposition and the structure of the Hecke rings of p-adic Chevalley groups*, Publ. Math. IHES **25** (1965) 5–48.

[KL] D. Kazhdan and G. Lusztig, *Proof of the Deligne-Langlands conjecture for Hecke algebras*, Inv. Math. **87** (1987) 153–215.

[L1] G. Lusztig, *Hecke algebras and Jantzen's generic decomposition patterns*, Adv. in Math. **37** (1980) 121–164.

[L2] G. Lusztig, *Some examples of square integrable representations of p-adic groups*, Trans. Amer. Math. Soc. **277** (1983) 623–653.

[L3] G. Lusztig, *Equivariant K-theory and representations of Hecke algebras*, Proc. Amer. Math. Soc. **94** (1985) 337–342.

[L4] G. Lusztig, *Equivariant K-theory and representations of Hecke algebras II*, Inv. Math. **80** (1985) 209–231.

[M] I. G. Macdonald, *Spherical functions on a group of p-adic type*, Ramanujan Institute Lecture Notes, Ramanujan Institute, Centre for Advanced Study in Mathematics, Madras (1971) 1–79.

[Mac] S. MacLane. *Homology*, Die Grundlehren der Mathematischen Wissenshaften in Einzeldarstellungen Vol. **114**, Academic Press, Springer-Verlag, Berlin-Göttingen-Heidelberg (1963) 1–422.

[PS] A. Pressley and G. Segal, *Loop Groups*, Clarendon Press, Oxford (1986)

[R] M. Ronan, *Lectures on buildings*, Perspectives in Mathematics, **7** Academic Press, Harcourt Brace Jovanovich, Boston-San Diego-New York-Berkeley-London-Sydney-Tokyo-Toronto (1989) 1–201.

[RS] M. Ronan and S. D. Smith, *Sheaves on buildings and modular representations of Chevalley groups*, J. Algebra **96** (1985) 319–346.

[S] L. Solomon, *A Mackey formula in the group ring of a Coxeter group*, J. Algebra **41** (1976) 255–264.

[T] J. Tits, *Two properties of Coxeter complexes*, appendix to *A Makey formula in the group ring of a Coxeter group* J. Algebra **41** (1976) 255–264.

Department of Mathematics
University of Illinois
273 Altgeld Hall
1409 W. Green St.
Urbana, Illinois 61801
U.S.A.

Poncelet Polygons and the Painlevé Equations

N.J. Hitchin

Dedicated to M.S.Narasimhan and C.S.Seshadri on the occasion of their 60th birthdays .

1 Introduction

The celebrated theorem of Narasimhan and Seshadri [13] relating stable vector bundles on a curve to unitary representations of its fundamental group has been the model for an enormous range of recent results intertwining algebraic geometry and topology. The object which mediates between the two areas in all of these generalizations is the notion of a *connection*, and existence theorems for various types of connection provide the means of establishing the theorems. In one sense, the motivation for this paper is to pass beyond the existence and demand more explicitness. What do the connections look like? Can we write them down? This question is our point of departure. The novelty of our presentation here is that the answer involves a journey which takes us backwards in time over two hundred years from the proof of Narasimhan and Seshadri's theorem in 1965.

For simplicity, instead of considering stable bundles on curves of higher genus we consider the analogous case of parabolically stable bundles, in the sense of Mehta and Seshadri [11], on the complex projective line $\mathbf{CP}^1$. Such a bundle consists of a vector bundle with a weighted flag structure at n marked points $a_1, \ldots, a_n$. The unitary connection that is associated with it is flat and has singularities at the points. In the generic case, the vector bundle itself is trivial, and the flat connection we are looking for can be written as a meromorphic $m \times m$ matrix-valued 1-form with a simple pole at each point a_i. The parabolic structure can easily be read off from the residues of the form. The other side of the equation is a representation of the fundamental group $\pi_1(\mathbf{CP}^1 \backslash \{a_1, \ldots, a_n\})$ in $U(m)$, the holonomy of the connection, and this presents more problems. Such questions occupied the attention of Fuchs, Klein and others in the last century under the alternative name of monodromy of ordinary differential equations. Now if we fix the holonomy, and ask for the corresponding 1-form for each set of distinct points $\{a_1, \ldots, a_n\} \subset \mathrm{CP}^1$, what in fact we are asking for is a solution of a differential equation, the so-called Schlesinger equation (1912) of isomonodromic deformation theory. To focus things even more, in the simple case where $m = 2$ and $n = 4$, an explicit form for the connection demands a knowledge of solutions to a single nonlinear second order differential equation. This equation, originally found in the context of isomonodromic deformations by R. Fuchs in 1907 [4], is nowadays called Painlevé's 6th equation

$$\begin{aligned}\frac{d^2y}{dx^2} &= 1/2\left(\frac{1}{y}+\frac{1}{y-1}+\frac{1}{y-x}\right)\left(\frac{dy}{dx}\right)^2-\left(\frac{1}{x}+\frac{1}{x-1}+\frac{1}{y-x}\right)\frac{dy}{dx}\\ &+ \frac{y(y-1)(y-x)}{x^2(x-1)^2}\left(\alpha+\beta\frac{x}{y^2}+\gamma\frac{x-1}{(y-1)^2}+\delta\frac{x(x-1)}{(y-x)^2}\right)\end{aligned}$$

and in the words of Painlevé, the general solutions of this equation are "transcendantes essentiellement nouvelles". That, on the face of it, would seem to be the end of the quest for explicitness – we are faced with the insuperable obstacle of Painlevé transcendants.

Notwithstanding Painlevé's statement, for certain values of the constants $\alpha, \beta, \gamma, \delta$, there do exist solutions to the equation which can be written down, and even solutions that are *algebraic*. One property of any solution to the above equation is that $y(x)$ can only have branch points at $x = 0, 1, \infty$. This is essentially the "Painlevé property", that there are no movable singularities. If we find an algebraic solution, then this means we have an algebraic curve with a map to $\mathbf{CP}^1$ with only three critical values. Such a curve has a numbr of special properties. On the one hand, it is defined by a subgroup of finite index in $\Gamma(2) \subset SL(2, \mathbf{Z})$, and also, by a well-known theorem of Weil, is defined over $\bar{\mathbf{Q}}$. In this paper we shall construct solutions by considering the case when the holonomy group Γ of the connection is *finite*. In that case the solution $y(x)$ to the Painlevé equation is algebraic.

Our approach here is to consider, for a finite subgroup Γ of $SL(2, \mathbf{C})$, the quotient space $SL(2, \mathbf{C})/\Gamma$ and an equivariant compactification Z. Thus Z is a smooth projective threefold with an action of $SL(2, \mathbf{C})$ and a dense open orbit. The Maurer-Cartan form defines a flat connection on $SL(2, \mathbf{C})/\Gamma$ with holonomy Γ, which extends to a meromorphic connection on Z. The idea is then to look for rational curves in Z such that the induced connection is of the required form. By construction the holonomy is Γ, and if we can find enough curves to vary the cross-ratio of the singular points $a_1, \ldots, a_4$, then we have a solution to the Painlevé equation. The question of finding and classifying such equivariant compactifications has been addressed by Umemura and Mukai [12], but here we focus on one particular case. We take Γ to be the binary dihedral group $\tilde{D}_k \subset SU(2)$. This might seem very restrictive within the context of parabolically stable bundles, but behind it there hides a very rich seam of algebraic geometry which has its origins further back in history than Painlevé.

In the case of the dihedral group, the construction of a suitable compactification is due to Schwarzenberger [16], who constructed a family of rank 2 vector bundles V_k over $\mathbf{CP}^2$. The threefold corresponding to the dihedral group D_k turns out to be the projectivized bundle $P(V_k)$. There are two types of relevant rational curves. Those which project to a line in $\mathbf{CP}^2$ yield the solution $y = \sqrt{x}$ to the Painlevé equation with coefficients $(\alpha, \beta, \gamma, \delta) = (1/8, -1/8, 1/2k^2, 1/2 - 1/2k^2)$. Those which project to a conic lead naturally to another problem, and this one goes back at least to 1746 (see [3]). It is the problem of *Poncelet polygons.* We seek conics B and C in the plane such that there is a k-sided polygon inscribed in C and circumscribed about B. Interest in this problem is still widespread. Poncelet polygons occur in questions of stable bundles on projective spaces [14] and more recently in the work of Barth and Michel [1]. In fact, we can use their approach to find the modular curve giving the algebraic solution $y(x)$ of the Painlevé equation corresponding to $\Gamma = \tilde{D}_k$. This satisfies Painlevé's equation with coefficients $(\alpha, \beta, \gamma, \delta) = (1/8, -1/8, 1/8, 3/8)$. It is essentially Cayley's solution in 1853 of the Poncelet problem which allows us to go further and produce explicit solutions. It is a method which fits in well with the isomonodromic approach.

There are a number of reasons why this is a fruitful area of study. One of them concerns solutions of Painlevé equations in general and their relation to integrable systems, another is the connection with self-dual Einstein metrics as discussed in [6]. In the latter context, the threefolds constructed are essentially twistor spaces, and the rational curves twistor lines, but we shall not pursue this line of approach here. Perhaps the most intriguing challenge is to find *any* explicit solutions to an equation to which Painlevé's remark refers.

The structure of the paper is as follows. In Section 2 we consider singular connections and the isomonodromic deformation problem, and in Section 3 see how equivariant compactifications give solutions to the problem. In Section 4 we look at the way the dihedral group fits in with the problem of Poncelet polygons. Sections 5 and 6 discuss the actual solutions of the Painlevé equation, especially for small values of k. Only there can we see in full explicitness the connection which, in the context of the theorem of Narasimhan and Seshadri, relates the parabolic structure and the representation of the fundamental group, however restricted this example may be. In the final section we discuss the modular curve which describes the solutions so constructed.

The author wishes to thank M.F. Atiyah and A. Beauville for useful conversations.

2 Singular connections

We introduce here the basic objects of our study – flat meromorphic connections with singularities of a specified type. For the most part we follow the exposition of Malgrange [10].

Definition 1 *Let Z be a complex manifold, Y a smooth hypersurface and E a holomorphic vector bundle over Z. Let ∇ be a flat holomorphic connection on E over $Z\backslash Y$ with connection form A in some local trivialization of E. Then on $U \subseteq Z$ we say that*

1. *∇ is meromorphic if A is meromorphic on U*
2. *∇ has a logarithmic singularity along Y if, in a local coordinate system $(z_1, \ldots, z_n)$ of Z, with Y given by $z_1 = 0$, A has the form*

$$A = A_1 \frac{dz_1}{z_1} + A_2 dz_2 + \ldots + A_n dz_n$$

where A_i is holomorphic on U.

One may easily check that the definition is independent of the choice of coordinates and local trivialization. The essential point about a logarithmic singularity is that the pole only occurs in the conormal direction to Y. In fact ∇ defines a holomorphic connection on E restricted to Y, with connection form

$$A_Y = \sum_{i=2}^{n} A_i(0, z_2, \ldots, z_n) dz_i.$$

If Z is 1-dimensional, then such a connection is just a meromorphic connection with simple poles. Flatness is automatic because the holomorphic curvature is a $(2,0)$ form which is identically zero in one dimension. If we take $Z = \mathrm{CP}^1$, $Y = \{a_1, \ldots, a_n, \infty\}$ and the bundle E to be trivial, then A is a matrix-valued meromorphic 1-form with simple poles at $z = a_1, \ldots, a_n, \infty$ and can thus be written as

$$A = \sum_{i=1}^{n} \frac{A_i dz}{z - a_i}$$

The holonomy of a flat connection on $Z\backslash Y$ is obtained by parallel translation around closed paths and defines, after fixing a base point b, a representation of the fundamental group

$$\rho : \pi_1(Z\backslash Y) \to GL(m, \mathbf{C}).$$

In one dimension, the holonomy may also be considered as the effect of analytic continuation of solutions to the system of ordinary differential equations

$$\frac{df}{dz} + \sum_{i=1}^{n} \frac{A_i f}{z - a_i} = 0$$

around closed paths through b. As such, one often uses the classical term *monodromy* rather than the differential geometric *holonomy*. Changing the basepoint to b' effects an overall conjugation (by the holonomy along a path from b to b') of the holonomy representation.

For the punctured projective line above, we obtain a representation of the group $\pi_1(S^2\backslash\{a_1, \ldots, a_n, \infty\})$. This is a free group on n generators, which can be taken as simple loops γ_i from b passing once around a_i. Moving b close to a_i, it is easy to see that $\rho(\gamma_i)$ is conjugate to

$$\exp(-2\pi i A_i).$$

There is also a singularity of A at infinity with residue A_∞. Since the sum of the residues of a differential is zero, we must have

$$A_\infty = -\sum_{i=1}^{n} A_i$$

and so $\rho(\gamma_\infty)$ is conjugate also to $\exp(-2\pi i A_\infty)$. In the fundamental group itself $\gamma_1\gamma_2 \cdots \gamma_n\gamma_\infty = 1$ so that in the holonomy representation

$$\rho(\gamma_1)\rho(\gamma_2) \ldots \rho(\gamma_\infty) = 1. \tag{1}$$

Thus the conjugacy classes of the residues A_i of the connection determine the conjugacy classes of $\rho(\gamma_i)$, and these must also satisfy (1).

This is partial information about the holonomy representation. However, the full holonomy group depends on the position of the poles a_i. The problem of particular interest to us here is the *isomonodromic deformation problem* – to determine the dependence of A_i on $a_1, \ldots, a_n$ in order that the holonomy representation should remain the same up to conjugation. All we have seen so far is that the conjugacy class of $\exp(2\pi i A_i)$ should remain constant.

One way of approaching the isomonodromic deformation problem, due to Malgrange, is via a universal deformation space. Let X_n denote the space of ordered distinct points $(a_1, \ldots, a_n) \in \mathbf{C}$, and $\tilde{X}_n$ its universal covering. It is well-known that this is a contractible space – the classifying space for the braid group on n strands. Now consider the divisor

$$Y_m = \{(z, a_1, \ldots, a_n) \in \mathbf{CP}^1 \times X_n : z \neq a_m\}$$

and let $\tilde{Y}_m$ be its inverse image in $\mathbf{CP}^1 \times \tilde{X}_n$. Furthermore, define $\tilde{Y}_\infty = \{(\infty, x) : x \in \tilde{X}_n\}$.

The projection onto the second factor $p : \mathbf{CP}^1 \times \tilde{X}_n \backslash \{\tilde{Y}_1 \cup \ldots \cup \tilde{Y}_n \cup \tilde{Y}_\infty\} \to \tilde{X}_n$ is a fibration with fibre $\mathbf{CP}^1 \backslash \{a_1, \ldots, a_n, \infty\}$, and the contractibility of $\tilde{X}_n$ implies from the exact homotopy sequence that the inclusion i of a fibre induces an isomorphism of fundamental groups

$$\pi_1(\mathbf{CP}^1 \backslash \{a_1^0, \ldots, a_n^0, \infty\}) \cong \pi_1(\mathbf{CP}^1 \times \tilde{X}_n \backslash \{\tilde{Y}_1 \cup \ldots \cup \tilde{Y}_n \cup \tilde{Y}_\infty\})$$

Thus a flat connection on $\mathbf{CP}^1 \backslash \{a_1^0, \ldots, a_n^0, \infty\}$ extends to a flat connection with the same holonomy on $\mathbf{CP}^1 \times \tilde{X}_n \backslash \{\tilde{Y}_1 \cup \ldots \cup \tilde{Y}_n \cup \tilde{Y}_\infty\}$. Malgrange's theorem asserts that this flat connection has logarithmic singularities along $\tilde{Y}_m$ and $\tilde{Y}_\infty$.

More precisely,

Theorem 1 (Malgrange [10]) *Let ∇^0 be a flat holomorphic connection on the vector bundle E^0 over $\mathbf{CP}^1 \backslash \{a_1^0, \ldots, a_n^0, \infty\}$, with logarithmic singularities at $a_1^0, \ldots, a_n^0$. Then there exists a holomorphic vector bundle E on $\mathbf{CP}^1 \times \tilde{X}_n$ with a flat connection ∇ with logarithmic singularities at $\tilde{Y}_1, \ldots, \tilde{Y}_n, \tilde{Y}_\infty$ and an isomorphism $j : i^*(E, \nabla) \to (E^0, \nabla^0)$. Furthermore, (E, ∇, j) is unique up to isomorphism.*

Now suppose that E^0 is holomorphically trivial. The vector bundle E will not necessarily be trivial on all fibres of the projection p, but for a dense open set $U \subseteq \tilde{X}_n$ it will be. Choose a basis $e_1^0, e_2^0, \ldots, e_m^0$ of the fibre of E^0 at $z = \infty$. Now since ∇ has a logarithmic singularity on $\tilde{Y}_\infty$, it induces a flat connection there, and since $\tilde{Y}_\infty \cong \tilde{X}_n$ is simply connected, by parallel translation we can unambiguously extend $e_1^0, e_2^0, \ldots, e_m^0$ to a trivialization of E over $\tilde{Y}_\infty$. Then since E is holomorphically trivial on each fibre over U, we can uniquely extend $e_1^0, e_2^0, \ldots, e_m^0$ along the fibres to obtain a trivialization $e_1, \ldots, e_m$ of E on $\mathbf{CP}^1 \times U$. It is easy to see that, relative to this trivialization, the connection form of ∇ can be written

$$A = \sum_{i=1}^{n} A_i \frac{dz - da_i}{z - a_i} \tag{2}$$

where A_i is a holomorphic function of $a_1, \ldots, a_n$.

The flatness of the connection can then be expressed as:

$$dA_i + \sum_{j \neq i} [A_i, A_j] \frac{da_i - da_j}{a_i - a_j} = 0$$

which is known as *Schlesinger's equation* [15].

The gauge freedom in this equation involves only the choice of the initial basis $e_1^0, e_2^0, \ldots, e_m^0$ and consists therefore of conjugation of the A_i by a constant matrix.

The case which interests us here is where the holonomy lies in $SL(2, \mathbf{C})$, (so that the A_i are trace-free 2×2 matrices), and where there are 3 marked points a_1, a_2, a_3 which, together with $z = \infty$, are the singular points of the connection. By a projective transformation we can make these points $0, 1, x$. Then

$$A(z) = \frac{A_1}{z} + \frac{A_2}{z-1} + \frac{A_3}{z-x}$$

and Schlesinger's equation becomes:

$$\begin{aligned} \frac{dA_1}{dx} &= \frac{[A_3, A_1]}{x} \\ \frac{dA_2}{dx} &= \frac{[A_3, A_2]}{x-1} \\ \frac{dA_3}{dx} &= \frac{-[A_3, A_1]}{x} - \frac{[A_3, A_2]}{x-1} \end{aligned} \qquad (3)$$

where the last equation is equivalent to

$$A_1 + A_2 + A_3 = -A_\infty = \text{const.}$$

The relationship with the Painlevé equation can best be seen by following [8]. Each entry of the matrix $A_{ij}(z)$ is of the form $q(z)/z(z-1)(z-x)$ for some quadratic polynomial q. Suppose that A_∞ is diagonalizable, and choose a basis such that

$$A_\infty = \begin{pmatrix} \lambda & 0 \\ 0 & -\lambda \end{pmatrix}$$

then A_{12} can be written

$$A_{12}(z) = \frac{k(z-y)}{z(z-1)(z-x)} \qquad (4)$$

for some $y \in \mathbf{CP}^1\backslash\{0,1,x,\infty\}$. If the $A_i(x)$ satisfy (3), then the function $y(x)$ satisfies the Painlevé equation

$$\begin{aligned}\frac{d^2y}{dx^2} &= 1/2\left(\frac{1}{y}+\frac{1}{y-1}+\frac{1}{y-x}\right)\left(\frac{dy}{dx}\right)^2-\left(\frac{1}{x}+\frac{1}{x-1}+\frac{1}{y-x}\right)\frac{dy}{dx} \\ &+ \frac{y(y-1)(y-x)}{x^2(x-1)^2}\left(\alpha+\beta\frac{x}{y^2}+\gamma\frac{x-1}{(y-1)^2}+\delta\frac{x(x-1)}{(y-x)^2}\right) \qquad (5)\end{aligned}$$

where

$$\begin{aligned}\alpha &= (2\lambda-1)^2/2 \\ \beta &= 2\det A_1^2 \\ \gamma &= -2\det A_2^2 \\ \delta &= (1+4\det A_3^2)/2 \qquad (6)\end{aligned}$$

For the formulae which reconstruct the connection from $y(x)$ we refer to [8], but essentially the entries of the A_i are rational functions of x, y and dy/dx. For our purposes it is useful to note the geometrical form of the definition of $y(x)$ given by (4):

Proposition 1 *The solution $y(x)$ to the Painlevé equation corresponding to an isomonodromic deformation $A(z)$ is the point $y \in \mathbf{CP}^1\backslash\{0,1,x,\infty\}$ at which $A(y)$ and A_∞ have a common eigenvector.*

Note that strictly speaking there are two Painlevé equations (with $\alpha = (\pm 2\lambda - 1)^2/2$) corresponding to the values of y with this property.

3 Equivariant compactifications

Consider the three-dimensional complex Lie group $SL(2,\mathbf{C})$ and the Lie algebra-valued 1-form

$$A = -(dg)g^{-1}.$$

The form A is the connection form for a trivial connection on the trivial bundle. It simply relates the trivializations of the principal frame bundle by left and right translation.

Now let Γ be a finite subgroup of $SL(2,\mathbf{C})$. Then $SL(2,\mathbf{C})/\Gamma$ is a non-compact complex manifold and since A is invariant under right translation, it descends to this quotient. Thus, on $SL(2,\mathbf{C})/\Gamma$, A defines a flat connection on the trivial rank 2 vector bundle. Its holonomy is tautologically Γ.

In this section, we shall consider an equivariant compactification of $SL(2,\mathbf{C})/\Gamma$, that is to say, a compact complex manifold Z on which $SL(2,\mathbf{C})$ acts with a dense open orbit with stabilizer conjugate to Γ. Let Z be such a compactification, then the action of the group embeds the Lie algebra $\underline{g}$ in the space of holomorphic vector fields on Z. Equivalently, we have a vector bundle homomorphism

$$\alpha : Z \times \underline{g} \to TZ$$

which is generically an isomorphism. It fails to be an isomorphism on the union of the lower dimensional orbits of $SL(2, \mathbf{C})$, and this is where $\wedge^3\alpha = 0$. But $\wedge^3\alpha \in H^0(Z, \mathrm{Hom}(\wedge^3\underline{g}, \wedge^3 T)) \cong H^0(Z, K^{-1})$ is a section of the anticanonical bundle, so the union of the orbits of dimension less than three form an *anticanonical divisor* Y, which may of course have several components or be singular.

On the open orbit $Z\backslash Y \cong SL(2, \mathbf{C})/\Gamma$, the action is equivalent to left multiplication, and the connection A above is given by

$$A = \alpha^{-1} : TZ \to Z \times \underline{g}.$$

It is clearly meromorphic on Z, but more is true:

Proposition 2 *If $\wedge^3\alpha$ vanishes non-degenerately on the divisor Y, then the connection $A = \alpha^{-1}$ has a logarithmic singularity along Y.*

This is a local statement, and so it can always be applied to the smooth part of Y even if there are singular points.

Proof. In local coordinates, α is represented by a holomorphic function $B(z)$ with values in the space of 3×3 matrices. The divisor Y is then the zero set of $\det B$. If $\det B$ has a non-degenerate zero at $p \in Y$, then its null-space is one-dimensional at p, so the kernel of α, the Lie algebra of the stabilizer of p, is one-dimensional. Thus the $SL(2, \mathbf{C})$ orbit through p is two-dimensional, and so Y is the orbit.

Now for any square matrix B, let $B^\vee$ denote the transpose of the matrix of cofactors. Then it is well-known that

$$BB^\vee = (\det B)I$$

Hence in local coordinates

$$A = \alpha^{-1} = \frac{B^\vee}{\det B}$$

and so A has a simple pole along Y. From Definition 1, we need to show that the residue is in the conormal direction. For this consider the invariant description of $B^\vee$. We have on Z

$$\wedge^2\alpha : \wedge^2\underline{g} \to \wedge^2 T$$

and using the identifications $\Lambda^2 \underline{g} \cong \underline{g}^*$ and $\Lambda^2 T \cong T^* \otimes \Lambda^3 T$, $B^\vee$ represents the dual map of $\Lambda^2\alpha$:

$$(\Lambda^2\alpha)^* : T \to \underline{g} \otimes \Lambda^3 T.$$

Now the image of α at p is the tangent space to the orbit Y at p by the definition of α. Thus the image of $\Lambda^2\alpha$ is $\Lambda^2 TY_p$ which means that $(\Lambda^2\alpha)^*$ annihilates TY, which is the required result.

Note that the kernel of $\Lambda^2\alpha$ is the set of two-vectors $v \wedge w$ where $w \in \underline{g}$ and v is in the Lie algebra of the stabilizer of p. Thus the residue at p of the connection A lies in the Lie algebra of the stabilizer.

Now suppose that P is a rational curve in Z which meets Y transversally at four points. Then the restriction of A to P defines a connection with logarithmic poles at the points and, from the map of fundamental groups

$$\pi_1(P\backslash\{a_1, \ldots, a_4\}) \to \pi_1(Z\backslash Y) \to \Gamma \to SL(2, \mathbf{C}),$$

its holonomy is contained in Γ. A deformation of P will define a nearby curve in the same homotopy class and hence the induced connection will have the same holonomy. To obtain isomonodromic deformations, we therefore need to study the deformation theory of such curves.

Proposition 3 *Let $P \subset Z$ be a rational curve meeting Y transversally at four points. Then P belongs to a smooth four-parameter family of rational curves on which the cross-ratio of the points is a nonconstant function.*

Proof. The proof is standard Kodaira-Spencer deformation theory. By hypothesis P meets the anticanonical divisor Y in four points, so the degree of K_Z on P is -4. Hence, if N is the normal bundle of $P \cong \mathbf{CP}^1$,

$$\deg N = -\deg K_Z + \deg K_P = 2$$

and so

$$N \cong \mathcal{O}(m) \oplus \mathcal{O}(2-m)$$

for some integer m. However, since C is transversal to the 2-dimensional orbit Y of $SL(2, \mathbf{C})$, the map α always maps *onto* the normal bundle to C. We therefore have a surjective homomorphism of holomorphic vector bundles

$$\beta : \mathcal{O} \otimes \underline{g} \to N$$

and this implies that $0 \leq m \leq 2$. As a consequence, $H^1(P, N) = 0$ and $H^0(P, N)$ is four-dimensional, so the existence of a smooth family follows from Kodaira [9].

Since β is surjective, its kernel is a line bundle of degree $-\deg N = -2$, so we have an exact sequence of sheaves:

$$\mathcal{O}(-2) \to \mathcal{O} \otimes \underline{g} \to N.$$

Under α, the kernel maps isomorphically to the sheaf of sections of the tangent bundle TP which vanish at the four points $P \cap Y$. From the long exact cohomology sequence we have

$$0 \to \underline{g} \to H^0(P, N) \xrightarrow{\delta} H^1(P, \mathcal{O}(-2)) \to 0$$

and since $H^0(P, N)$ is 4-dimensional and $\underline{g}$ is 3-dimensional, the map δ is surjective. But $\alpha\delta$ is the Kodaira-Spencer map for deformations of the four points on P, so since it is non-trivial, the cross-ratio is non-constant.

Example

As the reader may realize, the situation here is very similar to the study of twistor spaces and twistor lines, and indeed there is a differential geometric context for this (see [6], [7]). This is not the agenda for this paper, but it is a useful example to see the standard twistor space – $\mathbf{C}P^3$ and the straight lines in it – within the current context.

Let V be the 4-dimensional space of cubic polynomials

$$p(z) = c_0 + c_1 z + c_2 z^2 + c_3 z^3$$

and consider V as a representation space of $SL(2, \mathbf{C})$ under the action

$$p(z) \mapsto p\left(\frac{az+b}{cz+d}\right)(cz+d)^3.$$

This is the unique (up to isomorphism) 4-dimensional irreducible representation of $SL(2, \mathbf{C})$. Then $Z = P(V) = \mathbf{C}P^3$ is a compact threefold with an action of $SL(2, \mathbf{C})$ and moreover the open dense set of cubics with distinct roots is an orbit. This follows since given any two triples of distinct ordered points in $\mathbf{C}P^1$, there is a unique element of $PSL(2, \mathbf{C})$ which takes one to the other. However, the cubic polynomial determines an *unordered* triple of roots, and hence the stabilizer in $PSL(2, \mathbf{C})$ is the symmetric group S_3. Thinking of this as the symmetries of an equilateral triangle, the holonomy group $\Gamma \subset SL(2, \mathbf{C})$ of the connection $A = \alpha^{-1}$ is the binary dihedral group $\tilde{D}_3$. The lower-dimensional orbits consist firstly of the cubics with one repeated root, which is

2-dimensional, and those with a triple root, which constitute a rational normal curve in $\mathbf{CP}^3$. Together they form the discriminant divisor Y, the anticanonical divisor discussed above.

A generic line in $\mathbf{CP}^3$, generated by polynomials $p(z), q(z)$ meets Y at those values of t for which the discriminant of $tp(z) + q(z)$ vanishes, i.e. where

$$\begin{aligned} tp(z) + q(z) &= 0 \\ tp'(z) + q'(z) &= 0 \end{aligned}$$

have a common root. This occurs for $t = -q(\alpha)/p(\alpha)$ where α is a root of the quartic equation

$$p'(z)q(z) - p(z)q'(z) = 0$$

and so the line meets Y in four generically distinct points. Thus the 4-parameter family of lines in $\mathbf{CP}^3$ furnish an example of the above proposition.

As we remarked above, this is an example of an isomonodromic deformation, as would be any family of curves P in Proposition 3. It yields a solution of the Painlevé equation either by applying the arguments of Theorem 1 to the connection with logarithmic singularities on Z, or appealing to the universality of Malgrange's construction. We shall not derive the solution of the Painlevé equation here from $\mathbf{CP}^3$, since it will appear via a different compactification in the context of Poncelet polygons. There we shall also see how a straight line in $\mathbf{CP}^3$ defines a pair of conics with the Poncelet property for triangles.

4 Poncelet polygons and projective bundles

In this section we shall study a particular class of equivariant compactifications, originally due to Schwarzenberger [16]. Consider the complex surface $\mathbf{CP}^1 \times \mathbf{CP}^1$ and the holomorphic involution σ which interchanges the two factors. The quotient space is $\mathbf{CP}^2$. A profitable way of viewing this is as the map which assigns to a pair of complex numbers the coefficients of the quadratic polynomial which has them as roots. In affine coordinates we have the quotient map

$$\begin{aligned} \pi : \mathbf{CP}^1 \times \mathbf{CP}^1 &\rightarrow \mathbf{CP}^2 \\ (w, z) &\mapsto (-(w+z), wz). \end{aligned}$$

From this it is clear that π is a double covering branched over the image of the diagonal, which is the conic $B \subset \mathrm{CP}^2$ with equation $4y = x^2$. Moreover, the line $\{a\} \times \mathrm{CP}^1 \subset \mathrm{CP}^1 \times \mathrm{CP}^1$ maps to a line in CP^2 which meets B at the single point $\pi(a, a)$. The images of the two lines $\{a\} \times \mathrm{CP}^1$ and $\mathrm{CP}^1 \times \{b\}$ are therefore the two tangents to the conic B from the point $\pi(a, b)$.

Now let $\mathcal{O}(k, l)$ denote the unique holomorphic line bundle of bidegree (k, l) on $\mathrm{CP}^1 \times \mathrm{CP}^1$, and define the direct image sheaf $\pi_*\mathcal{O}(k, 0)$ on CP^2. This is a locally free sheaf, a rank 2 vector bundle V_k, and we may form the projective bundle $P(V_k)$, a complex 3-manifold which fibres over CP^2

$$p : P(V_k) \rightarrow \mathrm{CP}^2$$

with fibres CP^1.

Clearly the diagonal action of $SL(2, \mathrm{C})$ on $\mathrm{CP}^1 \times \mathrm{CP}^1$ induces an action on $P(V_k)$. Take a point $z \in P(V_k)$ and consider its stabilizer. If $p(z) \in \mathrm{CP}^2 \backslash B$, then $p(z) = \pi(a, b)$ where $a \neq b$. Consider the projective bundle pulled back to $\mathrm{CP}^1 \times \mathrm{CP}^1$. The point (a, b) is off the diagonal in $\mathrm{CP}^1 \times \mathrm{CP}^1$, so the fibre of $P(V_k) = P(\pi_*\mathcal{O}(k, 0))$ is

$$P(\mathcal{O}(k, 0)_a \oplus \mathcal{O}(k, 0)_b). \tag{7}$$

The stabilizer of (a, b) in $SL(2, \mathrm{C})$ is one-dimensional, and acts on $(u, v) \in \mathcal{O}(k, 0)_a \oplus \mathcal{O}(k, 0)_b$ as

$$(u, v) \mapsto (\lambda^k u, \lambda^{-k} v).$$

Thus, as long as $u \neq 0$ or $v \neq 0$, the stabilizer of the point represented by (u, v) in the fibre is finite. Thus the generic orbit is three-dimensional.

We have implicitly just defined the divisor Y of lower-dimensional orbits, but to be more precise, we have the inverse image of the branch locus

$$D_1 = \pi^{-1}(B)$$

as one component. The other arises from the direct image construction as follows.

Recall that by definition of the direct image, for any open set $U \subseteq \mathrm{CP}^2$,

$$H^0(U, V_k) \cong H^0(\pi^{-1}(U), \mathcal{O}(k, 0))$$

so that there is an evaluation map

$$\mathrm{ev} : H^0(\pi^{-1}(U), \pi^* V_k) \rightarrow H^0(\pi^{-1}(U), \mathcal{O}(k, 0)).$$

The kernel of this defines a distinguished line sub-bundle of $\pi^*(V_k)$ and thus a section of the pulled back projective bundle $P(V_k)$. This copy of $\mathrm{CP}^1 \times \mathrm{CP}^1$ in $P(V_k)$ is a divisor D_2.

Both divisors are components of the anticanonical divisor Y, and it remains to check the multiplicity. Now let U be the divisor class of the tautological line bundle over the projective bundle $P(V_k)$. The divisor D_2 is a section of $P(V_k)$ pulled back to $\mathbf{C}P^1 \times \mathbf{C}P^1$, and from its definition it is in the divisor class $p^*(-U) + \mathcal{O}(k, 0)$. Thus in $P(V_k)$,

$$D_2 \sim -2U + kH \tag{8}$$

where H is the divisor class of the pull-back by π of the hyperplane bundle on $\mathbf{C}P^2$. Clearly, since B is a conic,

$$D_1 \sim 2H. \tag{9}$$

Now from Grothendieck-Riemann-Roch applied to the projection π, we find $c_1(V_k) = (k-1)H$, from which it is easy to see that the canonical divisor class is

$$K \sim 2U - (k+2)H$$

so since $-K \sim -2U + (k+2)H \sim D_1 + D_2$, the multiplicity is 1 for each divisor, and we can take $Z = P(V_k)$ as an example of an equivariant compactification to which Proposition 2 applies.

The stabilizer of a point in $Z \backslash Y$ is in this case the binary dihedral group $\tilde{D}_k$, which is the inverse image in $SU(2)$ of the group of symmetries in $SO(3) \cong SU(2)/\pm 1$ of a regular plane polygon with k sides. Although this can be seen quite easily from the above description of the action, there is a direct way of viewing $Z \backslash Y = P(V_k) \backslash D_1 \cup D_2$ as the $SL(2, \mathbf{C})$ orbit of a plane polygon.

Note that a polygon centred on $0 \in \mathbf{C}^3$ is described by a non-null axis orthogonal to the plane of the polygon, and by k (if k is odd) or $k/2$ (if k is even) equally spaced axes through the origin in that plane. Now, given a point $z \in P(V_k) \backslash D_1 \cup D_2$, its projection $p(z) = x \in P(\mathbf{C}^3) \backslash B$ is a non-null direction in $\mathbf{C}^3$ which we take to be the axis. To find the other axes we use two facts:

- The map $s \mapsto s^k$ from $\mathcal{O}(1, 0)$ to $\mathcal{O}(k, 0)$ defines a rational map m_k : $P(V_1) \to P(V_k)$ of degree k.
- The projective bundle $P(V_2)$ is the projectivized tangent bundle $P(T)$ of $\mathbf{C}P^2$.

The first fact is a direct consequence of the definition of the direct image sheaf:

$$H^0(U, V_k) \cong H^0(\pi^{-1}(U), \mathcal{O}(k, 0))$$

for any open set $U \subseteq \mathbf{CP}^2$. The second can be found in [16].

Given these two facts, consider the set of points

$$m_2(m_k^{-1}(z)) \subset P(V_2).$$

Depending on the parity of k this consists of k or $k/2$ points in $P(T)$ all of which project to $x \in \mathbf{CP}^2$. In other words they are lines through x or, using the polarity with respect to the conic B, points on the polar line of x. Reverting to linear algebra, these are axes in the plane orthogonal to x.

We now need to apply Proposition 3 to this particular set of examples – to find rational curves which meet the divisor $Y = D_1 + D_2$ transversally in four points. Now if P is such a curve, then the intersection number $P.D_1 \leq 4$ so $p(P) = C$ is a plane curve of degree d which meets the branch conic B in $2d \leq 4$ points, hence $d = 1$ or 2. We consider the case $d = 2$ first. The curve C is a conic in $\mathbf{CP}^2$. The set of all conics forms a 5-parameter family, and we want to determine the 4-parameter family of conics which lift to $P(V_k)$.

Theorem 2 *A conic $C \subset \mathbf{CP}^2$ meeting B transversally lifts to $P(V_k)$ if and only if there exists a k-sided polygon inscribed in C and circumscribed about B.*

Proof. A lifting of C is a section of $P(V_k)$ over C, or equivalently a line subbundle $M \subset V_k$ over C. Since C is a conic, the hyperplane bundle H is of degree 2 on C, so we can write $M \cong H^{n/2}$ for some integer n. The inclusion $M \subset V_k$ thus defines a holomorphic section s of the vector bundle $V_k \otimes H^{-n/2}$ over C. But V_k is the direct image sheaf of $\mathcal{O}(k,0)$, so we have an isomorphism

$$H^0(C, V_k \otimes H^{-n/2}) \cong H^0(\tilde{C}, \mathcal{O}(k,0) \otimes \pi^*(H^{-n/2}))$$

where $\tilde{C} = \pi^{-1}(C) \subset \mathbf{CP}^1 \times \mathbf{CP}^1$ is the double covering of the conic C branched over its points of intersection with B. But it is easy to see that $\pi^*(H) \cong \mathcal{O}(1,1)$, so on $\tilde{C}$ we have a holomorphic section $\tilde{s}$ of $\mathcal{O}(k - n/2, -n/2)$.

We have more, though, for since the intersection number $-K.P = (D_1 + D_2).P = 4$ and $D_1.P = B.C = 4$, P lies in $P(V_k)\backslash D_2$, where D_2 was given as the kernel of the evaluation map. If the section $\tilde{s}$ vanishes anywhere, then the section of $P(V_k)$ will certainly meet D_2, thus $\tilde{s}$ is everywhere non-vanishing and $\mathcal{O}(k - n/2, -n/2)$ is the trivial bundle. In particular, its degree is zero on $\tilde{C}$. Now C is a conic, so $\tilde{C}$ is the divisor of a section of $\pi^*(H^2) \cong \mathcal{O}(2,2)$

and so the degree of the line bundle is $2k - 2n = 0$ and thus $n = k$. Hence a conic in CP^2 lifts to $P(V_k)$ if and only if it has the property that

$$\mathcal{O}(k/2, -k/2) \cong \mathcal{O} \quad \text{on} \quad \tilde{C}.$$

Now recall the Poncelet problem [3]: to find a polygon with k sides which is inscribed in a conic C and circumscribed about a conic B. The projection

$$\pi : \mathrm{CP}^1 \times \mathrm{CP}^1 \to \mathrm{CP}^2$$

we have already used is the correct setting for the problem.

Let (a, b) be a point in $\mathrm{CP}^1 \times \mathrm{CP}^1$ and consider the two lines $\{a\} \times \mathrm{CP}^1$ and $\mathrm{CP}^1 \times \{b\}$ passing through it. The first line is a divisor of the linear system $\mathcal{O}(1,0)$ and the second of $\mathcal{O}(0,1)$. As we have seen, their images in CP^2 are the two tangents to the branch conic B from the point $\pi(a,b)$. Now let C be the conic which contains the vertices of the Poncelet polygon, and let $P_1 = (a_1, b_1) \in \tilde{C} \subset \mathrm{CP}^1 \times \mathrm{CP}^1$ be a point lying over an initial vertex. The line $\{a_1\} \times \mathrm{CP}^1$ meets $\tilde{C} \sim \mathcal{O}(2,2)$ in two points generically, which are P_1 and a second point $P_2 = (a_1, b_2)$. The two points $\pi(P_1)$ and $\pi(P_2)$ lie on C, and the line joining them is $\pi(\{a_1\} \times \mathrm{CP}^1)$ which is tangent to B, and hence is a side of the polygon. The other side of the polygon through $\pi(P_2)$ is $\pi(\mathrm{CP}^1 \times \{b_2\})$ which meets the conic C at $\pi(P_3) = \pi(a_2, b_2)$. We carry on this procedure using the two lines through each point, to obtain $P_1, \ldots, P_{k+1}$. Since the Poncelet polygon is closed with k vertices, we have $\pi(P_{k+1}) = \pi(P_1)$.

Consider now the divisor classes $P_i + P_{i+1}$. We have

$$\begin{aligned} P_1 + P_2 &\sim \mathcal{O}(1,0) \\ P_2 + P_3 &\sim \mathcal{O}(0,1) \\ P_3 + P_4 &\sim \mathcal{O}(1,0) \\ \cdots & \quad \cdots \end{aligned}$$

and $P_k + P_{k+1} \sim \mathcal{O}(1,0)$ if k is odd and $\sim \mathcal{O}(0,1)$ if k is even.

In the odd situation, taking the alternating sum we obtain

$$P_1 + P_{k+1} \sim \mathcal{O}((k+1)/2, -(k-1)/2) \tag{10}$$

and since $\pi(P_{k+1}) = \pi(P_1)$, then $P_{k+1} = P_1$ or $\sigma(P_1)$. However, in the former case, we would have

$$P_k + P_1 \sim P_k + P_{k+1} \sim \mathcal{O}(1,0) \sim P_1 + P_2$$

and consequently $P_2 \sim P_k$ on the elliptic curve $\tilde{C}$ which implies $P_2 = P_k$. But $\pi(P_k)$ and $\pi(P_2)$ are different vertices of the polygon, so we must have $P_{k+1} = \sigma P_1$. This implies that the divisor $P_{k+1} + P_1 = \pi^{-1}(\pi(P_1))$ and so in the notation above

$$P_{k+1} + P_1 \sim H^{1/2} = \mathcal{O}(1/2, 1/2).$$

From (10) we therefore obtain the constraint on $\tilde{C}$

$$\mathcal{O}(k/2, -k/2) \sim \mathcal{O} \tag{11}$$

which is exactly the condition for the conic to lift to $P(V_k)$. A similar argument leads to the same condition for k even, where in this case $P_{k+1} = P_1$.

In the case that $d = 1$, C is a line, but the argument is very similar. Here $M \cong H^n$ for some n and on $\tilde{C}$ we have a section ξ of $\mathcal{O}(k-n, -n)$. This time, since $P.D_2 = 2$, the line bundle is of degree 2, so $k - 2n = 2$, and so a lifting is defined by a section of $\mathcal{O}(1 + k/2, 1 - k/2)$ on $\tilde{C}$.

Example

Let us now compare this interpretation with the equivariant compactification $\mathbf{C}P^3$ of $SL(2, \mathbf{C})/\tilde{D}_3$ discussed earlier. In the first place, consider the line bundle

$$\tilde{U} = U - 2H$$

on $P(V_3)$. Now since for any 2-dimensional vector space $V^* \cong V \otimes \Lambda^2 V^*$, $P(V_k) = P(V_k^*)$, but with different tautological bundles. The tautological bundle for $P(V_3^*)$ is actually $\tilde{U}$, and so there are canonical isomorphisms

$$\begin{aligned} H^0(P(V_3), -\tilde{U}) &\cong H^0(\mathbf{C}P^2, V_3) \cong H^0(\mathbf{C}P^1 \times \mathbf{C}P^1, \mathcal{O}(3,0)) \\ &\cong H^0(\mathbf{C}P^1, \mathcal{O}(3)) \cong \mathbf{C}^4. \end{aligned}$$

The linear system $|-\tilde{U}|$ therefore maps $P(V_3)$ equivariantly to $\mathbf{C}P^3$. Since $P.D_1 = 4$ and $P.D_2 = 0$, it follows from (8) and (9), that $P.H = 2$ and $P.U = 3$, and so $P.\tilde{U} = -1$, so under this mapping the curves P map to projective lines.

There is a more geometric way of seeing the relation of lines in $\mathbf{C}P^3$ to Poncelet triangles. Recall that we are viewing $\mathbf{C}P^3$ as the space of cubic polynomials, and $\mathbf{C}P^2$ as the space of quadratic polynomials. The quadratics with a fixed linear factor $z - \alpha$

describe, as we have seen, a line in $\mathbf{CP}^2$ which is tangent to the discriminant conic at the quadratic $(z-\alpha)^2$. Thus the three linear factors of a cubic $(z-\alpha),(z-\beta),(z-\gamma)$ describe the sides of a triangle circumscribing the conic B. Its vertices are the quadratic factors $(z-\alpha)(z-\beta),(z-\beta)(z-\gamma),(z-\gamma)(z-\alpha)$.

Now consider a straight line of cubics $p_t(z) = tp(z) + q(z)$ with roots $\alpha(t), \beta(t)$ and $\gamma(t)$. We have a 1-parameter family of triangles and

$$\begin{aligned} tp(\alpha) + q(\beta) &= 0 \\ tp(\beta) + q(\beta) &= 0. \end{aligned}$$

Now from these two equations

$$0 = p(\alpha)q(\beta) - p(\beta)q(\alpha) = (\alpha - \beta)r(\alpha, \beta)$$

where $r(\alpha,\beta)$ is a symmetric polynomial in α, β. It is in fact *quadratic* in $\alpha+\beta, \alpha\beta$ and thus defines a conic C in the plane.

Hence, as t varies, the vertices of the triangle lie on a fixed conic C, and we have a solution of the Poncelet problem for $k = 3$.

5 Solutions of Painlevé VI

To find more about the connection we have just defined on $Z = P(V_k)$ entails descending to local coordinates, which we do next.

Consider the projective bundle $P(V_k)$ pulled back to $\mathbf{CP}^1 \times \mathbf{CP}^1$. At a point off the diagonal $(a,b) \in \mathbf{CP}^1 \times \mathbf{CP}^1$, as in (7), the fibre is

$$P(\mathcal{O}(k,0)_a \oplus \mathcal{O}(k,0)_b) = P(\mathcal{O}(k,0) \oplus \mathcal{O}(0,k))_{(a,b)}$$

and away from the zero section of the second factor, this is isomorphic to

$$\mathcal{O}(k,-k)_{(a,b)}.$$

Now choose standard affine coordinates (w,z) in $\mathbf{CP}^1 \times \mathbf{CP}^1$. Since $K_{\mathbf{CP}^1} \cong \mathcal{O}(-2)$, we have corresponding local trivializations dw and dz of $\mathcal{O}(-2,0)$ and $\mathcal{O}(0,-2)$. These define a local trivialization $(dw)^{-k/2}(dz)^{k/2}$ of $\mathcal{O}(k,-k)$, and thus coordinates

$$(w,z,s) \mapsto s(dw)^{-k/2}(dz)^{k/2}_{(w,z)}.$$

Note that Z is the quotient of this space by the involution $(w, z, s) \mapsto (z, w, s^{-1})$. From this trivialization, the natural action of $SL(2, \mathbf{C})$ on differentials gives the action on Z:

$$(w, z, s) \mapsto \left(\frac{aw + b}{cw + d}, \frac{az + b}{cz + d}, \frac{(cz + d)^k}{(cw + d)^k} s \right).$$

Differentiating this expression at the identity gives the tangent vector (w', z', s') corresponding to a matrix

$$\begin{pmatrix} a' & b' \\ c' & -a' \end{pmatrix} \in \underline{g}$$

as

$$\begin{aligned} w' &= -c'w^2 + 2a'w + b' \\ z' &= -c'z^2 + 2a'z + b' \\ s' &= -kc'(w - z)s \end{aligned}$$

This is $\alpha(a', b', c') \in TZ_{(w,z,s)}$. Solving for (a', b', c') gives the entries of the matrix of 1-forms $A = \alpha^{-1}$ as

$$\begin{aligned} A_{11} &= \frac{dw - dz}{2(w - z)} - \frac{(w + z)ds}{2ks(w - z)} \\ A_{12} &= \frac{wdz - zdw}{(w - z)} + \frac{wzds}{ks(w - z)} \\ A_{21} &= -\frac{ds}{ks(w - z)} \end{aligned} \tag{12}$$

Proposition 4 *The residue of the connection at a singular point is conjugate to*

$$\begin{pmatrix} 1/4 & 0 \\ 0 & -1/4 \end{pmatrix} \text{ on } D_1 \quad \text{and} \quad \begin{pmatrix} 1/2k & 0 \\ 0 & -1/2k \end{pmatrix} \text{ on } D_2$$

Proof. In these coordinates, $s = 0$ is the equation of D_2. From (2), the residue of A at $s = 0$ is

$$\begin{pmatrix} -(w + z)/2k(w - z) & wz/k(w - z) \\ -1/k(w - z) & (w + z)/2k(w - z) \end{pmatrix} \tag{13}$$

which has determinant $-1/4k^2$ and therefore eigenvalues $\pm 1/2k$.

To find the residue at D_1, we need different coordinates, since the above ones are invalid on the diagonal. Take the affine coordinates $x = -(w +$

$z), y = wz$ on $\mathbf{CP}^2$. Since the holomorphic functions in w, z form a module over the symmetric functions generated by $1, w - z$ we can use these to give coordinates in the projectivized direct image $P(V_k)$, which are valid for $w = z$. We obtain an affine fibre coordinate t related to s above by

$$s = \frac{t + w - z}{t - w + z}.$$

Using this and local coordinates x and $u = (w - z)^2 = x^2 - 4y$ on $\mathbf{CP}^2$ the divisor D_1 is given by $u = 0$ and the residue here is

$$\begin{pmatrix} 1/4 + x/2kt & x/4 + x^2/4kt \\ -1/kt & -1/4 - x/2kt \end{pmatrix} \tag{14}$$

This has determinant $-1/16$ and hence eigenvalues $\pm 1/4$.

Remark. Exponentiating the residues we see that the holonomy of a small loop around the divisor D_1 or D_2 is conjugate to:

$$\begin{pmatrix} i & 0 \\ 0 & -i \end{pmatrix} \text{ on } D_1 \quad \text{and} \quad \begin{pmatrix} e^{i\pi/k} & 0 \\ 0 & e^{-i\pi/k} \end{pmatrix} \text{ on } D_2$$

In the dihedral group $D_k \subset SO(3)$ the conjugacy classes are those of a reflection in the plane and a rotation by $2\pi/k$.

These facts tell us something of the structure of the divisor D_1. Since we know that the residue of the meromorphic connection at a singular point lies in the Lie algebra of the stabilizer of the point, and this is here semisimple, the orbit is isomorphic to

$$SL(2, \mathbf{C})/\mathbf{C}^* \cong \mathbf{CP}^1 \times \mathbf{CP}^1 \backslash \Delta$$

where Δ is the diagonal. The projections onto the two factors must, by $SL(2, \mathbf{C})$ invariance, be the two eigenspaces of the residue corresponding to the eigenvalues $\pm 1/4$.

Now $D_1 = \pi^{-1}(B)$ is a projective bundle over the conic $B \cong \mathbf{CP}^1$. By invariance it must be one of the factors above. To see which, note that from (14), eigenvectors for the eigenvalues $1/4$ and $-1/4$ are respectively

$$\begin{pmatrix} x + kt \\ -2 \end{pmatrix} \quad \text{and} \quad \begin{pmatrix} x \\ -2 \end{pmatrix}$$

and so, from the choice of coordinates above, clearly the second represents the projection to B. Note, moreover, that on the diagonal $w = z$, the coordinate

$x = -(w+z) = -2z$, so that x is an affine parameter on $B \cong \Delta \cong \mathbf{C}P^1$. Furthermore, when $x = \infty$ the vector $\begin{pmatrix} 1 \\ 0 \end{pmatrix}$ is an eigenvector of the residue with eigenvalue $-1/4$.

Now let us use this information to determine the solution to the Painlevé equation corresponding to a rational curve $P \subset Z$. Recall that the curve $C = \pi(P)$ is a plane curve of degree d, where $d = 1$ or $d = 2$. As we have seen, when $d = 1$, any line is of this form, but when $d = 2$, the conic must circumscribe a Poncelet polygon.

By the $SL(2, \mathbf{C})$ action, we can assume that C meets the conic B at the point $x = \infty$. From the discussion above, if A_∞ is the residue of the connection at this point, then

$$A_\infty \begin{pmatrix} 1 \\ 0 \end{pmatrix} = -\frac{1}{4} \begin{pmatrix} 1 \\ 0 \end{pmatrix}$$

From Proposition 1, the solution of the Painlevé equation is the point y on the curve P at which $A(y)$ has this same eigenvector, i.e. where

$$A_{21}(y) = 0$$

Proposition 5 *A line in the plane defines a solution to Painlevé's sixth equation with coefficients* $(\alpha, \beta, \gamma, \delta) = (1/8, -1/8, 1/2k^2, 1/2 - 1/2k^2)$. *A Poncelet conic in the plane defines a solution to the Painlevé equation with coefficients* $(\alpha, \beta, \gamma, \delta) = (1/8, -1/8, 1/8, 3/8)$.

Proof. The residues on D_1 and D_2 are given by (14) and (13). The lifting of a line meets D_1 and D_2 in two points each, so using (6) (and taking account of the fact that the roles of the two basis vectors are interchanged), we obtain the first set of coefficients. The lifting of a Poncelet conic meets D_1 in four points, which gives the second set, again from (6).

Proposition 6 *The lifting of a line in* $\mathbf{C}P^2$ *to* $P(V_k)$ *defines the solution*

$$y = \sqrt{x}$$

of Painlevé VI with $(\alpha, \beta, \gamma, \delta) = (1/8, -1/8, 1/2k^2, 1/2 - 1/2k^2)$.

Proof. Taking the double covering $\tilde{C} \subset \mathbf{CP}^1 \times \mathbf{CP}^1$ of the line C and the coordinates w, z, s on the corresponding covering of Z, the lifted curve P is defined locally by a function s on the curve $\tilde{C}$. In fact, as we shall see next, s is a meromorphic function on $\tilde{C}$ with certain properties.

From the comments following Theorem 2, the lifting is given by a holomorphic section ξ of $\mathcal{O}(1+k/2, 1-k/2)$. This line bundle has degree 2 on $\tilde{C}$, and so ξ vanishes at twc points. Applying the involution σ, then $\sigma^*\xi$ is a section of $\mathcal{O}(1-k/2, 1+k/2)$. Considering ξ as a section of $\mathcal{O}(k,0)\otimes\mathcal{O}(1-k/2,1-k/2)$, the lifting of C to $P(V_k)$ is defined by $(\xi_{(a,b)}, \xi_{(b,a)})$, or in the coordinates w, z, s,

$$s(dw)^{-k/2}(dz)^{k/2} = \xi/\sigma^*\xi. \tag{15}$$

Since $dw^{-1/2}$ and $dz^{-1/2}$ are holomorphic sections of $\mathcal{O}(1,0)$ and $\mathcal{O}(0,1)$, it follows that on $\tilde{C}$, s is a meromorphic function. Now using the $SL(2,\mathbf{C})$ action, we may assume that the line C is given by $x = 0$, which means that $\tilde{C}$ has equation

$$w = -z$$

which defines an obvious trivialization of $\mathcal{O}(k,-k)$ and from which we deduce that s has two simple zeros at $(a_1, -a_1), (a_2, -a_2)$ and two poles at $(-a_1, a_1), (-a_2, a_2)$. Using z as an affine parameter on $\tilde{C}$, we obtain, up to a constant multiple,

$$s = \frac{(z+a_1)(z+a_2)}{(z-a_1)(z-a_2)}. \tag{16}$$

Now $y = wz$ is an affine parameter on the line C, which meets the conic B at $y = 0, \infty$. The lifting P meets the divisor D_2 where $y = -a_1^2$ and $y = -a_2^2$, so putting $a_1 = i$ and $a_2 = \sqrt{-x}$, then P is a projective line with a parametrization such that the singular points of the induced connection are $0, 1, x, \infty$, as required for the Painlevé equation.

It remains to determine the solution of the equation, which is given by $A_{21}(y) = 0$. But from (12), this is where $ds = 0$, and from (16) this is equivalent to

$$\frac{1}{z+a_1} - \frac{1}{z-a_1} + \frac{1}{z+a_2} - \frac{1}{z-a_2} = 0$$

which gives

$$y = -z^2 = -a_1 a_2 = \sqrt{x}$$

with the above choices of a_1, a_2.

Remarks.

1. By direct calculation, the function $y = \sqrt{x}$ solves Painlevé VI for any coefficients satisfying $\alpha + \beta = 0$ and $\gamma + \delta = 1/2$. From (6) this occurs when the residues are conjugate in pairs.

2. When $k = 2$, we obtain $(\alpha, \beta, \gamma, \delta) = (1/8, -1/8, 1/8, 3/8)$ which are the coefficients arising from Poncelet conics. We shall see the same solution appearing in the next section in the context of Poncelet quadrilaterals.

Naturally, the solutions corresponding to Poncelet conics are more complicated, and we shall give some explicitly in Section 6. Here we give the general algebraic procedure for obtaining them.

In the case of a conic C in $\mathbf{CP}^2$, we have a section ξ, in fact a trivialization, of the bundle $\mathcal{O}(k/2, -k/2)$ on the elliptic curve $\tilde{C}$. As in (15), we still define the lifting by

$$s(dw)^{-k/2}(dz)^{k/2} = \xi/\sigma^*\xi$$

but in this case ξ is non-vanishing. The section $(dw)^{-k/2}$ vanishes to order k at $\infty \in \mathbf{CP}^1$, and so at the two points $(\infty, \infty), (\infty, b)$ where $\tilde{C}$ meets $\{\infty\} \times \mathbf{CP}^1$. Similarly $(dz)^{-k/2}$ vanishes at $(\infty, \infty), (b, \infty)$. The meromorphic function s can be regarded as a map of curves

$$s : \tilde{C} \to \mathbf{CP}^1.$$

It follows that s is a meromorphic function on $\tilde{C}$ with a zero of order k at (b, ∞), a pole of order k at (∞, b) and no other zeros or poles.

The derivative ds is invariantly defined as a section of $K_{\tilde{C}} \otimes s^* K^{-1}_{\mathbf{CP}^1} \cong s^*\mathcal{O}(2)$ (since $\tilde{C}$ is an elliptic curve and hence has trivial canonical bundle). In particular ds vanishes with total multiplicity $2k$. But since (∞, a) and (a, ∞) are branch points of order k, ds has a zero of order $k - 1$ at each of these points, leaving two extra points as the remaining zeros. Since the involution σ takes s to s^{-1}, these points are paired by the involution, and give a single point $y \in \mathbf{CP}^1$ which is our solution to the Painlevé equation.

Fortunately Cayley's solution in 1853 to the Poncelet problem gives us the means to find y algebraically. A useful modern account of this is given by Griffiths and Harris in [5], but the following description I owe to M. F. Atiyah.

Suppose the elliptic curve $\tilde{C}$ is described as a cubic in $\mathbf{CP}^2$ given by $v^2 = h(u)$, where $h(u)$ is a cubic polynomial and $h(0) = c_0^2 \neq 0$. We shall find the condition on the coefficients of h in order that there should exist a polynomial $g(v, u)$ of degree $(n-1)$ (a section of $\mathcal{O}(n-1)$) on the curve with a zero of order $(2n-1)$ at $(v, u) = (c_0, 0)$ and a pole of order $(n-2)$ at

$u = \infty$. Given such a polynomial,

$$s(v,u) = \frac{g(v,u)}{g(-v,u)}$$

is the function on the curve (for $k = (2n-1)$) used above, and the zeros of its derivative define the solution $y(x)$ to the Painlevé equation. A very similar procedure deals with the case of even k.

To find g, expand $\sqrt{h(u)}$ as a power series in z, making a choice c_0 of square root of $h(0)$:

$$v = c_0 + c_1 u + c_2 u^2 + \ldots$$

and then put

$$v_m = c_0 + c_1 u + \ldots + c_{m-1} u^{m-1}.$$

Now clearly $v - v_n$ has a zero of order n on the curve at $u = 0$, as do other functions constructed from the v_m:

$$\begin{aligned}
v - v_n &= c_n u^n + \ldots + c_{2n-2} u^{2n-2} + \ldots \\
u(v - v_{n-1}) &= c_{n-1} u^n + \ldots + c_{2n-3} u^{2n-2} + \ldots \\
u^2(v - v_{n-2}) &= c_{n-2} u^n + \ldots + \ldots \\
\ldots &= \ldots \\
u^{n-2}(v - v_2) &= c_2 u^n + \ldots + c_n u^{2n-2} + \ldots
\end{aligned}$$

We can then find $n-1$ coefficients $\lambda_0, \lambda_1, \ldots, \lambda_{n-2}$ such that

$$g(v,u) \equiv \lambda_0(v - v_n) + \lambda_1 u(v - v_{n-1}) + \ldots + \lambda_{n-2} u^{n-2}(v - v_2)$$

vanishes at $u = 0$ to order $2n-1$ if and only if

$$\det M = 0 \quad \text{where} \quad M = \begin{bmatrix} c_n & c_{n-1} & \ldots & c_2 \\ c_{n+1} & c_n & \ldots & c_3 \\ \ldots & \ldots & \ldots & \ldots \\ c_{2n-2} & c_{2n-3} & \ldots & c_n \end{bmatrix} \tag{17}$$

This is Cayley's form of the Poncelet constraint.

If (17) holds, $g(v,u)$ is a polynomial of degree $n-1$ which, upon inspection, vanishes with multiplicity $n-2$ at the inflexion point at infinity of the curve. Its total intersection number with the cubic $\tilde{C}$ is $3(n-1) = (2n-1) + (n-2)$, so there are no more zeros. Thus the condition $\det M = 0$ is necessary and sufficient for the construction of the required function s with a zero of order k at $(v,u) = (c_0, 0)$ and a pole of order k at $(v,u) = (-c_0, 0)$.

In the case of a pair of conics in the plane, defined by symmetric matrices B and C, the constraint is on the cubic $h(u) = \det(B + uC)$ in order for the conics to satisfy the Poncelet condition.

Note that

$$g(v,u) = p(u)v + q(u)$$

where p and q are polynomials of degree $n-2$ and $n-1$ respectively. Thus

$$s = \frac{p(u)v + q(u)}{-p(u)v + q(u)}$$

and ds vanishes if

$$pqv' + (p'q - pq')v = 0.$$

Using $v^2 = h(u)$, this is equivalent to

$$r(u) \equiv p(u)q(u)h'(u) + 2(p'(u)q(u) - p(u)q'(u))h(u) = 0$$

This is a polynomial in u of degree $2n-1$, which by construction vanishes to order $k-1 = 2n-2$ at $u = 0$. It is thus of the form

$$r(u) = au^{2n-2}(u - b)$$

and so y, the solution to the Painlevé equation which corresponds to a zero of ds, is defined in terms of the ratio of the two highest coefficients of $r(u)$. Since the solution to the Painlevé equation has singularities at the four points $0, 1, x, \infty$, a Möbius transformation gives the variable x, and the solution $y(x)$, as:

$$x = \frac{e_3 - e_1}{e_2 - e_1} \qquad y = \frac{b - e_1}{e_2 - e_1} \tag{18}$$

where e_1, e_2 and e_3 are the roots of $h(u) = 0$.

To calculate b explicitly is easy. Putting $p(u) = p_0 + p_1 u + \ldots + p_{n-2}u^{n-2}$ and $q(u) = q_0 + q_1 u + \ldots q_{n-1}u^{n-1}$ and looking at the coefficients of $r(u)$, we find

$$\begin{aligned} b &= \frac{p_{n-3}}{p_{n-2}} - 3\frac{q_{n-2}}{q_{n-1}} \\ &= \frac{\lambda_{n-3}}{\lambda_{n-2}} - 3\frac{\lambda_0 c_{n-2} + \lambda_1 c_{n-3} + \ldots + \lambda_{n-2}c_0}{\lambda_0 c_{n-1} + \lambda_1 c_{n-2} + \ldots + \lambda_{n-2}c_1} \end{aligned}$$

from the definition of $p(u)$ and $q(u)$. Now the coefficients λ_i are the entries of a column vector λ such that $M\lambda = 0$. Thus in the generic case where the rank of M is $n-3$, these are given by cofactors of M. We can then write

$$b=-\frac{\begin{vmatrix} c_n & \dots & c_4 & c_2 \\ c_{n+1} & \dots & c_5 & c_3 \\ \dots & \dots & \dots & \dots \\ c_{2n-3} & \dots & c_{n+1} & c_{n-1} \end{vmatrix}}{\begin{vmatrix} c_n & c_{n-1} & \dots & c_3 \\ c_{n+1} & c_n & \dots & c_4 \\ \dots & \dots & \dots & \dots \\ c_{2n-3} & c_{2n-4} & \dots & c_n \end{vmatrix}}-3\frac{\begin{vmatrix} c_n & c_{n-1} & \dots & c_2 \\ \dots & \dots & \dots & \dots \\ c_{2n-3} & c_{2n-4} & \dots & c_{n-1} \\ c_{n-2} & c_{n-3} & \dots & c_0 \end{vmatrix}}{\begin{vmatrix} c_n & c_{n-1} & \dots & c_2 \\ \dots & \dots & \dots & \dots \\ c_{2n-3} & c_{2n-4} & \dots & c_{n-1} \\ c_{n-1} & c_{n-2} & \dots & c_1 \end{vmatrix}} \tag{19}$$

This effectively gives us a concrete form for the solution of the Painlevé equation for k odd, though we shall try to be more explicit in special cases in the next section. When k is even, a similar analysis can be applied. Very briefly, if $k=2n$, and $n \geq 2$, then the vanishing of

$$\begin{vmatrix} c_{n+1} & c_n & \dots & c_3 \\ c_{n+2} & c_{n+1} & \dots & c_4 \\ \dots & \dots & \dots & \dots \\ c_{2n-1} & \dots & \dots & c_{n+1} \end{vmatrix}$$

is the condition for the existence of λ_i so that

$$g(v,u)=\lambda_0(v-v_{n+1})+\lambda_1 u(v-v_n)+\ldots+\lambda_{n-2}u^3(v-v_3)$$

has a zero of order $2n$ at $u=0$. The rest follows in a similar manner to the above.

6 Explicit solutions

We shall now calculate explicit solutions to Painlevé VI with coefficients $(\alpha,\beta,\gamma,\delta)=(1/8,-1/8,1/8,3/8)$ for small values of k. Clearly, from the interpretation in terms of Poncelet polygons, we must have $k \geq 3$. The discussion in the previous section shows that we need to perform calculations with the coefficients of the cubic $h(u)=\det(B+uC)$ where B and C are symmetric matrices representing the conics we have denoted with the same symbol. For convenience, we take the cubic

$$\begin{aligned} h(u) &= (1+(x_1+x_2)u)(1+(x_2+x_0)u)(1+(x_0+x_1)u) \\ &= 1+2s_1u+(s_1^2+s_2)u^2+(s_1s_2-s_3)u^3 \end{aligned} \tag{20}$$

where s_i is the ith elementary symmetric function in x_0, x_1, x_2.

6.1 Solution for $k = 3$

For $k = 3$ the Poncelet constraint from (17) is $c_2 = 0$, which is $s_2 = 0$ for the above cubic, and can therefore be written

$$\frac{1}{x_0} + \frac{1}{x_1} + \frac{1}{x_2} = 0.$$

This is the equation in homogeneous coordinates (x_0, x_1, x_2) for a plane conic which can clearly be parametrized rationally by

$$x_0 = \frac{1}{1+s} \qquad x_1 = \frac{-1}{s} \qquad x_2 = -1 \tag{21}$$

The polynomial g is just $g(v, u) = v - (1 + s_1 u)$, and this gives

$$r(u) = s_1 s_3 u^3 + 3 s_3 u^2$$

and hence $b = -3/s_1$. Substituting the parametrization (21) in (18) and using the fact that the roots of $h(u) = 0$ are $u = -1/(x_1 + x_2)$ etc, gives the solution $y(x)$ to the Painlevé equation as

$$y = \frac{s^2(s+2)}{(s^2+s+1)} \quad \text{where} \quad x = \frac{s^3(s+2)}{2s+1}$$

6.2 Solution for $k = 4$

The Poncelet constraint here is $c_3 = 0$, which in the formalism above is $s_3 = 0$, that is

$$x_0 = 0 \qquad x_1 = 0 \qquad x_2 = 0.$$

In CP^2 this consists of three lines. Take the component $x_0 = 0$ and parametrize it by

$$x_1 = 1 \qquad x_2 = s.$$

Now the polynomial g is given by $g(v, u) = v - (1 + s_1 u + \frac{1}{2} s_2 u^2)$, which yields

$$r(u) = \frac{1}{2} s_1 s_2^2 u^4 + s_2^2 u^3$$

and hence $b = -2/s_1$. Substituting the parametrization, we obtain $x = s^2$, $y = s$, thus the solution to the Painlevé equation is

$$y(x) = \sqrt{x}.$$

Remark. In Proposition 6, we saw that the same solution $y = \sqrt{x}$ with coefficients $(\alpha, \beta, \gamma, \delta) = (1/8, -1/8, 1/8, 3/8)$ arises from taking a curve P in $P(V_2)$ with $P.D_1 = P.D_2 = 2$, and hence has holonomy in $\tilde{D}_2$, which is the quaternion group $\{\pm 1, \pm i, \pm j, \pm k\}$, a proper subgroup of $\tilde{D}_4$. Recall also that from [16], $P(V_2) \cong P(T)$, the projectivized tangent bundle of CP^2. In geometrical terms, a point of $P(T)$ consists of a point in CP^2 together with a line passing through it, or equivalently a line with a distinguished point. Thus there is a projection also to the dual projective plane CP^{2*}. In other words

$$P(T) = \{(p, L) \in \mathrm{CP}^2 \times \mathrm{CP}^{2*} : p \in L\}$$

with projections onto the two factors. In the terminology above, the two corresponding hyperplane divisor classes are H and $H - U$. Now the curve P which defines the solution to the Painlevé equation satisfies $P.H = P.(H - U) = 1$. It follows easily that P is obtained by taking a point $q \in \mathrm{CP}^2$ and a skew line M. The set

$$P = \{(p, L) \in P(T) : q \in L \quad \text{and} \quad p = L \cap M\}$$

describes the rational curve $P \subset P(T)$.

According to our formula, this curve must correspond to a Poncelet conic. In fact, let (p, L) be a point of P, and let ℓ denote the pole of the line L with respect to the conic B. The line ℓp has pole q, and as L varies in the pencil of lines through p, the point q describes a conic C. If L is a tangent through p to the conic B, then q is the point of contact $B \cap L$. Let q_1 and q_2 be the two points of contact of the tangents through p, then the conic C passes through p, q_1 and q_2. But pq_1pq_2 is then a degenerate Poncelet quadrilateral, and by Poncelet's theorem [3], if there is a Poncelet polygon through one point of C, then there exists one through each point.

6.3 Solution for $k = 5$

From (17), the Poncelet constraint is

$$\det \begin{bmatrix} c_3 & c_2 \\ c_4 & c_3 \end{bmatrix} = 0$$

which in terms of the symmetric functions s_i is

$$4s_3^2 + s_2^2 - 4s_1s_2s_3 = 0. \tag{22}$$

Now from (19),

$$b = -\frac{c_2}{c_3} - 3\frac{(c_3 - c_1c_2)}{(c_1c_3 - c_2^2)}$$

$$= \frac{s_2}{s_3} - 6\frac{(s_3 + s_1 s_2)}{(2s_1 s_3 + s_2^2)}$$

and using (22) this becomes

$$b = -20\frac{s_2 s_3}{(4s_3^2 + 3s_2^3)}.$$

It is convenient to introduce coordinates u, v by setting

$$x_0 = 1 \qquad u = \frac{1}{x_1} + \frac{1}{x_2} \qquad v = \frac{1}{x_1 x_2}$$

and then the constraint (22) becomes

$$v = \frac{(1+u)(1-u)^2}{4u} \tag{23}$$

and

$$b = -5\frac{(u+1)(u-1)^2}{(3u(u+1)^2 + (u-1)^2)}$$

Now $1/x_1$ and $1/x_2$ are the roots of the quadratic $z^2 - uz + v = 0$, and so, using (23),

$$\frac{1}{x_1}, \frac{1}{x_2} = \frac{1}{2}(u \pm \sqrt{1 + u - u^{-1}})$$

and now putting

$$w^2 = 1 + u + u^{-1} \tag{24}$$

we finally obtain the solution of the Painlevé equation as

$$y = \frac{(u-w+2)(u+w)}{(u+w-2)(u-w)}\left(1 - \frac{20u^2}{(3u(u+1)^2 + (u-1)^2)}\right)$$

$$\text{where} \quad x = \frac{(u-w-2)(u-w+2)(u+w)^2}{(u+w-2)(u+w+2)(u-w)^2} \quad \text{and} \quad w^2 = 1 + u + u^{-1}$$

Note that (24) is the equation of an elliptic curve, so that x and y are meromorphic functions on the curve. It is a special elliptic curve, in fact under the Cremona transformation $x_i \mapsto 1/x_i$, the equation (22) transforms into the plane cubic

$$s_1^3 - 4s_1 s_2 + 4s_3 = 0$$

and the symmetric group S_3 clearly acts as automorphisms of the curve.

The study of the Poncelet constraints for small values of k is undertaken in [1], and the reader will find that, apart from $k = 6$ and $k = 8$, the formulae rapidly become more complicated. We shall only consider now these two further cases.

6.4 Solution for $k = 6$

Here, from [1], we find that the constraint factorizes

$$(x_0x_1 + x_1x_2 + x_2x_0)(-x_0x_1 + x_1x_2 + x_2x_0)(x_0x_1 - x_1x_2 + x_2x_0)$$
$$(x_0x_1 + x_1x_2 - x_2x_0) = 0.$$

The first factor represents the case $k = 3$ embedded in $k = 6$, by thinking of a repeated Poncelet triangle as a hexagon. We choose instead the third factor, which can be written as

$$\frac{1}{x_0} = \frac{1}{x_1} + \frac{1}{x_2}.$$

This is a conic, and we rationally parametrize it by setting

$$x_0 = \frac{1}{1+s} \qquad x_1 = \frac{1}{s} \qquad x_2 = 1.$$

After some calculation, this gives a solution to the Painlevé equations

$$y = \frac{s(1+s+s^2)}{(2s+1)}$$
$$\text{where} \quad x = \frac{s^3(s+2)}{(2s+1)}$$

6.5 Solution for $k = 8$

Here, again referring to [1], we find the constraint equation splits into components

$$(-x_0^2x_1^2 + x_1^2x_2^2 + x_2^2x_0^2)(x_0^2x_1^2 - x_1^2x_2^2 + x_2^2x_0^2)(x_0^2x_1^2 + x_1^2x_2^2 - x_2^2x_0^2) = 0.$$

one of which is given by the equation

$$\frac{1}{x_0^2} = \frac{1}{x_1^2} + \frac{1}{x_2^2}$$

and, parametrizing this conic rationally in the usual way with

$$x_0 = 1 \qquad x_1 = \frac{1+s^2}{2s} \qquad x_2 = \frac{1+s^2}{1-s^2}$$

one may obtain the solution to the Painlevé equation as

$$y = \frac{4s(3s^2 - 2s + 1)}{(1+s)(1-s)^3(s^2+2s+3)}$$
$$\text{where} \quad x = \left(\frac{2s}{1-s^2}\right)^4.$$

7 Painlevé curves

The examples above show that there do indeed exist algebraic solutions to Painlevé's sixth equation, for certain values of the coefficients, despite the general description of solutions to these equations as "Painlevé transcendants". In general, an algebraic solution is given by a polynomial equation

$$R(x,y) = 0$$

which defines an algebraic curve. So far, we have only seen explicit examples where this curve is rational or elliptic, but higher genus curves certainly do occur. We make the following definition:

Definition 2 *A Painlevé curve is the normalization of an algebraic curve $R(x,y) = 0$ which solves Painlevé's sixth equation (5) for some values of the coefficients $(\alpha, \beta, \gamma, \delta)$.*

Just as the elliptic curve above corresponding to the solution for $k = 5$ was special, so are Painlevé curves in general. The equation (5) was in fact found not by Painlevé, but by R. Fuchs [4], but nevertheless falls into the Painlevé classification by its characteristic property that its solutions have no "movable singular points". What this means is that the branch points or essential singularities of solutions $y(x)$ are independent of the constants of integration. In the case of Painlevé VI, these points occur only at $x = 0, 1, \infty$. Now if X is a Painlevé curve, x and y are meromorphic functions on X, and so there are no essential singularities. The function

$$x : X \to \mathbf{CP}^1$$

is thus a map with branch points only at $x = 0, 1$ and ∞.

Such curves have remarkable properties. In the first place, it follows from Weil's rigidity theorem [17], that X is defined over the algebraic closure $\bar{\mathbf{Q}}$ of the rationals (from Belyi's theorem [2] this actually characterizes curves with such functions). Secondly, by uniformizing $\mathbf{CP}^1 \backslash \{0, 1, \infty\}$,

$$X \cong \overline{H/\Gamma}$$

where H is the upper half-plane and Γ is a subgroup of finite index in the principal congruence subgroup $\Gamma(2) \subset SL(2, \mathbf{Z})$. Thus, in some manner, each algebraic solution of Painlevé VI gives rise to a problem involving elliptic curves.

Coincidentally, the investigation of the Poncelet problem by Barth and Michel in [1] proceeds by studying a modular curve. This is a curve which occurs in parametrizing elliptic curves with

- a level-2 structure
- a primitive element of order k

In our model of the Poncelet problem, the elliptic curve is $\tilde{C}$, the level-2 structure identifies the elements of order two (or equivalently an ordering of the branch points of π), and the Poncelet constraint (11) selects the line bundle $\mathcal{O}(1/2,-1/2)$ of order k on $\tilde{C}$, or equivalently a point of order k on the curve, the zero of the function s in Section 5. Choosing a primitive element avoids recapturing a solution for smaller k.

As described by Barth and Michel, the stabilizer of a primitive element of order k is

$$\Gamma_{00}(k) = \left\{ \begin{pmatrix} a & b \\ c & d \end{pmatrix} \in SL(2,\mathbf{Z}) : a \equiv d \equiv 1(k) \quad \text{and} \quad c \equiv 0(k) \right\} \qquad (25)$$

Now if k is odd, matrices $A \in \Gamma_{00}(k)$ can be chosen such that A mod 2 is any element of $SL(2,\mathbf{Z_2})$. Thus $\Gamma_{00}(k)$ acts transitively on the level-2 structures. Consider in this case the modular curve

$$X_{00}(k,2) = \overline{H/\Gamma_{00}(k) \cap \Gamma(2)} \qquad (26)$$

Since $-I \in \Gamma_{00}(k)$ acts trivially on H, this is a curve parametrizing opposite pairs of primitive elements of order k, and level-2 structures.

When k is even, however, only the two matrices

$$\begin{pmatrix} 1 & 0 \\ 0 & 1 \end{pmatrix} \quad \begin{pmatrix} 1 & 1 \\ 0 & 1 \end{pmatrix}$$

are obtained by reducing mod 2 from $\Gamma_{00}(k)$. Thus the group has three orbits on the level-2 structures, and the curve $X_{00}(k,2)$ has three components each given by (26).

There is another curve in the picture: the curve Π_k defined by the Cayley constraint (17), with the c_i symmetric functions of (x_0, x_1, x_2) as defined by (20). This is a plane curve in homogeneous coordinates (x_0, x_1, x_2). Barth and Michel show that a birational image of $X_{00}(k,2)$ lies as a union of components of Π_k. In the examples of Section 6, we have already seen this curve, connected for $k = 3,5$ but with different components for $k = 4,6,8$.

In the algebraic construction of y in Section 5, it is clear from (19) that x and y are meromorphic functions on Π_k, and so the Painlevé curve X_k is a rational image of the modular curve. In fact, we have the following

Proposition 7 *The Painlevé curve X_k defined by Poncelet polygons is birationally equivalent to the modular curve $\overline{H/\Gamma_{00}(k) \cap \Gamma(2)}$.*

Proof. Let Y_k be the modular curve, then we already have a map $f : Y_k \to X_k$ as described above. We shall define an inverse on the complement of a finite set.

Let the Painlevé curve be defined by the equation

$$R(x, y) = 0$$

and suppose (x, y) is a point on the curve such that $\partial R/\partial y \neq 0$. Then

$$\frac{dy}{dx} = -\frac{\partial R/\partial x}{\partial R/\partial y}$$

is finite, and thus we can recover the connection matrix $A(z)$ on $\mathbf{CP}^1\backslash\{0, 1, x, \infty\}$, its coefficients being rational in $x, y, dy/dx$, by using the formulae for defining the connection from the solution of the Painlevé equation as in [8] (cf Section 2).

Now pull the connection back to the elliptic curve E which is the double covering of $\mathbf{CP}^1$ branched over the four points. The fundamental group of the punctured elliptic curve consists of the words of even length in the generators $\gamma_1, \gamma_2, \gamma_3$ of $\pi_1(\mathbf{CP}^1\backslash\{0, 1, x, \infty\})$. But under the holonomy representation into $SO(3) = SU(2)/\pm 1$ these generators map to reflections in a plane. Thus the even words map to the rotations in the dihedral group, the cyclic group $\mathbf{Z}_k$. Around a singular point in $\mathbf{CP}^1$, the holonomy is conjugate to

$$\begin{pmatrix} i & 0 \\ 0 & -i \end{pmatrix}$$

and so on the double covering branched around the point, the holonomy is

$$\begin{pmatrix} -1 & 0 \\ 0 & -1 \end{pmatrix}$$

and hence the identity in $SO(3)$.

Thus the holonomy is that of a smooth connection, and so defines an element of $\mathrm{Hom}(\pi_1(E), \mathbf{Z}_k)$. This is a flat line bundle of order k, and through the constructions in Sections 4 and 5 is the same bundle as $\mathcal{O}(1/2, -1/2)$. We have actually made a choice here, since holonomy is determined up to conjugation. There is a rotation in $SO(3)$ which takes the generating rotation of the cyclic group $\mathbf{Z}_k$ to its inverse. Thus x, y defines a pair of $k-$torsion points on the elliptic curve, and hence a single point of the modular curve $Y_k = X_{00}(k, 2)$.

Remark. A straightforward consideration of the branching over $0, 1, \infty$ leads to the formula

$$g = \frac{1}{4}(p-3)^2$$

for the genus g of $X_{00}(p,2)$ when p is prime (see [1]). Thus the Painlevé curve can have arbitrarily large genus.

References

[1] W. Barth & J. Michel, *Modular curves and Poncelet polygons,* Math. Ann. **295** (1993) 25–49.

[2] G. V. Belyi, *On Galois extensions of a maximal cyclotomic field,* Math. USSR Izvestiya, **14** (1980) 247–256.

[3] H. J. M. Bos, C. Kers, F. Oort & D. W. Raven, *Poncelet's closure theorem,* Exposition. Math. **5** (1987) 289–364.

[4] R. Fuchs, *Über lineare homogene Differentialgleichungen zweiter Ordnung mit drei im Endlichen gelegene wesentlich singulären Stellen,* Math. Ann. **63** (1907) 301–321.

[5] Ph. Griffiths & J. Harris, *On Cayley's explicit solution to Poncelet's porism,* Enseign. Math. **24** (1978) 31–40.

[6] N. J. Hitchin, *A new family of Einstein metrics,* Proceedings of conference in honour of E. Calabi, Pisa (1993), Academic Press (to appear)

[7] N. J. Hitchin, *Twistor spaces and isomonodromic deformations,* (in preparation)

[8] M. Jimbo and T. Miwa, *Monodromy preserving deformation of linear ordinary differential equations with rational coefficients II,* Physica , **2D** (1981) 407–448.

[9] K. Kodaira, *A theorem of completeness of characteristic systems for analytic families of compact submanifolds of complex manifolds,* Ann. of Math. **75** (1962) 146–162.

[10] B. Malgrange, *Sur les deformations isomonodromiques I. Singularités régulières,* in Mathématique et Physique, Séminaire de l'Ecole Normale Supérieure 1979–1982, Progress in Mathematics **37**, Birkhäuser, Boston (1983) 401–426.

[11] V. B. Mehta and C. S. Seshadri, *Moduli of vector bundles on curves with parabolic structure,* Math. Ann. **248** (1980) 205–239.

[12] S. Mukai and H. Umemura, *Minimal rational 3-folds,* in Algebraic Geometry (Proceedings, Tokyo/Kyoto 1982), Lecture Notes in Mathematics No.1016, Springer, Heidelberg (1983) 490–518.

[13] M. S. Narasimhan and C. S. Seshadri, *Stable and unitary bundles on a compact Riemann surface,* Ann. of Math. **82** (1965) 540–567.

[14] M. S. Narasimhan and G. Trautmann, *Compactification of $M_{\mathbf{P}_3}(0,2)$ and Poncelet pairs of conics,* Pacific J. Math. **145** (1990) 255–365.

[15] L. Schlesinger, *Über eine Klasse von differentialsystemen beliebiger Ordnung mit festen Kritischer Punkten,* J. für Math. **141** (1912) 96–145.

[16] R. L. E. Schwarzenberger, *Vector bundles on the projective plane,* Proc. London Math. Soc. **11** (1961) 623–640.

[17] A. Weil, *The field of definition of a variety,* American J. Math. **78** (1956) 509–524.

Mathematics Institute
University of Warwick
Coventry CV4 7AL
U.K.

Scalar conservation laws with boundary condition

K.T. Joseph

1 Introduction

Many of the balance laws in Physics are conservation laws. We consider scalar conservation laws in a single space variable,

(1.1) $$u_t + f(u)_x = 0.$$

On the flux function $f(u)$, we assume either

(H_1) $$f''(u) > 0 \text{ and' } \lim_{|u|\to\infty} \frac{f(u)}{|u|} = \infty$$

or

(H_2) $$f(u) = log\left[ae^u + be^{-u}\right],$$

where a and b are positive constants such that $a + b = 1$. An important special case is the Burgers equations i.e., when $f(u) = \frac{u^2}{2}$.

Initial value problem for (1.1) is to find $u(x,t)$ satisfying (1.1) and the initial data

(1.2) $$u(x,0) = u_0(x).$$

It is well known that (see Lax [8]) solution of (1.1) in the classical sense develop singularities after a finite time, no matter how smooth the initial data $u_0(x)$ is and cannot be continued as a regular solution. They can be continued however as a solution in weak sense. However, weak solutions of (1.1) are not determined uniquely by their initial values. Therefore, some additional priciple is needed for prefering the physical solution to others. One such condition is (See Lax [9]),

(1.3) $$u(x+0,t) \leq u(x-0,t).$$

This condition is called entropy condition.

Existence and uniqueness of weak solution of (1.1) and (1.2) satisfying the entropy condition (1.3) is well known (see Hopf [2], Lax [9], Olenik [14], Kruskov [7] And Quinn [13]). Hopf [2] derived an explicit explicit for the solution when $f(u) = \frac{u^2}{2}$ and Lax [9] extended this formula for general convex $f(u)$.

Let $f^*(u)$ be the convex dual of $f(u)$ i.e.,

(1.4) $$f^*(u) = \max_{\theta}[u\theta - f(\theta)],$$

(1.5) $$U_0(x) = \int_0^x u_0(y)dy$$

and

(1.6) $$U(x,t) = \min_{-\infty<y<\infty}\left[u_0(y) + tf^*\left(\frac{x-y}{t}\right)\right].$$

For each fixed (x,t), there may be several minimisers $y_0(x,t)$ for (1.6), define

(1.7) $$y_0^+(x,t) = \max\{y_0(x,t)\}, y_0^-(x,t) = \min\{y_0(x,t)\}.$$

Lax [9] proved that, for each fixed $t > 0, y_0^-(.,t)$ and $y_0^+(.,t)$ are left continuous and right continuous respectively, and both are continuous except on a common denumerable set of points of x and at the points of continuity

$$y_0^+(x,t) = y_0^-(x,t).$$

Define

(1.8) $$u(x,t) = (f^*)'\left(\frac{x - y_0(x,t)}{t}\right)$$

and

(1.9) $$u(x\pm,t) = (f^*)'\left(\frac{x - y_0^{\pm}(x,t)}{t}\right).$$

Clearly $u(x,t)$ is well defined a.e. (x,t) and $u(x\pm,t)$ is well defined for all (x,t). Lax [9] proved the following theorem.

Theorem 1. *$u(x,t)$ defined by (1.8) and (1.9) is the weak solution of (1.1) and (1.2) which satisfies the entropy condition (1.3).*

Let us consider the mixed initial boundary value problem for (1.1) in $x > 0, t > 0$. We prescribe the initial data

(1.10) $$u(x,0) = u_0(x), x > 0.$$

It follows from the work of Bardos et al [1] that we really cannot impose such a boundary condition

(1.11) $$u(0,t) = \lambda(t)$$

at $x = 0$, arbitrarily and hope to have a solution. Bardos et al studied this problem in several space variables by vanishing viscosity method. For one space variable their formulation is as follows.

$$\sup_{k \in I(u(0+,t),\lambda(t))} \{sgn(u(0+,t)-k)(f(u(0+,t))-f(k))\} = 0 \text{ a.e.} t > 0 \tag{1.12}$$

where the closed interval $I(u,\lambda)$ is defined by $I(u,\lambda) = [\min(u,\lambda), \max(u,\lambda)]$. Under the assumption $f''(u) > 0$, (1.12) is equivalent to saying (see Lefloch [10])

$$\begin{cases} either \\ u(0+,t) = \lambda^+(t) \\ \quad or \\ f'(u(0+t)) \leq 0 \text{ and } f(u(0+,t)) \geq f(\lambda^+(t)) \end{cases} \tag{1.13}$$

where

$$\lambda^+(t) = \max\{\lambda(t), u^*\}, \tag{1.14}$$

and u^* is the solution of $f'(u) = 0$. There exists only the solution u^*, because of the strict convexity of $f(u)$.

Definition. Let $u_0(x)$ be in $L^\infty(0,\infty)$ and $\lambda(t)$ is continuous, by a solution of (1.1)-(1.10) and (1.11) we mean a solution in the sense of Bardos-Leroux and Nedelec. That is bounded measurable function $u(x,t)$ in $x \geq 0, t > 0$ such that

$$\int_0^\infty \int_0^\infty (u\phi_t + f(u)\phi_x) dx dt = 0 \tag{1.15}$$

for all test functions $\phi(x,t) \in C_0^\infty(0,\infty) \times (0,\infty)), u(x,0)$ satisfies (1.10) a.e. $x > 0, u(x+,t)$ and $u(x-,t)$ satisfy (1.3) for $x > 0$ and $u(0+,t)$ satisfies (1.13).

Bardos et al [1] proved the existence and uniqueness of solution of (1.1), (1.10) and (1.11) (see also Lefloch [10]). We are interested in extending Lax formula (1.8) for the case of initial boudary value problem. Lefloch [10] derived a formula for the solution, which contains solution of a variational inequality. This variational inequality is not solvable explicitly. In a series of papers Joseph [3], Joseph and Veerappa Gowda [4,5], an explicit formula is derived for the solution of (1.1), (1.10) and (1.11), generalizing theorem 1. The case of two boundaries is considered in [6].

Before the statement of our main theorem we introduce some notations. For each fixed $(x,y,t), x \geq 0, y \geq 0, t > 0$ and $\alpha > 0, \mathcal{C}_\alpha(x,y,t)$ denotes the following class of paths $(\beta(s), s)$ in the quarter plane

$$D = \{(z,s) : z \geq 0,\ s \geq 0\}.$$

Each path connects the point $(y, 0)$ to (x, t) and is of the form

$$z = \beta(s)$$

where $\beta(s)$ is a piecewise linear function with one line or three straight lines of possible shapes shown in Figure, where the absolute value of slope of each straight line is $\leq \alpha$.

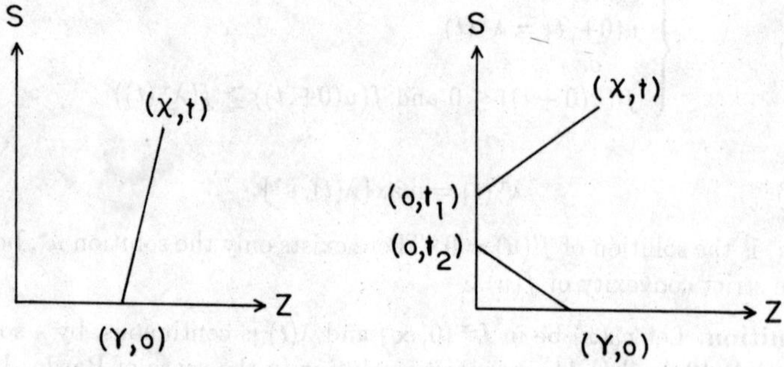

Let $u_0(x) \varepsilon L^{\infty}(0, \infty)$ and $\lambda(t)$ is continuous. Let $\lambda^+(t)$ be defined by (1.14) and let

$$\alpha = \begin{cases} \infty & \text{if } f(u) \text{ satisfies } (H_1) \\ 1 & \text{if } f(u) \text{ satisfies } (H_2)\ . \end{cases} \tag{1.16}$$

For each fixed $(x, y, t), x \geq 0, y \geq 0$, define

$$\mathrm{Q}(\mathrm{x}, \mathrm{y}; \mathrm{t}) = \min_{\beta \varepsilon \mathcal{C}_\alpha(\mathrm{x},\mathrm{y},\mathrm{t})} \left[- \int_{\{s:\beta(s)=0\}} \mathrm{f}(\lambda^+(\mathrm{s}))\mathrm{ds} + \int_{\{s:\beta(s)>0\}} \mathrm{f}^\star\left(\frac{\mathrm{ds}}{\partial \mathrm{s}}\right) \mathrm{ds}\right] \tag{1.17}$$

It can be shown that $Q(x, y, t)$ is Lipschitz continous function. Let

$$Q_1(x, y, t) = \frac{\partial}{\partial x} Q(x, y, t) \tag{1.18}$$

and

$$U(x, t) = \min_{0 \leq y < \infty} \left[\int_0^y u_0(z) dz + Q(x, y, t) \right] . \tag{1.19}$$

For each fixed (x, t) there may be several minimisers for (1.19). Define

$$y_0^-(x, t) = \min\{y_0(x, t)\}, y_0^+(x, t) = \max\{y_0(x, t)\} \tag{1.20}$$

It can be shown that (see [5]), for each fixed $t > 0, y_0^-(.,t)$ and $y_0^+(.,t)$ are left continuous and right continuous respectively, and both are continuous except on a comman denumerable set of points of x and at the points of continuity

$$y_0^+(x,t) = y_0^-(x,t)$$

Define

$$u(x,t) = Q_1(x, y_0(x,t), t) \text{ and} \tag{1.21}$$

$$u(x\pm,t) = Q_1(x, y_0^{\pm}(x,t), t)\ . \tag{1.22}$$

Clearly $u(x,t)$ is well defined a.e $x > 0, t > 0$ and $u(x\pm,t)$ is well defined for all $x > 0,\ t > 0$. Our main result is the following theorem.

Theorem 2. *Let $u(x,t)$ be given by (1.21) and (1.22), then it is the solution of (1.1), (1.10), (1.13) and (1.3).*

Here we remark that viscosity solution of Hamilton Jacobi equation with Neumann boundary condition (see Lions [11, 12]).

$$\begin{cases} U_t + f(U_x) = 0 \\ U(x,0) = U_0(x) \\ \text{“}U_x(0,t) = \lambda(t)\text{”} \end{cases} \tag{1.23}$$

is closely related to our problem. In fact the proof of theorem 2 shows the following result.

Theorem 3. *The function $U(x,t)$ defined by*

$$U(x,t) = \min_{0 \le y < \infty} [U_0(y) + Q(x,y,t)] \tag{1.24}$$

is a viscosity solution of (1.23).

Here $Q(x,y,t)$ is defined the same way as (1.17).

We arrived at these results, by first working out two examples namely the Burgers equation [3] and the Lax's equation [4].

2 Burgers Equation

In this section, we consider the Burgers equation in $x > 0, t > 0$

$$\begin{cases} u_t + \left(\frac{u^2}{2}\right)_x = \frac{\varepsilon}{2} u_{xx}, \\ u(x,0) = u_0(x), \\ u(0,t) = \lambda(t)\ . \end{cases} \tag{2.1}$$

Let us define

$$U^\varepsilon(x,t) = -\int_x^\infty u(y,t)dy,\ U_0(y) = -\int_0^\infty u_0(y)dy\ . \tag{2.2}$$

Then (2.1) becomes

$$(2.3)\qquad \begin{cases} U_t + \left(\frac{U_x^2}{2}\right) = \frac{\varepsilon}{2} U_{xx}, \\ U(x,0) = U_0(x), \\ U_x(0,t) = \lambda(t)\,. \end{cases}$$

By the Hopf-Cole transformation

$$(2.4)\qquad V = e^{-\frac{1}{\varepsilon}U,}$$

We can linearize the problem (2.3)

$$(2.5)\qquad \begin{cases} V_t = \frac{\varepsilon}{2} V_{xx}, \\ \varepsilon V_x(0,t) + \lambda(t) V(0,t) = 0, \\ V(x,0) = e^{-\frac{1}{\varepsilon}U_0(x)}. \end{cases}$$

When $\lambda(t) = \lambda$, a constant, the solution of (2.5) can be explicitly written down:

$$(2.6)\qquad V^\varepsilon(x,t) = \frac{1}{(2\pi t\varepsilon)^{1/2}} \Big[\int_0^\infty e^{-\frac{1}{\varepsilon}\left[U_0(y) + \frac{(x-y)^2}{2t}\right]} dy +$$

$$+ \int_0^\infty e^{-\frac{1}{\varepsilon}\left[U_0(y) + \frac{(x+y)^2}{2t}\right]} dy \Big] + \frac{2(\lambda/\varepsilon)}{(2\pi t\varepsilon)^{1/2}} \int_0^\infty \int_y^\infty e^{-\frac{1}{\varepsilon}\left[\lambda(y-z) + U_0(y) + \frac{(y+z)^2}{2t}\right]} dz dy.$$

From (2.4), we have

$$(2.7)\qquad U^\varepsilon(x,t) = -\varepsilon \log(V^\varepsilon(x,t))\,.$$

Substituting (2.6) in (2.7), we have an explicit formula for $U^\varepsilon(x,t)$. Using the method of stationary phase one can show that the $\lim_{\varepsilon\to 0} U^\varepsilon(x,t) = U(x,t)$ exists and is given by (1.24). The general variable boundary data $\lambda(t)$ can be treated by using a comparison theorem and some elementary convex analysis. The details can be found in [3].

3 Lax's equation

In this section, we consider the Lax's equation

$$u_t + \left(\log\left[ae^u + be^{-u}\right]_x\right) = 0\ , \tag{3.1}$$

in $x > 0, t > 0$. In the case of no boundary Lax [9] studied this example by a difference scheme. We consider the case with boundary. Let

$$u_k^n \simeq u(k\Delta, n\Delta), k = 0, 1, 2, \ldots, n = 0, 1, 2, \ldots \tag{3.2}$$

Δ being the mesh size. Let $u^\Delta(x,t)$ be the approximate solution defined by

$$\begin{cases} u_k^{n+1} = u_k^n + \left[g\left(u_{k-1}^n, u_k^n\right) - g\left(u_k^n, u_{k+}^n\right)\right] \\ u_k^0 = u_0(k\Delta) \\ u_0^n = \lambda(n\Delta) \end{cases} \tag{3.3}$$

where the numerical flux $g(u,v)$ is given by

$$g(u,v) = \log\left[ae^u + be^{-v}\right]\ . \tag{3.4}$$

Let

$$U_k^n = -\sum_{j=k}^{\infty} u_j^n, U_k^k = -\sum_{j=k}^{\infty} u_0(j\Delta),\ U_k^n = \log V_k^n\ , \tag{3.5}$$

then from (3.3), we get

$$\begin{cases} V_k^{n+1} = aV_{k+1}^n + bV_{k-1}^n & , n = 1, 2, \ldots k = 1, 2, 3, \ldots \\ V_k{}^0 = e^{-U_k^0} & , k = 0, 1, 2, \\ V_0^n = e^{\lambda(n)} V_1^n & , n = 1, 2, \ldots \end{cases} \tag{3.6}$$

When $\lambda = \lambda(t)$, a constant, an explicit formula can be obtained for the solution of (3.6) namely

$$V_k^n = \begin{cases} e^{(n-k)\lambda} \sum_{q=0}^{n} \binom{n}{q} a^q b^{n-q} V_{2n-2q}^0 \\ \qquad\qquad + \sum_{j=0}^{n-k-1} e^{j\lambda} S_{n,k+j}, & \text{if } n > k \\ \sum_{q=0}^{n} \binom{n}{q} a^q b^{n-q} V_{n+k-2q}^0, & \text{if } n \le k \end{cases} \tag{3.7}$$

where

$$S_{n,k} = \sum_{q=0}^{r} C_{q,k}^{\,n} a^q b^{n-q} \left(V^0_{n+k-2q} - e^{\lambda} V^0_{n+k+1-2q}\right)$$

$$r = \begin{cases} \frac{n+k-1}{2} & \text{if } (n+k) \text{ is odd} \\ \frac{n+k-2}{2} & \text{if } (n+k) \text{ is even} \end{cases}$$

$$C_{q,k}^{\,n} = \begin{cases} \binom{n}{q} \quad , & \text{if} \quad q \leq k-1 \\ \binom{n}{q} - \binom{n}{q-k}, & \text{if } q \geq k \end{cases}$$

and

$$\binom{n}{j} = \frac{n!}{j!(n-j)!}$$

By retracing the transformation (3.5), and using (3.7), we get an explicit formula for $U^{\Delta}(x,t) = -\int_x^{\infty} u^{\Delta}(y,t)dy$. Using Stirling's asymptotic formula one can study the limit, $\lim_{\Delta \to 0} U^{\Delta}(x,t) = U(x,t)$ and show that $U(x,t)$ is given by (1.24). As in the case of Burgers equation here again one can prove a comparison theorem and with the help of this comparison theorem general variable $\lambda(t)$ can be treated. The details are omitted and can be found in [4].

4 Proof of Main theorem

The examples of last two section suggest the formula for the case of general convex $f(u)$.

Following Lax [9], we introduce

(4.1)
$$u_N(x,t) = \frac{\int_0^{\infty} Q_1(x,y,t) e^{-N[U_0(y)+Q(x,y,t)]} dy}{\int_0^{\infty} e^{-N[U_0(y)+Q(x,y,t)]} dy}$$

(4.2)
$$f_N(x,t) = \frac{\int_0^{\infty} f\left(Q_1(x,y,t)\right) e^{-N[U_0(y)+Q(x,y,t)]} dy}{\int_0^{\infty} e^{-N[U_0(y)+Q(x,y,t)]} dy}$$

(4.3)
$$V_N(x,t) = \int_0^{\infty} e^{-N[U_0(y)+Q(x,y,t)]} dy \ .$$

(4.4)
$$U_N(x,t) = -\frac{1}{N} \log V_N \ .$$

where $U_0(x), Q(x,y,t), Q_1(x,y,t)$ are defined in Sec 1. It is clear that

(4.5)
$$\begin{cases} \lim_{N\to\infty} u_N(x,t) = Q_1(x, y_0(x,t), t) & \text{a.e } (x,t) \\ \lim_{N\to\infty} f_N(x,t) = f\left[Q_1(x, y_0(x,t), t)\right] & \text{a.e } (x,t) \end{cases}$$

where $y_0(x,t)$ minimises (1.19) and

$$\lim_{N\to\infty} U_N(x,t) = U(x,t)\ , \tag{4.6}$$

and

$$\frac{\partial U}{\partial x} = Q_1(x, y_0(x,t), t). \tag{4.7}$$

It can be shown that

$$\int_0^\infty \int_0^\infty (u_N \phi_t + f_N) dx dt = 0 \tag{4.8}$$

for all test function $\phi \in C_0{}^\infty(0,\infty) \times (0,\infty))$.

Let $N \to \infty$ and use (4.5) we get from (4.8)

$$u(x,t) = Q_1(x, y_0(x,t), t) \tag{4.9}$$

solves (1.1) in distribution. It can be shown that $u(x,t)$ defined by (4.9) satisfies initial condition (1.10) boundary condition (1.13) and entropy condition (1.3). The details can be found in [5].

References

[1] Bardos, C., Leroux, A.Y. and Nedelec, J.C., *First order quasilinear equations with boundary conditions*, Comm. Partial Differential equations **4** (1979) 1017–1034.

[2] Hopf, E., *The partial differential equation* $u_t u u_x = \mu U_{xx}$, Comm. Pure and Appl. Math **13** (1950) 201–230.

[3] Joseph, K.T., *Burgers equation in the quarter plane, a formula for the weak limit*, Comm. Pure and Appl. Math **41** (1988) 133–149.

[4] Joseph, K.T. and Veerappa Gowda, G.D., *Explicit-formula for the limit of a difference approximation*, Duke Math. J. **61** (1990) 369–393.

[5] Joseph, K.T. and Veerappa Gowda, G.D., *Explicit-formula for the solution of convex conservation laws with boundary condition*, Duke Math. J. **62** (1991) 401–416.

[6] Joseph, K.T. and Veerappa Gowda, G.D., *Solution of convex conservation laws in a strip*, Proc. Ind. Acad. Sci, **102**, 1, (1992) 29–47.

[7] Kruskov, S. *First order quasilinear equations with several space variables*, Math. USSR Sb, **10** (1970) 217–273.

[8] Lax, P.D., *Weak solutions of non linear hyperbolic equations and their numerical computation*, Comm. Pure Appl. Math. **7** (1954) 159–193.

[9] Lax, P.D : *Hyperbolic Systems of Conservation laws II*, Comm. Pure Appl. Math. **10** (1957) 537–566.

[10] Lefloch, P., *Explicit-formula for Scalar non-linear Conservation laws with boundary condition*, Math Methods in Appl. Sci **10** (1988) 265–287.

[11] Lions, P.L., *Generalized solutions of Hamilton-Jacobi equation*, Pitman Lecture Notes, Pitman, London (1982).

[12] Lions, P.L., *Neumann type boundary conditions for Hamilton-Jacobi equations*, Duke Math. J. **52** (1985) 793–820.

[13] Quinn, B., *Solutions with shocks : an example of L^1-contraction semi-group*, Comm. Pure Appl. Math. **24** (1971) 125–132.

[14] Oleinik, O., *Discontinuous solutions of non-linear differential equation*, Usp. Mat. Nauk, (N.S), **12** (1957) 3–73; English transl. in AMS Transl. Ser. 2, **26** (1957) 95–172.

TIFR Centre
P.B. No.1234
IISc. Campus
Bangalore 560 012

Bases for Quantum Demazure modules-I

V. Lakshmibai *

(Dedicated to Professors M.S. Narasimhan and C.S. Seshadri on their 60th birthdays)

1 Introduction

Let $\mathfrak{g}$ be a semi-simple Lie algebra over $\mathbb{Q}$ of rank n. Let U be the quantized enveloping algebra of $\mathfrak{g}$ as constructed by Drinfeld (cf [D]) and Jimbo (cf [J]). This is an algebra over $\mathbb{Q}(v)$ (v being a parameter) which specializes to $U(\mathfrak{g})$ for $v = 1, U(g)$ being the universal enveloping algebra of $\mathfrak{g}$. This algebra has generators $E_i, F_i, K_i, 1 \leq i \leq n$, which satisfy *the quantum Chevalley and Serre relations* (cf [L1]). Let $A = \mathbb{Z}[v, v^{-1}]$ and $U_A^{\pm}$ be the A-subalgebra of U generated by $E_i^{(r)}$ (resp. $F_i^{(r)}$), $1 \leq i \leq n, r \in \mathbb{Z}^+$, (here $E_i^{(r)}, F_i^{(r)}$ are the quantum divided powers (cf [J])). Let $d = (d_1, .., d_n) \in (\mathbb{Z}^+)^n$ and V_d be the simple U-module with a non-zero vector e such that $E_i e = 0, K_i e = v^{-d_i} e$ (recall that V_d is unique up to isomorphism). Let us denote V_d by just V. Let W be the Weyl group of $\mathfrak{g}$.

Let $w \in W$ and let $w = s_{i_1} \cdots s_{i_r}$ be a reduced expression for w. Let $U_{w,A}^-$ denote the A-submodule of U spanned by $F_{i_1}^{(a_1)} \cdots F_{i_r}^{(a_r)}, a_i \in \mathbb{Z}^+$. We observe ([L4]) that $U_{w,A}^-$ depends only on w and not on the reduced expression chosen. For $w \in W$, let $V_{w,A} = U_{w,A}^- e$. We shall refer to $V_{w,A}$ as the *Quantum Demazure module associated with* w. Let w_0 be the unique element in W of maximal length. Then $V_{w_0,A}$ is simply $U_A^- e$. In the sequel, we shall denote $V_{w_0,A}$ by just V_A.

Let $g = s\ell\,(\mathbf{3})$. In this paper, we construct an A-basis for V_A, which is compatible with $\{V_{w,A}, w \in W\}$. The construction is done using the configuration of Schubert varieties in the Flag variety G/B. Let $Id = \mu_0 < \mu_1 < \mu_2 < \mu_3 = w_0$ be a chain in $W (= S_3)$. Let $\mu_{i-1} = s_{\beta_i} \mu_i$ for some positive

*Partially supported by NSF Grant DMS 9103129 and also by the Faculty Development Fund of Northeastern University.

root β_i. Let $n(\mu_{i-1}, \mu_i) = (\mu_{i-1}(\lambda), \beta_i^*)$, where $\lambda = d_1\omega_1 + d_2\omega_2, \omega_1$ and ω_2 being the fundamental weights of $s\ell(3)$. Let us denote $n(\mu_{i-1}, \mu_i)$ by just n_i. Let $C = \{(\mu_0, \mu_1, \mu_2, \mu_3; m_1, m_2, m_3) : 1 \geq \frac{m_1}{n_1} \geq \frac{m_2}{n_2} \geq \frac{m_3}{n_3} \geq 0\}$. Given $c = (\mu_0, \mu_1, \mu_2, \mu_3; m_1, m_2, m_3)$, let $\tau_c = \mu_r$, where r is the largest integer such that $m_r \neq 0$. Given two elements

$$c_1 = \{\mu_0, \mu_1, \mu_2, \mu_3; m_1, m_2, m_3\}, c_2 = \{\lambda_0, \lambda_1, \lambda_2, \lambda_3; p_1, p_2, p_3\}$$

in C, let us denote $\frac{m_i}{n(\mu_{i-1},\mu_i)}$ (resp. $\frac{p_i}{n(\lambda_{i-1},\lambda_i)}$) by just a_i (resp. b_i). We say $c_1 \sim c_2$, if

(1) $a_i = b_i$
(2) either
(a) $a_1 = a_2 > a_3, \mu_2 = \lambda_2$
or
(b) $a_1 > a_2 = a_3, \mu_1 = \lambda_1$
or
(c) $a_1 = a_2 = a_3$

We shall denote $C/\sim$ by $\overline{C}$, For $\theta \in \overline{C}$, we shall denote $\tau_\theta = \tau_c, c$ being a representative for θ (note that τ_θ is well-defined). We have (Theorems 6.7 and 7.2)

Theorem. *V_A has an A-basis $B_d = \{v_\theta, \theta \in \overline{C}\}$ where $v_\theta = D_\theta e, D_\theta$ being a monomial in F_i's of the form $F_{i_1}^{(n_1)} \cdots F_{i_r}^{(n_r)}$. Further, for $w \in W, \{v_\theta \mid w \geq \tau_\theta\}$ is an A-basis for $V_{w,A}$.*

Let $\mathcal{B}_d$ denote Lusztig's canonical basis for V_A (cf [L2]). It turns out that the transition matrix from B_d to $\mathcal{B}_d$ is upper triangular. We also give a conjectural A-basis B_d for V_A for $\mathfrak{g}$ of other types. An element in B_d is again of the form $F_{i_1}^{(n_1} \cdots F_{i_r}^{(n_r)} e$. We conjecture that the transition matrix from B_d to $\mathcal{B}_d$ is upper triangular. The sections are organized as follows. In §2, we recall some results pertaining to the configuration of Schubert varieties in G/B. In §3, we describe a conjectural A-basis for V_A (and also for $V_{w,A}$). In §4, we study $\overline{C}$ in detail and construct B_d, for $G = SL(3)$. In §5, we establish a bijection between B_d and $\mathcal{B}_d$. In §6 and §7, we prove the results for $G = SL(3)$. In §8, an Appendix, we have explicitly established a bijection between the elements of $\overline{C}$ (the indexing set for B_d) and the classical standard Young tableaux on $SL(3)$ of type (d_1, d_2).

The author would like to express her gratitude to SPIC Science Foundation for the hospitality extended to her during her stay (Jan-Feb, 1992) there, when research pertaining to this paper was carried out.

2 Preliminaries

Let G be a semi-simple, simply connected Chevalley group defined over a field k. Let T be a maximal k-split torus, B a Borel subgroup, $B \supset T$. Let W be the Weyl group, and R the root system of G relative to T. Let R^+ (resp. S) be the system of positive (resp. simple) roots of G relative to B. For $w \in W$, let $X(w) = \overline{BwB}$ (mod B) be the Schubert variety in G/B associated with w.

2.1 Definition. Let $X(\varphi)$ be a Schubert divisor in $X(\tau)$. We say that $X(\varphi)$ is a *moving divisor in* $X(\tau)$ *moved by the simple root* α, if $\varphi = s_\alpha \tau$.

2.2 Lemma (cf [L-S]) . *Let* $X(\varphi)$ *be a moving divisor in* $X(\tau)$ *moved by* α. *Let* $X(w)$ *be a Schubert subvariety in* $X(\tau)$. *Then either*

(i) $X(w) \subseteq X(\varphi)$ *or*

(ii) $X(w) = X(s_\alpha w')$ *for some* $X(w') \subseteq X(\varphi)$.

2.3 Definition. Let λ be a dominant integral weight of G. Let $X(w)$ be a divisor in $X(\tau)$. Let $w = s_\beta \tau$, for some $\beta \in R^+$. We define $m_\lambda(w,\tau)$ as the non-negative integer $m_\lambda(w,\tau) = (w(\lambda), \beta^*)(= -(\tau(\lambda), \beta^*))$, and call it the λ *multiplicity* of $X(w)$ in $X(\tau)$. (Here (,) is a W-invariant scalar product on $Hom(T, G_m)$).

2.4 Lemma. *Let* $X(\varphi)$ *be a moving divisor in* $X(\tau)$ *moved by* α. *Let* $X(w)$ *be a divisor in* $X(\tau)$. *Let* $w = s_\beta \tau, \theta = s_\alpha w$, $\theta = s_\gamma \varphi$, *where* $\beta, \gamma \in R^+$ *(note that* $\gamma = s_\alpha(\beta)$*). Let* $m = m_\lambda(w,\tau), s = m_\lambda(\theta, w), p = m_\lambda(\varphi,\tau), r = (\beta, \alpha^*)$. *Then*

(a) $m_\lambda(\theta,\varphi) = m$

(b) $p = s + mr$

Proof.

(a) $m_\lambda(\theta,\varphi) = (\theta(\lambda), \gamma^*) = (\theta(\lambda), (s_\alpha(\beta))^*) = (s_\alpha\theta(\lambda), \beta^*) = (w(\lambda), \beta^*) = m$

(b) Now $s_\beta s_\alpha \theta(\lambda) = s_\alpha s_\gamma \theta(\lambda)$ implies that $s\alpha + m\beta = m\gamma + p\alpha$. Hence $(p-s)\alpha = m\beta - m\gamma = m\beta - m(\beta - r\alpha) = mr\alpha$. Hence $p = s + mr$.

2.5 Lemma. *Let $X(\theta)$ be a divisor in $X(\tau)$. Let $\theta = s_\beta\tau$, where $\beta \in R^+$. Let β be non-simple, say $\beta = \Sigma c_i\alpha_i, \alpha_i \in S$. Then for at least one i with $c_i \neq 0$, we have $(\tau(\lambda), \alpha_i^*) < 0$.*

Proof. Let $m = m_\lambda(\theta, \tau)$. Then $(\tau(\lambda), \beta^*) = -m < 0$. The assertion follows from this.

2.6 Lexicographic shellability

Given a finite partially ordered set H which is graded (i.e., which has an unique maximal and an unique minimal element and in which all maximal chains m i.e., maximal totally ordered subsets of H, have the same length), the *lexicographic shellability* of H (cf [B-W]) consists in labelling the maximal chains $\underline{m}$ in H, say $\lambda(\underline{m}) = (\lambda_1(\underline{m}), \lambda_2(\underline{m}), \cdots, \lambda_r(\underline{m}))$ (here r is the length of any maximal chain in H), where $\lambda_i(\underline{m})$ belong to some partially ordered set Ω, in such a way that the following two conditions hold:

(L1) If two maximal chains $\underline{m}$ and $\underline{m}'$ coincide along their first d edges, for some $d, 1 \leq d \leq r$, then $\lambda_i(\underline{m}) = \lambda_i(\underline{m}'), 1 \leq i \leq d$.

(L2) For any interval $[x, y]$ $(= \tau \in H : x \leq \tau \leq y)$, together with a chain $\underline{c}$ going down from the unique maximal element in H to y, there is a unique maximal chain m_0 in $[x, y]$ whose label is increasing (i.e. $\lambda_1(m_o) \leq \lambda_2(m_0) \leq \cdots \leq \lambda_t(m_0), t$ being the length of any maximal chain in $[x, y]$) and if $\underline{m}$ is any other maximal chain in $[x, y]$, then $\lambda(\underline{m_0})$ is lexicographically $< \lambda(m)$ (here, the label for any chain $\underline{m}$ in $[x, y]$ is induced from the maximal chain of H consisting of $\underline{c}$, followed by $\underline{m_0}$, followed by an arbitrary path from x to the unique minimal element of H.)

2.7 Theorem. (cf [B-W]) *the Bruhat order of a Coxeter group is lexicographic shellable.*

2.8 Labelling of maximal chains in $[x.y]$ for $H = W$

We fix a reduced expression for w_0, the element of the maximal length in W and label the maximal chains in W as in [B-W] (with respect to this fixed reduced expression for w_0). Let $\underline{m}$ be a maximal chain in $[x, y]$. Let $\underline{c}$ be the (unique) chain from w_0 to y whose label is increasing. We take the label for $\underline{m}$ as that induced by maximal chain in W consisting of $\underline{c}$, followed by $\underline{m}$, followed by an arbitrary path from x to Id.

3 A conjectural Bruhat-order compatible A-basis for V_A

3.1 Let $\mathfrak{g} = Lie(G)$. Let U, A, $U_A^{\pm}, E_i, F_i, K_i, V_d, V_A, V_{w,A}$ etc. be as in §1. Let $\lambda = \Sigma d_i\omega_i, \omega_i$ being the fundamental weights of G. We shall index the set of simple roots of G as in [B]. Let $\underline{c} = \{\mu_0, \mu_1, \cdots, \mu_r\}$ be a chain in W, i.e., $\ell(\mu_i) = \ell(\mu_{i-1}) + 1$ (if $d_t = 0$ for $t = i_1, \cdots, i_s$, then we shall work with W^Q, the set of minimal representatives of W_Q in W, W_Q being the subgroup of W generated by the set of simple reflections $\{s_t, t = i_1..., i_s\}$). Let $\mu_{i-1} = s_{\beta_i}\mu_i, \beta_i \in R^+$. Let $m_\lambda(\mu_{i-1}, \mu_i) = m_i$. We set $\ell(\underline{c}) = r$, and call it the *length* of $\underline{c}$.

3.2 Definition. A chain $\underline{c}$ is called *simple* (resp. *non-simple* if all (resp. some) $\beta_i's$ are simple (resp. non-simple).

3.3 Definition. By a *weighted* chain, we shall mean $(\underline{c}, \underline{n})$ where $\underline{c} = \{\mu_0, \cdots, \mu_r\}$ is a chain and $\underline{n} = \{n_1, \cdots, n_r\}, n_i \in \mathbb{Z}^+$.

3.4 Definition. A weighted chain $(\underline{c}, \underline{c}, \underline{n})$ is said to be *admissible* if $1 \geq \frac{n_1}{m_1} \geq \frac{n_2}{m_2} \geq \cdots \geq \frac{n_r}{m_r} \geq 0$.

3.5 Lemma. *Let $X(\varphi)$ be a moving divisor in $X(\tau)$ moved by the simple root α. Let $\underline{c} = \{\mu_0, \mu_1, \cdots \mu_r = \tau\}$ be a chain, and let $\mu_{i-1} = s_\beta\mu_i, \beta_i \in R^+$. Further let $\beta_r \neq \alpha$, and $\beta_i \in S, i \neq r$ (note that we allow β_r to be non-simple). Then either*

(1) $\beta_i = \alpha$ *, for some i,* (or)

(2) $\beta_i \neq \alpha, 1 \leq i \leq r$ *, in which case $\mu_i > s_\alpha\mu_i$, and $s_\alpha\mu_i < \varphi, 0 \leq i \leq r$*

Proof. Let $\beta_i \neq \alpha, 1 \leq i \leq r$. We shall now show that $\mu_i > s_\alpha\mu_i, 0 \leq i \leq r$. For $i = r$, this is clear (since $s_\alpha\mu_r = \varphi < \tau$). For $i = r - 1$, this follows from Lemma 2.2. We have

$$(\mu_{r-2}(\lambda), \alpha^*) = (\mu_{r-1}(\lambda) + m_{r-1}\beta_{r-1}, \alpha^*) < 0,$$

since $(\mu_{r-1}(\lambda), \alpha^*) < 0$, and $(\beta_{r-1}, \alpha^*) \leq 0$. (note that $\mu_{r-1} > s_\alpha\mu_{r-1} \Rightarrow (\mu_{r-1}(\lambda), \alpha^*) < 0$, and that $(\beta_{r-1}, \alpha^*) \leq 0$, since $\beta_i \in S, 1 \leq i \leq r-1$). Hence $\mu_{r-2} > s_\alpha\mu_{r-2}$. In a similar way one concludes $\mu_i > s_\alpha\mu_i$. The assertion that $s_\alpha\mu_i$ is $< \varphi$ follows from Lemma 2.2.

3.6 Let $(\underline{c}, \underline{n})$ be an admissible weighted chain. With notations and assumptions as in Lemma 3.5, we define an admissible weighted chain $(s_\alpha(\underline{c}), s_\alpha(\underline{n}))$ as follows:

Case 1. Let $\beta_i \neq \alpha, 1 \leq i \leq r$. We set

$$s_\alpha(\underline{c}) = \{s_\alpha\mu_0, s_\alpha\mu_1, \cdots, s_\alpha\mu_r = \varphi\}$$
$$s_\alpha(\underline{n}) = \underline{n}$$

(Note that $m_\lambda(s_\alpha\mu_{i-1}, s_\alpha\mu_i) = m_\lambda(\mu_{i-1}, \mu_i)$ (Lemma 2.4, (a)) and hence $(s_\alpha(c), s_\alpha(n))$ is admissible).

Case 2. Let $\beta_i = \alpha$, from some $i, 1 \leq i \leq r$. Let t be the largest integer, $1 \leq t \leq r$ such that $\beta_t = \alpha$. We set

$$s_\alpha(\underline{c}) = \{\mu_0, \mu, \cdots, \mu_{t-1} = s_\alpha\mu_t, s_\alpha\mu_{t+1}, \cdots, s_\alpha\mu_r = \varphi\}$$

and we define $s_\alpha(\underline{n}) = \{n'_i, \cdots, n'_{r-1}\}$ by

$$n'_i = n_i, 1 \leq i \leq t-1$$
$$n'_i = n_{i+1}, t \leq i \leq r-1$$

(note that $1 \geq \frac{n'_1}{m'_1} \geq \cdots \geq \frac{n'_{r-1}}{m'_{r-1}} \geq 0$, by the same considerations as in Case 1 (here $m'_i = m_i, 1 \leq i \leq t-1, m'_i\ (= (s_\alpha\mu_i, s_\alpha\mu_{i+1})) = m_{i+1}, t \leq i \leq r-1$)).

3.7 With notations and assumptions as in 3.6, we denote by $(\underline{c}_\alpha, \underline{n}_\alpha)$ the admissible weighted chain, where

$$\underline{c}_\alpha = (s_\alpha(\underline{c}), \mu_r), n_\alpha = (s_\alpha(\underline{n}), n'_r)$$

and n'_r is given as follows:

Let

$$k = \begin{cases} 0, & \text{if case 1 holds} \\ t, & \text{if case 2 holds} \end{cases}$$

For $i \geq k$, let $\gamma_i = s_\alpha(\beta_i)$ (note that $s_\alpha\mu_{i-1} = s_{\gamma_i}s_\alpha\mu_i$).

Let

$$x = \begin{cases} m(s_\alpha\mu_0, \mu_0), & \text{if case 1 holds} \\ n_t, & \text{if case 2 holds} \end{cases}$$

$$y = \begin{cases} \sum n_i(\beta_i, \alpha^*) \\ \{i > k \mid \gamma_i \neq \beta_i\} \end{cases}$$

$$n'_r = x + y$$

Then n'_r is given by

$$n'_r = x + y$$

3.8 Let $(\underline{c}, \underline{n})$ be an admissible weighted chain. Further, let $\underline{c}$ be not simple. To $(\underline{c}, \underline{n})$, we attach a canonical (not necessarily admissible) weighted chain $(\delta(\underline{c}), \delta(\underline{n}))$ with $\delta(\underline{c})$ simple, as follows: We preserve the above notation for $\underline{c}$. We do the construction using induction on $r (= \ell(\underline{c}))$.

Starting point of induction Let $\underline{c} = \{(\tau, w)\}$, where $X(w)$ is a divisor in $X(\tau), w = s_\beta \tau, \beta \in R^+, \beta$ non-simple. (We refer to this situation as $X(w)$ *being a non-moving divisor in* $X(\tau)$). Let $n = (n_1)$, where $n_1 \leq m$ $(= m_\lambda(w, \tau))$. By induction on dim $X(\tau)$, we may suppose that $X(\tau)$ is of least dimension such that $X(\tau)$ has a non-moving divisor. Let $\beta = \Sigma c_i \alpha_i$. Then at least for one i with $c_i \neq 0$, we have $(\tau(\lambda), \alpha_i^*) < 0$ (Lemma 2.5). Let t be the least integer with $c_t \neq 0$ such that $(\tau(\lambda), \alpha_t^*) < 0$ (the indexing of the simple roots being as in [B]). Denote α_t by just α. Let $s_\alpha w = \theta, \varphi = s_\gamma \theta$. Then $\gamma = s_\alpha(\beta)$. Further by our assumption on dim $X(\tau), \gamma \in S$. Hence we obtain $(\beta, \alpha^*) > 0$, say $(\beta, \alpha^*) = r$, and $\beta = \gamma + r\alpha$. We set

$$(\delta(\underline{c}), \delta(\underline{n})) = \{(\theta, \varphi, \tau); (p_1, p_2)\}$$

where $p_1 = n_1, p_2 = n_1 + a$ with $a = m_\lambda(\theta, w)$ (Lemma 2.4).

Let now $\ell(\underline{c}) > 1$. Let $\underline{c} = \{\mu_0, \mu_1, ..., \mu_r\}, \mu_{i-1} = s_\beta \mu_i, 1 \leq i \leq r$. We may suppose that $\beta_i \in S, 1 \leq i < r$. For, otherwise, if i is the least integer such that β_i is non-simple we may work with $(\Delta(\underline{n}), \Delta(\underline{c}))$, (where $\Delta(\underline{c})$ is the chain $\delta(\mu_0, \cdots, \mu_i)$ followed by $\{\mu_{i+1}, \cdots, \mu_r\}$, and $\Delta(\underline{n})$ is the sequence $\delta(n_1, \cdots, n_i)$ followed by $(n_{i+1}, \cdots, n_r)$) and use induction on $\#\{t, 1 \leq t \leq r : \beta_t$ is non-simple$\}$. Let us denote β_r by just β. Let $\beta = \Sigma c_i \alpha_i$. Since $(\tau(\lambda), \beta^*) < 0$, we have (Lemma 2.5), for at least one t with $c_t \neq 0, (\tau(\lambda), \alpha_t^*) < 0$. Let i be the least integer such that $c_i \neq 0$ and $(\tau(\lambda), \alpha_i^*) < 0$. Let us denote α_i by just α. Let $\varphi = s_\alpha \tau$. We set (3.7)

$$\delta(\underline{c}) = \underline{c}_\alpha, \delta(\underline{n}) = \underline{n}_\alpha$$

3.9 Given a simple weighted chain $(\underline{c}, \underline{n})$ (not necessarily admissible) we set

$$v_{\underline{c},\underline{n}} = F_{i_r}^{(n_r)} \cdots F_{i_1}^{(n_1)} e_\mu$$

where $\underline{c} = \{\mu = \mu_0, ..., \mu_r\}, \underline{n} = \{n_1, ..., n_r\}, \beta_t = \alpha_{i_t}, 1 \leq t \leq r$, and e_μ is the extremal weight vector associated to μ. (Note that it $\tau_0 = Id < \tau_1 < \cdots < \tau_r = \mu$ is any simple chain from Id to μ, and $\tau_{i-1} = s_\beta \tau_i, \beta_i \in S, 1 \leq i \leq r$, then $e_\mu = F_{\beta_r}^{(n_r)} ... F_{\beta_1}^{(n_1)} e$, where $n_i = m_\lambda(\tau_{i-1}, \tau_i)$ (2.3))

3.10 Let $(\underline{c}, \underline{n})$ be admissible. Let us denote the unequal values in $\{\frac{n_1}{m_1}, \cdots, \frac{n_r}{m_r}\}$ by $a_1, \cdots, a_s$ so that $1 \geq a_1 > a_2 > \cdots > a_s \geq 0$. Let $i_0, \cdots, i_s$ be defined by

$$i_0 = 0, i_s = r, \frac{n_j}{m_j} = a_t, i_{t-1} + 1 \leq j \leq i_t$$

We set

$$D_{\underline{c},\underline{n}} = \{(a_1, \cdots, a_s); (\mu_{i_0}, \cdots, \mu_{i_s})\}$$

3.11 Definition. Let $(\underline{c}, \underline{n}), (\underline{c}', \underline{n}')$ be two admissible weighted chains. Let $D_{\underline{c},\underline{n}} = \{(a_1, \cdots, a_s); (\mu_{i_0}, \cdots, \mu_{i_s})\}$, and $D_{\underline{c}',\underline{n}'} = \{(a'_1, \cdots, a'_\ell); (\tau_{j_0}, \cdots, \tau_{j_\ell})\}$. We say $(\underline{c}, \underline{n}) \sim (\underline{c}', \underline{n}')$, if $s = t$, and $a_t = a'_t, i_t = j_t, \mu_{i_t} = \tau_{j_t}, 0 \leq t \leq s$

3.12 Given $\mu, \tau \in W, \mu < \tau$, we shall label the chain in $[\mu, \tau]$ as in 2.8. Let $C =$ {all admissible weighted chains}, and $\overline{C} = C/\sim$. Given $x \in \overline{C}$, let

$$S_x = \{(\underline{c}, \underline{n}) \in C : (\underline{c}, \underline{n}) \text{ is a representative of } x \textit{ and } \underline{c} \text{ is simple}\}$$

$$N_x = \{(\underline{c}, \underline{n}) \in C : (\underline{c}, \underline{n}) \text{ is a representative of } x \textit{ and } \underline{c} \text{ is non} - \text{simple}\}$$

Let us define $x_{\min} \in C$ as follows.

Case 1. $S_x \neq \phi$. We set

$$x_{\min} = (\underline{c}_0, \underline{n}_0)$$

where $(\underline{c}_0, \underline{n}_0)$ is lexicographically the least in S_x.

Case 2. $S_x = \phi$. In this case we set

$$x_{\min} = (\underline{c}_0, \underline{n}_0)$$

where $(\underline{c}_0, \underline{n}_0)$ is lexicographically the least in N_x.

3.13 Let $x \in \overline{C}$. Let $x_{min} = (\underline{c}_0, \underline{n}_0)$. Define $v_x \in V_A$ as follows:

Case 1. $S_x \neq \phi$. We set

$$v_x = v_{\underline{c}_0, \underline{n}_0}$$

Case 2. $S_x = \phi$. We set

$$v_x = v_{\delta(\underline{c}_0), \delta(\underline{n}_0)}$$

3.14 Let $x \in \overline{C}$. Let $(\underline{c}, \underline{n})$ be a representative of x. Let $\underline{c} = (\mu_0, \cdots, \mu_r)$. We define $\tau_x = \mu_s$, where s is the largest integer such that $n_s \neq 0$ (Note that τ_x is well-defined).

3.15 Conjectures.

(1) $\{v_x, x \in \overline{C}\}$ is an A-basis for V_A

(2) $\{v_x : w \geq \tau_x\}$ is an A-basis for $V_{A,w}$

(3) Let B_d denote $\{v_x, x \in \overline{C}\}$, and $\mathcal{B}_d$, Lusztig's canonical basis for V_A ([L2]). The transition matrix from B_d to $\mathcal{B}_d$ is upper triangular.

4 The case G = $SL(3)$

4.1 For the rest of the paper we shall suppose that G = $SL(3)$. Let us denote the elements of W by $\{\tau_i, \varphi_j, i = 0,1,2,3, j = 1,2\}$, where $\tau_0 = Id, \tau_1 = s_1, \tau_2 = s_2s_1, \tau_3 = s_1s_2s_1, \varphi_1 = s_2, \varphi_2 = s_1s_2$. We shall label the maximal chains in W with respect to the reduced expression $s_1s_2s_1$ of w_0 ([B-W]).

4.2 Let $d = (d_1, d_2)$ and $\lambda = d_1\omega_1 + d_2\omega_2$. We shall suppose that d_1, d_2 are both non-zero (If $d_1 = 0$ for instance, then we work with $\{Id, s_2, s_1s_2\}$, the set of minimal representatives of W_P (= $\{s_1, Id\}$) in W). Also, for simplicity of notation, we shall denote d_1 by m and d_2 by n.

4.3 Given a pair (φ, τ) such that $X(\varphi)$ is a divisor in $X(\tau)$, let us denote $m_\lambda(\varphi, \tau)$ by just $m(\varphi, \tau)$. We have

$$m(\varphi.\tau) = \begin{cases} m, & \text{if } (\varphi,\tau) = (\tau_0, \tau_1), (\varphi_1, \tau_2) \text{ or } (\varphi_2, t_3) \\ m+n, & \text{if } (\varphi,\tau) = (\tau_1, \tau_2) \text{ or } (\varphi_1, \varphi_2) \\ n, & \text{if } (\varphi,\tau) = (\tau_0, \varphi_1), (\tau_1, \varphi_2) \text{ or } (\tau_2, \tau_3) \end{cases}$$

4.4 We shall denote an admissible weighted chain by $\underline{c} = (\mu_0, \mu_1, \mu_2, \mu_3), \underline{n} = (n_1, n_2, n_3)$, where if $n_i = 0, i$ being the least such integer, then $\underline{c}$ is to be understood as the chain $(\mu_0, \cdots, \mu_{i-1})$. For instance, if $n_3 = 0$ and n_1, n_2 are non-zero then $\underline{c} = (\mu_0, \mu_1, \mu_2)$. If $n_1 = 0 = n_2 = n_3$, we shall call $\underline{c}$ a *trivial chain* consisting of just μ_0.

4.5 We have four types of admissible weighted chains given as follows.

$$\begin{aligned}
&\text{Type I}: && \{(\tau_0, \tau_1, \tau_2, \tau_3), (n_1, n_2, n_3) : 1 \geq \tfrac{n_1}{m} \geq \tfrac{n_2}{m+n} \geq \tfrac{n_3}{n} \geq 0\} \\
&\text{Type II}: && \{(\tau_0, \varphi_1, \tau_2, \tau_3), (n_1, n_2, n_3) : 1 \geq \tfrac{n_1}{n} \geq \tfrac{n_2}{m} \geq \tfrac{n_3}{n} \geq 0\} \\
&\text{Type III}: && \{(\tau_0, \tau_1, \varphi_2, \tau_3), (n_1, n_2, n_3) : 1 \geq \tfrac{n_1}{m} \geq \tfrac{n_2}{n} \geq \tfrac{n_3}{m} \geq 0\} \\
&\text{Type IV}: && \{(\tau_0, \varphi_1, \varphi_2, \tau_3), (n_1, n_2, n_3) : 1 \geq \tfrac{n_1}{n} \geq \tfrac{n_2}{m+n} \geq \tfrac{n_3}{m} \geq 0\}
\end{aligned}$$

4.6 Given

$$\Delta_1 = ((\mu_0,\mu_1,\mu_2,\mu_3),(n_1,n_2,n_3)), \Delta_2 = ((\lambda_0,\lambda_1,\lambda_2,\lambda_3),(p_1,p_2,p_3))$$

in C, let use denote $\frac{n_i}{m(\mu_{i-1},\mu_i)}$ (resp. $\frac{p_i}{m(\lambda_{i-1},\lambda_i)}$) by just a_i (resp. b_i). We have $\Delta_1 \sim \Delta_2$, if

(1) $a_i = b_i$

(2)

(a) $a_1 = a_2 > a_3, \mu_2 = \lambda_2$, or

(b) $a_1 > a_2 = a_3, \mu_1 = \lambda_1$, or

(c) $a_1 = a_2 = a_3$

We note that if (a) holds, then $\lambda_2 = \tau_2$ or φ_2, and the equivalence can hold only between elements of either Type I and II, or Type III and IV respectively. If (b) holds, then $\lambda_1 = \tau_1$ or φ_1 , and the equivalence can hold only between elements of either Type I and III or Type II and IV respectively.

4.7 Let $x \in \overline{C}$. Let $x_{\min} = (\underline{c},\underline{n})$ (3.12). Let $\underline{c} = (\mu_0,\mu_1,\mu_2,\mu_3), \underline{n} = (n_1,n_2,n_3)$, and $a_i, 1 \le i \le 3$ as above. Note that if in 4.5, 2(a) holds, and $\mu_2 = \tau_2(resp.\varphi_2)$, then $(\underline{c},\underline{n})$ is of Type I (resp Type IV). If 2(b) holds, and $\mu_1 = \tau_1(resp.\varphi_1)$, then $(\underline{c},\underline{n})$ is of Type I (resp. type IV). If 2(c) holds, then $(\underline{c},\underline{n})$ is of Type I.

4.8 With notation as in 4.7, the element $v_x \in V_A$ (3.13) may be expressed explicitly as

$$v_x = \begin{cases} F_1^{(n_3)} F_2^{(n_2)} F_1^{(n_1)} e, & \text{if } \underline{c} \text{ is of Type I} \\ F_1^{(n_3)} F_2^{(n_1+n_2)} F_1^{(n_2)} e, & \text{if } \underline{c} \text{ is of Type II} \\ F_2^{(n_3)} F_1^{(n_1+n_2)} F_2^{(n_2)} e, & \text{if } \underline{c} \text{ is of Type III} \\ F_2^{(n_3)} F_1^{(n_2)} F_2^{(n_1)} e, & \text{if } \underline{c} \text{ is of Type IV} \end{cases}$$

(note that v_x is extremal $\iff n_i$ is either 0 or $= m_i$, where $m_i = m_\lambda(\mu_{i-1},\mu_i)$). We shall denote $\{v_x, x \in \overline{C}\}$ by B_d.

4.9 Lusztig's canonical basis for V_A. An element in V_A of the form $F_1^{(p)} F_2^{(q)} F_1^{(r)} e, q \ge p + r$ or $F_2^{(u)} F_1^{(t)} F_2^{(s)} e, t \ge u + s$ will be referred to as a *Lusztig* element or just a *L-element*. We have, $F_1^{(r)} e \ne 0 \iff r \le m$. Let $r \le m$; then $F_2^{(q)} F_1^{(r)} e \ne 0 \iff q \le r + n$ (using the relation

$$F_2^{(q)} F_1^{(r)} = \sum_{j=0}^{\min(q,r)} v^{-j-(q-j)(r-j)} F_1^{(r-j)} F_{\alpha_1+\alpha_2}^{(j)} F_2^{(q-j)})$$

Let now, $r \le m, p+r \le q \le r+n$; then $F_1^{(p)}F_2^{(q)}F_1^{(r)}e \ne 0$ (by $U_q(\mathbf{s}\ell_2)$-theory, since $K_1(F_2^{(q)}F_1^{(r)}e) = v^{-a}F_2^{(q)}F_1^{(r)}e$ (where $a = m+q-2r$, and $p \le a$). Thus an L-element of the form $F_1^{(p)}F_2^{(q)}F_1^{(r)}e$ (resp. $F_2^{(u)}F_1^{(t)}F_2^{(s)}e$) is non-zero if and only if $r \le m$, and $q \le r+n$ (resp. $s \le n, t \le s+m$). Hence if $\mathcal{B}_d$ denotes Lusztig's canonical basis for V_A, then

$$\mathcal{B}_d = \left\{ \begin{array}{ll} F_1^{(p)}F_1^{(q)}F_1^{(r)}e, p+r \le q \le r+n, r \le m, & \text{and} \\ F_2^{(u)}F_1^{(t)}F_2^{(s)}e, u+s \le t \le s+m, s \le n & \end{array} \right\}$$

(here one notes that the if $q = p+r$, then $F_1^{(p)}F_2^{(q)}F_1^{(r)}e = F_2^{(r)}F_1^{(q)}F_2^{(p)}e$) Let $p,q,r,p',q',r' \in \mathbb{Z}^+$. Let

$$L = \left\{ \begin{array}{ll} (p,q,r), r \le m, p+r \le q \le r+n, & \text{and} \\ (p'.q'.r'), r' \le n, p'+r' \le q' \le r'+m & \end{array} \right\}$$

where (p,q,r) is identified with (p',q',r'), if $q = q' = r+p, p = r', r = p'$ (note that L is an indexing set for $\mathcal{B}_d$).

5 A bijection between L and $\overline{C}$

5.1 Lemma. *Let (p,q,r) in L be such that $r \le m, p+r \le q \le r+n$. Further let $\frac{r}{m}, \frac{p}{n}, \frac{q-r}{n}$ be all distinct. Then precisely one of the following holds*

(1) $1 \ge \frac{r}{m} \ge \frac{q}{m+n} \ge \frac{p}{n} \ge 0$

(2) $1 \ge \frac{q-r}{n} \ge \frac{r}{m} \ge \frac{p}{n} \ge 0$

(3) $1 \ge \frac{q-r}{n} \ge \frac{p+r}{m+n} \ge \frac{r}{m} \ge 0$

Proof. We first observe that under the hypothesis that $\frac{r}{m}, \frac{p}{n}, \frac{q-r}{n}$ are distinct, the three cases are mutually exclusive. We now distinguish the following two cases.

Case 1. $\frac{r}{m} < \frac{q}{m+n}$

This implies that $\frac{q-r}{n} > \frac{r}{m}$. In this case $\frac{p+r}{m+n} < \frac{q-r}{n}$, necessarily; for, $\frac{r+p}{m+n} \ge \frac{q-r}{n}$ would imply $\frac{p+2r-q}{m}$ $(\ge \frac{q-r}{n}) > \frac{r}{m}$, which is not possible, since $q \ge p+r$. Hence either

(a) $1 \ge \frac{q-r}{n} \ge \frac{p+r}{m+n} \ge \frac{r}{m} \ge 0$, in which case (3) holds or

(b) $1 \ge \frac{q-r}{n} \ge \frac{r}{m} \ge \frac{p+r}{m+n} \ge 0$, in which case (2) holds.

Case 2. $\frac{r}{m} \ge \frac{q}{m+n}$ The hypothesis that $q \ge p+r$ implies that $\frac{q}{m+n} \ge \frac{p}{n}$. Thus in this case (1) holds.

5.2 Let (p,q,r) in L be such that $r \leq m, p+r \leq q \leq r+n$. We now define an element $\theta(p,q,r)$ in C. Let us denote $\theta(p,q,r)=(\underline{c},\underline{n})$, where $\underline{c}$ and $\underline{n}$ are given as follows.

(a) Let $\frac{r}{m}, \frac{p}{n}, \frac{q-r}{n}$ be all distinct.

We set

$$\underline{c}=\begin{cases}(\tau_0,\tau_1,\tau_2,\tau_3), & \text{if (1) of 5.1 holds}\\(\tau_0,\varphi_1,\tau_2,\tau_3), & \text{if (2) of 5.1 holds}\\(\tau_0,\varphi_1,\tau_2,\tau_3), & \text{if (3) of 5.1 holds}\end{cases}$$

$$\underline{n}=\begin{cases}(r,q,p), & \text{if (1) of 5.1 holds}\\(q-r,r,p), & \text{if (2) of 5.1 holds}\\(q-r,p+r,r), & \text{if (3) of 5.1 holds}\end{cases}$$

(b) Let $\frac{r}{m}=\frac{q-r}{n}\neq\frac{p}{n}$.

Then (3) of 5.1 cannot hold, and the cases (1) and (2) of 5.1 coincide. We set

$$\underline{c}=(\tau_0,\tau_1,\tau_2,\tau_3), \underline{n}=(r,q,p)$$

(c) Let $\frac{r}{m}=\frac{p}{n}\neq\frac{q-r}{n}$.

Then (1) of 5.1 cannot hold, and (2) and (3) of 5.1 coincide. We set

$$\underline{c}=(\tau_0,\varphi_1,\varphi_2,\tau_3),\ \underline{n}=(q-r,p+r,r)$$

(d) Let $\frac{p}{n}=\frac{q-r}{n}\neq\frac{r}{m}$.

This implies that $q=r+p$. In this case (2) of 5.1 cannot hold, and (1) and (3) are mutually exclusive. We set

$$\underline{c}=\begin{cases}(\tau_0,\tau_1,\tau_2,\tau_3), & \text{if (1) of 5.1 holds}\\(\tau_0,\varphi_1,\varphi_2,\tau_3), & \text{if (3) of 5.1 holds}\end{cases}$$

$$\underline{n}=\begin{cases}(r,q,p), & \text{if (1) of 5.1 holds}\\(q-r,p+r,r), & \text{if (3) of 5.1 holds}\end{cases}$$

(e) Let $\frac{r}{m}=\frac{p}{n}=\frac{q-r}{n}$

This implies that $\frac{q}{m+n}=\frac{r}{m}=\frac{p+r}{n+m}$. Then all three cases of 5.1 coincide, and we set

$$\underline{c}=(\tau_0,\tau_1,\tau_2,\tau_3),\ \underline{n}=(r,q,p)$$

5.3 Remark. We observe that in all of the cases (a) through (e) above, $(\underline{c},\underline{n})$ is of type I, II or IV. Also, if $x=(\underline{c},\underline{n})$, then it is easily seen that $x_{\min}=(\underline{c},\underline{n})$.

5.4 Lemma. *Let (p', q', r') in L be such that $r' \leq n, p' + r' \leq q' \leq r' + m$. Further, let $\frac{r'}{n}, \frac{q'-r'}{m}, \frac{p'}{m}$ be all distinct. Then precisely one of the following holds.*

(1) $1 \geq \frac{r'}{n} \geq \frac{q'}{m+n} \geq \frac{p'}{m} \geq 0$

(2) $1 \geq \frac{q'-r'}{m} \geq \frac{r'}{n} \geq \frac{p'}{m} \geq 0$

(3) $1 \geq \frac{q'-r'}{m} \geq \frac{p'+r'}{m+n} \geq \frac{r'}{n} \geq 0$

The proof is similar to that of Lemma 5.1.

5.5 Let (p', q', r') in L be such that $r' \leq n, p' + r' \leq q' \leq r' + m$. We now define an element $\theta(p', q', r')$ in C. Let us denote $\theta(p', q', r') = (\underline{c}', \underline{n}')$, where $\underline{c}'$ and $\underline{n}'$ are given as follows.

(a) Let $\frac{r'}{m}, \frac{p'}{m}, \frac{q'-r'}{m}$, be all distinct. We set

$$\underline{c}' = \begin{cases} (\tau_0, \varphi_1, \varphi_2, \tau_3), & \text{if (1) of 5.4 holds} \\ (\tau_0, \tau_1, \varphi_2, \tau_3), & \text{if (2) of 5.4 holds} \\ (\tau_0, \tau_1, \tau_2, \tau_3), & \text{if (3) of 5.4 holds} \end{cases}$$

$$\underline{n}' = \begin{cases} (r', q', p'), & \text{if (1) of 5.4 holds} \\ (q' - r', r', p'), & \text{if (2) of 5.4 holds} \\ (q' - r', p' + r', r'), & \text{if (3) of 5.4 holds} \end{cases}$$

(b) Let $\frac{r'}{n} = \frac{q'-r'}{m} \neq \frac{p'}{m}$. We set

$$\underline{c}' = (\tau_0, \varphi_1, \varphi_2, \tau_3), \underline{n}' = (r', q', p')$$

(the discussion being as in 5.2 (b))

(c) Let $\frac{r'}{n} = \frac{p'}{m} \neq \frac{q'-r'}{n}$. We set

$$\underline{c}' = (\tau_0, \tau_1, \tau_2, \tau_3), n' = (q' - r', p' + r', r')$$

(d) Let $\frac{p'}{m} = \frac{q'-r'}{m} \neq \frac{r'}{n}$. In this case, (2) of 5.4 cannot hold and (1) and (3) of 5.4 are mutually exclusive.

We set

$$\underline{c}' = \begin{cases} (\tau_0, \varphi_1, \varphi_2, \tau_3), & \text{if (1) of 5.4 holds} \\ (\tau_0, \tau_1, \tau_2, \tau_3), & \text{if (3) of 5.4 holds} \end{cases}$$

$$\underline{n}' = \begin{cases} (r', q', p'), & \text{if (1) of 5.4 holds} \\ (q' - r', p' + r', r'), & \text{if (3) of 5.4 holds} \end{cases}$$

(e) Let $\frac{r'}{n} = \frac{p'}{m} = \frac{q'-r'}{m}$. This implies that $\frac{r'}{n} = \frac{p'+r'}{m+n} = \frac{q'}{m+n}$. Then all the three cases of 5.4 coincide and we set

$$\underline{c}' = (\tau_0, \tau_1, \tau_2, \tau_3), \underline{n}' = (q' - r', p' + r', r')$$

5.6 Remark. In all of the cases (a) through (e) above, $(\underline{c}', \underline{n}')$ is of Type I, III or IV. Also if $y = (\overline{c', n'})$, then $y_{\min} = (\underline{c}', \underline{n}')$.

5.7 Let

$$\begin{aligned} L_1 &= \{(p,q,r) \in L : r \le m, p+r \le q \le r+n\} \\ L_2 &= \{(p',q',r') \in L : r' \le n, p'+r' \le q' \le r'+m\} \end{aligned}$$

An element (p',q',r') in L_2 will be identified with the element (p,q,r) in L_1, if $r' = p, p' = r, q' = q = r+p = r'+p'$. When this happens, we shall express it as $(p',q',r') \sim (p,q,r)$. Let θ be as in 5.2 (resp. 5.5). We observe that if $(p',q',r') \sim (p,q,r)$, then $\theta(p',q',r') = \theta(p,q,r)$. To see this, let $\theta(p,q,r) = (\underline{c}, \underline{n})$. Then since $\frac{q-r}{n} = \frac{p}{n}$, only (d) or (e) of 5.2 can hold. If (e) of 5.2 holds, then (e) of 5.5 also holds. We have (5.2 (e), 5.5 (e)),

$$\underline{c} = (\tau_0, \tau_1, \tau_2, \tau_3), \underline{n} = (r, q, p)$$

$$\underline{c}' = (\tau_0, \tau_1, \tau_2, \tau_3), \underline{n}' = (q'-r', p'+r', r') \ (= (r,q,p))$$

Thus $(\underline{c}, \underline{n}) = (\underline{c}', \underline{n}')$.

Let then (d) of 5.2 hold. We distinguish the following two cases.

Case 1. $\frac{r}{m} > \frac{p}{n}$

This implies that $1 \ge \frac{r}{m} > \frac{q}{m+n} > \frac{p}{n} \ge 0$, and $1 \ge \frac{q'-r'}{m} > \frac{r'+p'}{m+n} > \frac{r'}{n} \ge 0$. Hence we get (5.2(d), 5.5(d))

$$\underline{c} = (\tau_0, \tau_1, \tau_2, \tau_3), \underline{n} = (r, q, p)$$

$$\underline{c}' = (\tau_0, \tau_1, \tau_2, \tau_3), \underline{n}' = (q'-r', r'+p', r') (= (r,q,p))$$

Thus $\underline{c}, \underline{n} = (\underline{c}', \underline{n}')$

Case 2. $\frac{p}{n} > \frac{r}{m}$.

This implies that $1 \ge \frac{p}{n} > \frac{p+r}{m+n} > \frac{r}{m} \ge 0$, and $1 \ge \frac{r'}{n} > \frac{q'}{m+n} > \frac{p'}{m} \ge 0$. Hence we get (5.2(d), 5.5(d))

$$\underline{c} = (\tau_0, \varphi_1, \varphi_2, \tau_3), \underline{n} = (q-r, p+r, r) \ (= (r', q', p'))$$

$$\underline{c}' = (\tau_0, \varphi_1, \varphi_2, \tau_3), \underline{n}' = (r', q', p')$$

Thus $(\underline{c}, \underline{n}) = (\underline{c}', \underline{n}')$.

In view of the discussion above we obtain a map $\theta : L \longrightarrow C$, which induces a map $\overline{\theta} : L \longrightarrow \overline{C}$ in an obvious way (as above, we identify an element (p',q',r') of L where $q' = p'+r', r' \le n, q' \le r'+m$, with the element (r',q',p')).

5.8 We now define a map $\overline{\psi} : \overline{C} \longrightarrow L$. Let $x \in \overline{C}$. Let $x_{\min} = (\underline{c}, \underline{n})$. We distinguish the following cases. Let $\underline{n} = (a, b, c)$. (We follow the convention in 4.4)

Case 1. $(\underline{c}, \underline{n})$ is of Type I.

This implies $1 \geq \frac{a}{m} \geq \frac{b}{m+n} \geq \frac{c}{n} \geq 0$. We have $b \geq c$ (since $\frac{b}{m+n} \geq \frac{c}{n}$)) We set

$$\overline{\psi}(x) = \begin{cases} (c, b, a), & \text{if } \ b \geq a + c \\ (b - c, a + c, c) & \text{if } \ b < a + c \end{cases}$$

(note that $\frac{a}{m} \geq \frac{b}{m+n} \Rightarrow \frac{a}{m} \geq \frac{b-a}{m}$ and hence $b \leq a + n$, since $\frac{a}{m} \leq 1$. Also, $c \leq n$, since $\frac{c}{n} \leq 1$. Thus $(c, b, a) \in L_1$, (in the case $b \geq a + c$) and $(b - c, a + c, c) \in L_2$ (in the case $b < a + c$).

Case 2. $(\underline{c}, \underline{n})$ is of Type II. This implies that $1 \geq \frac{a}{n} \geq \frac{b}{m} \geq \frac{c}{n} \geq 0$. Then $(c, a + b, b) \in L_1$, and we set

$$\overline{\psi}(x) = (c, a + b, b)$$

Case 3. $\underline{c}, \underline{n}$ is of Type III . This implies that $1 \geq \frac{a}{m} \geq \frac{b}{n} \geq \frac{c}{m} \geq 0$ Then $(c, a + b, b) \in L_2$, and we set

$$\overline{\psi}(x) = (c, a + b, b)$$

Case 4. $(\underline{c}, \underline{n})$ is of Type IV.
This implies that $1 \geq \frac{a}{n} \geq \frac{b}{n+m} \geq \frac{c}{m} \geq 0$. Then we set (as in Case 1)

$$\overline{\psi}(x) = \begin{cases} (c, b, a), & \text{if } \ \text{b} \geq a + c \\ (b - c, c + a, c), & \text{if } \ b < a + c \end{cases}$$

(note that $(c, b, a) \in L_2$ (if $b \geq a + c$) and $(b - c, c + a, c) \in L_1$ (if $b < a + c$)). It is easily checked that $\overline{\psi} o \overline{\theta} = Id_L$ and $\overline{\theta} o \overline{\psi} = Id_{\overline{C}}$. Thus we obtain

5.9 Theorem. *The map $\theta : L \rightarrow C$ is a bijection*

6. An A-basis for V_A

6.1 For $r \in \mathbb{Z}, s \in \mathbb{Z}^+$, we set

$$\begin{bmatrix} r \\ s \end{bmatrix} = \frac{[r][r-1]\cdots[r+1-s]}{[s]\cdots[1]}$$

where for $m \in \mathbb{Z}, [m] = \frac{\nu^m - \nu^{-m}}{\nu - \nu^{-1}}$. Let $\alpha = a_i, a' = \alpha_j$, where $a_{ij} = -1$. Let

$$(1) \qquad F_{\alpha+\alpha'} = \nu F_\alpha F_{\alpha'} - F_{\alpha'} F_\alpha.$$

We have ([L1]),

$$(2) \qquad F_\alpha^{(M)} F_{\alpha'} = \nu^{-1} F_{\alpha+\alpha'} F_\alpha^{(M-1)} + \nu^{-M} F_{\alpha'} F_\alpha^{(M)}$$

$$(3) \qquad F_\alpha F_{\alpha'}^{(N)} = v^{-1} F_{\alpha'}^{(N-1)} F_{\alpha+\alpha'} + \nu^{-N} F_{\alpha'}^{(N)} F_\alpha$$

Hence we obtain (using (1))

$$(4) \qquad F_{\alpha'}^{(N-1)} F_\alpha F_{\alpha'} = [N-1] F_{\alpha'}^{(N)} F_\alpha + F_\alpha F_{\alpha'}^{(N)}$$

$$(5) \qquad F_\alpha F_{\alpha'} F_{\alpha'}^{(M-1)} = [M-1] F_{\alpha'} F_\alpha^{(M)} + F_\alpha^{(M)} F_{\alpha'}$$

6.2 Lemma. *Let α, α' be as above. For $t, u, v \in \mathbb{N}$, we have,*

$$F_\alpha^{(t)} F_{\alpha'}^{(u)} F_\alpha^{(v)} = \sum_{j=t-k}^{t} \binom{t+v-u}{j} F_{\alpha'}^{(u-(t-j))} F_\alpha^{(v+t)} F_{\alpha'}^{(t-j)}$$

where $k = min(u,t)$. This is proved by induction on t using (2) - (5) above.

6.3 Let $\mathcal{B}_d$ be the canonical A-basis for V_A as constructed in [L2]. Then we have ([L3]), $\mathcal{B}_d = \mathcal{B}_1 \cup \mathcal{B}_2$, where

$$\begin{aligned} \mathcal{B}_1 &= \{F_1^{(c)} F_2^{(b)} F_1^{(a)} e, (a,b,c) \in L_1\} \\ \mathcal{B}_2 &= \{F_2^{(c')} F_1^{(b')} F_2^{(a')} e, (a',b',c') \in L_2\} \end{aligned}$$

(here we identify $F_2^{(c')} F_1^{(b')} F_2^{(a')} e$, with $F_2^{(c)} F_1^{(b)} F_2^{(a)} e$ if $((a,b,c) \sim (a',b',\underline{c}'))$. Note that if $(a',b',c') \in L_2$, and $b' = c' + a'$, then $F_2^{(c')} F_1^{(b')} F_2^{(a')} = F_1^{(a')} F_2^{(b')} F_1^{(c')}$.

6.4 Let $x \in \overline{C}, x_{\min} = (\underline{c}, \underline{n})$, and $\underline{n} = (a, b, c)$. We have (4.8)

$$v_x = \begin{cases} F_1^{(c)} F_2^{(b)} F_1^{(a)} e, & \text{if } (\underline{c}, \underline{n}) \text{ if of Type I} \\ F_1^{(c)} F_2^{(a+b)} F_1^{(b)} e, & \text{if } (\underline{c}, \underline{n}) \text{ if of Type II} \\ F_2^{(c)} F_1^{(a+b)} F_2^{(b)} e, & \text{if } (\underline{c}, \underline{n}) \text{ if of Type III} \\ F_2^{(c)} F_1^{(b)} F_2^{(a)} e, & \text{if } (\underline{c}, \underline{n}) \text{ if of Type IV} \end{cases}$$

6.5 Let us take an indexing I of L such that

(1) If $(p,q,r), (a,q,b)$ are in L_1 with $a > p$, then (p,q,r) preceeds (a,q,b).

(2) if $(p',q',r'), (a',q',b')$ are in L_2, with $a' > p'$, then (p',q',r') preceeds (a',q',b').

Then via the bijection $\overline{\psi} : \overline{C} \longrightarrow L$, we obtain an indexing J of $\overline{C}$ induced by I. Let M be the matrix expressing the elements in B_d as A-linear combinations of the elements in $\mathcal{B}_d$, for the indexing J of B_d (resp. I of $\mathcal{B}_d$).

6.6 Theorem. M is upper triangular with diagonal entries equal to 1.

Proof. Let $x \in \overline{C}, x_{\min} = (\underline{c}, \underline{n}), \underline{n} = (a,b,c)$. We may suppose that $c \neq 0$; for, if $c = 0$, then $v_x \in \mathcal{B}_d$ obviously. If $(\underline{c}, \underline{n})$ is of Type II or III, then $v_x \in \mathcal{B}_d$ clearly. We now distinguish the following two cases:

Case 1. $(\underline{c}, \underline{n})$ is of Type I.
We have $v_x = F_1^{(c)} F_2^{(b)} F_1^{(a)} e$. Hence, if $b \geq a + c$, then $v_x \in \mathcal{B}_d$. Let then $b < a + c$. We have (Lemma 6.2, with $\alpha = \alpha_1, \alpha' = \alpha_2, t = c, u = b, v = a$),

$$(*) \qquad v_x = \sum_{j=0}^{c} \begin{bmatrix} c+a-b \\ j \end{bmatrix} F_2^{(b-(c-j))} F_1^{(a+c)} F_2^{(c-j)} e$$

(Note that $(\underline{c}, \underline{n})$ being of Type I, we have $1 \geq \frac{a}{m} \geq \frac{b}{m+n} \geq \frac{c}{n} \geq 0$. Hence $b > c$, and $\min(c,b) = c$). Now, on R.H.S. of $(*)$, each term $F_2^{(b-\ell)} F_1^{(a+c)} F_2^{(\ell)} e$ is in L_2, since $\ell \leq c \leq n, a+c \leq m+c$ (as $a \leq m$), and $a+c > b-\ell+\ell\,(= b)$. Also, the term $F_2^{(b-c)} F_1^{(a+c)} F_2^{(c)} e$ occurs with coefficient 1. Now, the element $(\underline{c}, \underline{n})$ in $\overline{C}$ corresponds to the element $(b-c, a+c, c)$ in L (under the indexing J (resp. I) for $\overline{C}$ (resp. L)). Also, it is clear that all the other terms (on the R.H.S. of (*)) succeed $F_2^{(b-c)} F_1^{(a+c)} F_2^{(c)}$ (in the indexing I for $\mathcal{B}_d$).

Case 2. $(\underline{c}, \underline{n})$ is of Type IV.

The discussion is exactly similar to that of Case 1.

6.7 Theorem. *B_d is an A-basis for V_A.*

Proof. *This follows from Theorem 6.6, since $\mathcal{B}_d$ is an A-basis for V_A.*

7 Basis for quantum Demazure modules

7.1 Let $x \in \overline{C}, x_{\min} = (\underline{c}, \underline{n}), \underline{n} = (a_1, a_2, a_3)$. Let τ_x be as in 3.14. Then τ_x is given as follows. If $a_1 = a_2 = a_3 = 0$, then $\tau_x = Id$. Let r be the largest integer ≤ 3 such that $a_r \neq 0$.

(1) $r = 1$. This implies that $\underline{c}$ is of Type I or IV. We have $\tau_x = \tau_1$ $(resp. \varphi_1)$ if $\underline{c}$ is of Type I (resp. IV).

(2) $r = 2$. This implies that $\tau_x = \tau_2$, if $\underline{c}$ is of Type I or II and $\tau_x = \varphi_2$, if $\underline{c}$ is of Type III or IV

(3) $r = 3$. This implies that $\tau_x = \tau_3$.

7.2 Theorem. *Let $w \in W$. Let $B_w = \{v_x : \tau_x \leq w\}$. Then B_w be an A-basis for $V_{w,A}$.*

Proof. Let $X(\varphi)$ be a moving divisor in $X(w)$, moved by α. Then we see easily that

$$(*) \qquad V_{w,A} = U^-_{\alpha,A} V_{\varphi,A}$$

where $U^-_{\alpha,A}$ is the A-submodule of U generated by $F^{(r)}_\alpha, r \in \mathbb{Z}^+$. For $w = \tau_0$ $(= Id)$, the result is clear. For $w = \tau_3$, the result follows from Theorem 6.7.

(1) Let $w = \tau_1$. Then (*) implies that $\{F^{(r)}_1 e, r \in \mathbb{Z}^+\}$ generates $V_{w,A}$. Now $F^{(r)}_1 e = 0$, for $r > m$. Hence $\{F^{(r)}_1 e, 0 \leq r \leq m\}$ is an a A-basis for $V_{w,A}$, while B_w is precisely $\{F^{(r)}_1 e, 0 \leq r \leq m\}$.

(2) Let $w = \varphi_1$. The proof is similar as in (1).

(3) Let $w = \tau_2$ and $\varphi = \tau_1$. Then, we have (in view of $(*)$), $\{F_2^{(q)}v, v \in B_\varphi\}$ generates $V_{w,A}$. We have $F_2^{(q)}F_1^{(r)}e = 0$, if $q > r + n$ (4.9). Hence $\{F_2^{(q)}F_1^{(r)}e, r \leq m, q \leq r+n\}$ generates $V_{w,A}$ as an A-module. Now, if $\frac{r}{m} \geq \frac{q}{m+n}$, then $F_2^{(q)}F_1^{(r)}e = v_x$, where $x = \overline{(\underline{c}, \underline{n})}, \underline{c} = (\tau_0, \tau_1, \tau_2), \underline{n} = (r, q)$; if $\frac{q}{m+n} > \frac{r}{m}$, then $F_2^{(q)}F_1^{(r)}e = v_x$, where $x = \overline{(\underline{c}, \underline{n})}, \underline{c} = (\tau_0, \varphi_1, \tau_2), \underline{n} = (q - r, r)$ (6.4). Hence we see that B_w generates $V_{w,A}$. The linear independence of B_w follows from Theorem 6.7 (since $B_w \subseteq B_d$).

(4) Let $w = \varphi_2$. The proof is similar to that in (3).

8 Appendix

We have used the results of [L2] mainly to prove that $\#\overline{C} = dim V_d$, (§ 5). We can get around proving $\#\overline{C} = dim(V_d)$, by showing that $\#\overline{C} = \#$ {standard Young tableaux on $SL(3)$ of type (m, n)}. We can then prove the results of §6, § 7 in the same spirit as in [L-S]. In this Appendix, we establish a bijection between C and {standard Young tableaux on $SL(3)$ of type (m, n)}.

8.1 Let $G = SL(3)$. Let $P_1 = \{Id, s_1\}, P_2 = \{Id, s_2\}$. Let us denote the set of minimal representatives of W_{P_1} (resp. W_{P_2}) in W by $\Theta = \{\Theta_1, \Theta_2, \Theta_3\}$ (resp. $\Lambda = \{\lambda_1, \lambda_2, \lambda_3\}$). Then Θ (resp. Λ) is totally ordered (under the Bruhat order $\succeq$). Let $\theta_3 \succeq \theta_2 \succeq \theta_1; \lambda_3 \succeq \lambda_2 \succeq \lambda_1$. Let $X = \{\theta_3, \theta_2, \theta_1, \lambda_3, \lambda_2, \lambda_1\}$. We have a partial order $\geq$ on X given as follows. Let $x, y \in X$. Then $x \geq y$, if either both $x, y \in \Theta$ (resp. Λ), and $x \succeq y$, or x (resp. y) $\in \Theta(resp. \Lambda)$, and $(x, y) \neq (\theta_1, \lambda_3)$. A classical standard Young tableau on G of type (m, n) can be denoted as

$$\tau_{11}\tau_{12}\cdots\tau_{1m}\tau_{21}\tau_{22}\cdots\tau_{2n}$$

where τ_{1j} (resp. τ_{2k}) $\in \Theta$(resp. Λ) and

$$\tau_{11} \geq \tau_{12} \geq \cdots \geq \tau_{1m} \geq \tau_{21} \geq \tau_{22} \geq \cdots \geq \tau_{2n}$$

Let Y = {standard Young tableaux of type (m, n)}

8.2 Let $a \in Y$, say $a = \tau_{11} \cdots \tau_{1m}\tau_{21} \cdots \tau_{2n}$. We define the integers $r_a, q_a, p_a, u_a, t_a, s_a$ as follows.

$$\begin{aligned} r_a &= \#\{\tau_{1j} : \tau_{1j} = \theta_3\} \\ p_a &= \#\{\tau_{1j}, \tau_{2k} : \tau_{1j} = \theta_2, \tau_{2k} = \lambda_3\} \\ q_a - r_a &= \#\{\tau_{2k} : \tau_{2k} = \lambda_3 \text{ or } \lambda_2\} \\ u_a &= r_a \\ t_a &= r_a + p_a \\ s_a &= q_a - r_a \end{aligned}$$

Note that $r_a \leq m, q_a \leq r_a + n, r_a + p_a \leq q_a - r_a + m$. Note also that r_a, p_a, q_a (resp. u_a, t_a, s_a) completely determine "a".

8.3 The map $f : Y \longrightarrow \overline{C}$.

For $a \in Y$, we define $f(a) = \overline{(\underline{c}_a, \underline{n}_a)}$ as follows. For simplicity of notation let us drop off the suffix 'a' is in $r_a, \cdots, s_a, \underline{c}_a, \underline{n}_a$. (We follow the convention in 4.4 while denoting a chain $(\underline{c}, \underline{n})$). We first observe that

$$\frac{r}{m}, \frac{q}{m+n}, \frac{q-r}{n}, \frac{p+2r-q}{m}, \frac{p+r}{m+n}, \frac{s}{n}, \frac{t-s}{m+n}, \frac{u+s}{m+n}, \frac{t}{m+n},$$

are all ≤ 1. We now distinguish the following cases.

Case 1. $1 \geq \frac{r}{m} \geq \frac{q}{m+n} \geq \frac{p}{n} \geq 0$

We set $\overline{c} = (\tau_0, \tau_1, \tau_2, \tau_3), \underline{n} = (r, q, p)$

Case 2. $\frac{q}{m+n} > \frac{r}{m}$

Now $\frac{q}{m+n} > \frac{r}{m} \Longleftrightarrow \frac{q-r}{n} > \frac{r}{m}$

We divide this case into the following cases.

Case 2(a) $1 \geq \frac{q-r}{n} > \frac{p+r}{m+n} \geq \frac{r}{m} \geq 0.$

This is equivalent to

$$1 \geq \frac{s}{n} \geq \frac{t}{m+n} \geq \frac{u}{m} \geq 0$$

We set $\underline{c} = (\tau_0, \varphi_1, \varphi_2, \varphi_3), \underline{n} = (s, t, u).$

Case 2(b) $1 \geq \frac{p+r}{m+n} \geq \frac{q-r}{n} \geq \frac{r}{m} \geq 0$

This is equivalent to

$$1 \geq \frac{p+2r-q}{m} \geq \frac{q-r}{n} \geq \frac{r}{m} \geq 0$$

We set $\underline{c} = (\tau_0, \varphi_1, \tau_2, \tau_3), n = (p+2r-q, q-r, r)$

Case 2(c) $1 \geq \frac{q-r}{n} \geq \frac{r}{m} \geq \frac{p+r}{m+n} \geq 0$

This is equivalent to

$$1 \geq \frac{q-r}{n} \geq \frac{r}{m} \geq \frac{p}{n} \geq 0$$

We set $\underline{c} = (\tau_0, \tau_1, \varphi_2, \tau_3), \underline{n} = (q-r, r, p)$

Case 3. $\frac{q}{m+n} \leq \frac{r}{m}, \frac{q}{m+n} < \frac{p}{n}$

Now, $\frac{q}{m+n}, \leq \frac{r}{m} \iff \frac{q-r}{n} < \frac{q}{m+n}$, and $\frac{q}{m+n} < \frac{p}{n} \iff \frac{q-p}{m} < \frac{q}{m+n} < \frac{p}{n}$ Hence in this case, we have $\frac{q-r}{n} \leq \frac{q}{m+n} < \frac{p}{n}$. This implies that $q < r+p$, i.e., $u+s < t$.

We divide this case into the following subcases.

Case 3(a) $1 \geq \frac{s}{n} \geq \frac{t}{m+n} \geq \frac{u}{m} \geq 0$

We set $\underline{c} = (\tau_0, \varphi_1, \varphi_2, \tau_3), \underline{n} = (s, t, u)$

Case 3(b) $\frac{t}{m+n} \geq \frac{s}{n}$. Now $\frac{t}{m+n} \geq \frac{s}{n} \iff \frac{t-s}{m} \geq \frac{s}{n}$. The condition $u+s < t$ implies $\frac{u+t}{m+n} < \frac{t-s}{m}$ (for, otherwise, $\frac{u+s}{m+n} \geq \frac{t-s}{m} \geq \frac{s}{n} \Rightarrow \frac{u+s-(t-s)}{n} \geq \frac{t-s}{m} \geq \frac{s}{n} \Rightarrow$ $u+s \geq t$, which is not true) Hence, either,

(1) $$1 \geq \frac{t-s}{m} \geq \frac{u+s}{m+n} \geq \frac{s}{n} \geq 0 (\text{or})$$

(2) $$1 \geq \frac{t-s}{m} \geq \frac{s}{n} \geq \frac{u+s}{m+n} \geq 0$$

Now (1) is equivalent to

$$1 \geq \frac{p+2r-q}{m} \geq \frac{q}{m+n} \geq \frac{q-r}{n} \geq 0$$

and we set $\underline{c} = (\tau_0, \tau_1, \tau_2, \tau_3), \underline{n} = (p+2r-q, q, q-r)$, if (1) holds.

Similarly, (2) is equivalent to

$$1 \geq \frac{t-s}{m} \geq \frac{s}{n} \geq \frac{u}{m} \geq 0$$

and we set $\underline{c} = (\tau_0, \tau_1, \varphi_2, \tau_3), \underline{n} = (t-s, s, u)$, if (2) holds.

8.4 The map $g : \overline{C} \longrightarrow Y$.

Let $x \in \overline{C}, x_{\min} = (\underline{c}, \underline{n}), \underline{n} = (a, b, c)$. Set $g(x) = a$, where $r\ (= r_a), q\ (= q_a), p\ (= p_a)$ are given as follows.

$$(r, q, p) = \begin{cases} (a, b, c), & \text{if } (\underline{c}, \underline{n}) \text{ is of Type I} \\ (b, a+b, c), & \text{if } (\overline{c}, \overline{n}) \text{ is of Type II} \end{cases}$$

$$(s, t, u) = \begin{cases} (b, a+b, c), & \text{if } (\underline{c}, \underline{n}) \text{ is of Type III} \\ (a, b, c), & \text{if } (\underline{c}, \underline{n}) \text{ is of Type IV} \end{cases}$$

(where, recall that $u = r, t = r + p, s = q - r$). It is easily checked that $gof = Id_Y, fog = Id_{\overline{C}}$.

Thus we obtain

8.5 Theorem. *The map $f : Y \longrightarrow \overline{C}$ is a bijection.*

References

[B] N. Bourbaki, *Groupes et algebres de Lie semisimples, Chapitres* IV, V et VI., Bourbaki (1961/62).

[B-W] A. Björner and M. Wachs, *Bruhat order of Coxeter groups and shellability,* Adv. Math. **43** (1982) 87–100.

[D] V.G. Drinfeld, *Hopf algebras and the Yang-Baxter equation,* Soviet Math. Dokl. **32** (1985) 254–258.

[J] M. Jimbo, *A q-difference analogue of $U(g)$ and the Yang-Baxter equation,* Lett. Math. Phys. **10** (1985) 63–69.

[L1] G. Lusztig, *Finite dimensional Hopf algebras arising from quantized universal enveloping algebras,* Jour. Amer. Math. Soc. **3** (1990) 257–296.

[L2] G. Lusztig, *Canonical baases arising from quantized enveloping algebras,* J. Amer. Math. Soc. **3** (1990) 447–498; II, Progress of Theor. Physics (to appear).

[L3] G. Lusztig, *Introduction to quantized enveloping algebras* (preprint).

[L4] G. Lusztig, *Problems on Canonical Bases* (preprint).

[LS] V. Lakshmibai and C.S. Seshadri, *Geometry of G/P – V*, J. Algebra, **100** (1986) 462–557.

Northeastern University
Boston, MA 02115
U.S.A.

An Appendix to Bases for Quantum Demazure modules-I

Let g be a symmetrizable Kac-Moody Lie algebra, and U be the quantized enveloping algebra of g as constructed by Drinfeld (cf [D]) and Jimbo (cf [J]). This is an algebra over $\mathbb{Q}(v)$ (v being a parameter) which specializes to $U(g)$ for $v = 1, U(g)$ being the universal enveloping algebra of g. This algebra has generators E_i, $F_i, K_i, 1 \leq i \leq n$, which satisfy "the quantum Chevalley and Serre relations". Let $U^{\pm}$ be the $\mathbb{Q}(v)$-sub algebra of U genrated by $E_i(respF_i), 1 \leq i \leq n$. Let $A = \mathbb{Z}[v, v^{-1}]$ and $U_A^{\pm}$ be the A-subalgebra of U generated by $E_i^{(r)}$ (resp. $F_i^{(r)}$), $1 \leq i \leq n, r \in \mathbb{Z}^+$, (here $E_i^{(r)}, F_i^{(r)}$ are the quantum divided powers of (cf [J])). Let λ be a dominant, integral weight and V_λ the associated simple U-module. Let us fix a highest weight vector e in V_λ and denote $V_A = U_A e \; (= U_A^- e)$. Let W be the Weyl group of g. For $w \in W$, let e_w be the corresponding extremal weight vector in V_λ of weight $w(\lambda)$. Let $V_w = U^+ e_w, V_{w,A} = U_A^+ e_w$. In [La] (see also [LS]), we proposed a conjecture (which we recall below) towards the construction of an $A-$basis for V_A compatible with $\{V_{w,A}, w \in W\}$. This conjecture consists of two parts. The first part gives a (conjectural) character formula for the U^+-module V_w in terms of certain weighted chains in W. The second part gives a conjectural A-basis B_λ for V_A, compatible with $\{V_{w,A}, w \in W\}$. We now state the conjecture:

Part I I_λ, an indexing set for B_λ

Let $\lambda = \Sigma d_i \omega_i, \omega_i$ being the fundamental weights.

Admissible weighted λ-chains

Let $\underline{c} = \{\mu_0, \mu_1, ..., \mu_r]\}$ be a λ-chain in W, i.e., $\mu_i > \mu_{i-1}, \ell(\mu_i) = \ell(\mu_{i-1}) + 1$ (if $d_t = 0$ for $t = i_1, \cdots, i_s$, then we shall work with W^Q, the set of minimal representatives of W_Q in W, W_Q being the Weyl group of the parabolic subgroup Q, where $S_Q = (\{\alpha_t, t = i_1, \cdots, i_s\})$.

Let $\mu_{i-1} = s_i\mu_i$, where β_i is some positive real root. Let $(\mu_{i-1}(\lambda), \beta^*) = m_i$.

A 1. Definition. A λ-chain $\underline{c}$ is called *simple* if all β_i' s are simple.

A 2. Definition. m By a *weighted* λ *chain* we shall mean $(\underline{c}, \underline{n})$ where $\underline{c} = \{\mu_0, \cdots, \mu_r\}$ is a chain and $\underline{n} = \{n_1, ..., n_r\}, n_i \in \mathbb{Z}^+$.

A 3. Definition. A weighted λ-chain $(\underline{c}, \underline{n})$ is said to be *admissible* if $1 \geq \frac{n_1}{m_1} \geq \frac{n_2}{m_2} \geq \cdots \geq \frac{n_r}{m_r} \geq 0$.

Let $(\underline{c}, \underline{n})$ be admissible. Let us denote the unequal values in $\{\frac{n_1}{m_1}, \cdots, \frac{n_r}{m_r}\}$ by $a_1, \cdots, a_s$ so that $1 \geq a_1 > a_2 > \cdots > a_s \geq 0$. Let $i_0, \cdots, i_s$ be defined by

$$i_0 = 0, i_s = r, \frac{n_j}{m_j} = a_t, i_{t-1} + 1 \leq j \leq i_t.$$

We set

$$D_{\underline{c},\underline{n}} = \{(a_1, ..., a_s); (\mu_{i_0}, ..., \mu_{i_s})\}$$

A 4. Definition. Let $(\underline{c}, \underline{n}), (\underline{c}', \underline{n}')$ be two admissible weighted λ-chains. Let $D_{\underline{c},\underline{n}} = \{(a_1, ..., a_s); (\mu_{i_0}, \cdots, \mu_{i_s})\}$, and $D_{\underline{c}',\underline{n}'} = \{(a_1', \cdots, a_\ell'); (\tau_{j_0}, ..., \tau_{j_\ell})\}$. We say $(\underline{c}, \underline{n}) \sim (\underline{c}', \underline{n}')$, if $s = t$, and $a_t = a_t', i_t = j_t, \mu_{i_t} = \tau_{j_t}, 1 \leq t \leq s$.

Let C_λ = { all admissible weighted λ -chains}, and $I_\lambda = C_\lambda / \sim$. Let $\pi \in I_\lambda$, and let $(\underline{c}, \underline{n})$ be a representative of π. With notations as above, we set

$$\tau(\pi) = \mu_{i_s}, v(\pi) = \sum_{t=0}^{s} (a_t - a_{t+1})\mu_{i_t}(\lambda)$$

where $a_o = 1$ and $a_{s+1} = 0$ (note that $\tau(\pi)$ and $v(\pi)$ depend only on π and not on the representative chosen). For $w \in W$, let

$$I_\lambda(w) = \{\pi \in I_\lambda \mid w \geq \tau(\pi)\}.$$

A 5. Conjecture.

$$\text{Char } V_w = \sum_{\pi \in I_\lambda(w)} e^{v(\pi)}$$

Part II: An A-basis for V_A compatible with $\{V_{w,A}, w \in W\}$

Let $\pi, (\underline{c}, \underline{n})$ etc. be as in Part I. To $(\underline{c}, \underline{n})$ there corresponds a canonical (not necessarily admissible) weighted chain $(\delta(\underline{c}), \delta(\underline{n}))$ with $\delta(\underline{c})$ *simple* (cf [La], 3.8). Let $\delta(\underline{c}) = \{\theta = \tau_o, \cdots, \tau_r\}, \underline{n} = \{n_1, \cdots, n_r\}, \beta_t = \alpha_{i_t}, 1 \leq t \leq r$ (note that β_t's are simple). We set

$$v_{\underline{c},\underline{n}} = F_{i_r}^{(n_r)} \cdots F_{i_1}^{(n_1)} e_\theta$$

A 6. Conjecture. For each $\pi \in I_\lambda$, choose a representative $(\underline{c}, \underline{n})$ for π. Then $\{v_{\underline{c},\underline{n}} : w \geq \tau(\pi)\}$ is an A-basis for $V_{w,A}$.

In [Li], Littelmann proves Conjecture 1, and as a consequence gives a Littlewood-Richardshon type "decomposition rule" for a symmetrizable Kac-Moody lie algebra g, and a "restriction rule" for a Levi subalgebra L of g which we state below.

Let θ be a dominant integral weight and let $\pi \in I_\theta$. Let $(\underline{c}, \underline{n})$ be a representative of π and let $D_{\underline{c},\underline{n}}$ be as above. Let us denote

$$p(\pi,\theta) = \{\sum_{k=t}^{s} (a_k - a_{k+l})\mu_{i_k}(\theta), 0 \leq t \leq s\}$$

A 7. Definition. Let λ, θ be two dominant, integral weights. Let $\pi \in I_\theta$. Then π is said to be λ-dominant if $\lambda + p(\pi, \theta)$ is contained in the dominant Weyl chamber.

A 8. Difinition. Let L be a Levi subalgebra of g, and let $\pi \in I_\lambda$. Then π is said to be *L-dominant* if $p(\pi, \lambda)$ is contained in the dominant Weyl chamber .of L.

Decomposition rule. ([Li]) Let λ, μ be two dominant integral weights. Let $I(\lambda, \mu) = \{\pi \in I_\mu \mid \pi \text{ is } \lambda\text{-dominant}\}$. Then

$$V_\lambda \otimes V_\mu = \bigoplus_{\pi \in I(\lambda,\mu)} V_{\lambda+v(\pi)}$$

Restriction rule. ([Li]) Let L be a Levi subalgebra of g. Let $I(\lambda, L) = \{\pi \in I_\lambda : \pi \text{ is L-dominant}\}$. Then

$$\text{res}_L\, V_\lambda = \bigoplus_{\pi \in I(\lambda,L)} U_{v(\pi)}$$

(here, for an integral weight θ contained in the dominant Weyl chamber of L, U_θ denotes the corresponding simple highest weight module of L)

In [Li], Littelmann introduces operators e_α, f_α on I_λ, (for α simple), and associates an oriented, colored (by the simple roots) graph $G(V_\lambda)$ with I_λ as the set of vertices, and $\pi \xrightarrow{\alpha} \pi'$ if $\pi' = f_\alpha(\pi)$. He conjectures that $G(V_\lambda)$ is the crystal graph of V_λ as constructed by Kashiwara ([K]).

Using the decomposition rule, Littelmann gives in [Li] a new (and simple) proof of the Parthasarathy-Ranga Rao- Varadarajan conjecture.

References

[D] V.G. Drinfeld, *Hopf algebras and the Yang-Baxter equation,* Soviet Math. Dokl. **32** (1985) 254–258.

[J] M. Jimbo, *A q-difference analogue of $U(g)$ and the Yang-Baxter equation,* Lett. Math. Phys. **10** (1985) 63–69.

[K] M. Kashiwara, *Crystalizing the q-analogue of Universal Enveloping Algebras,* Commun. Math. Phys. **133** (1990) 249–260.

[La] V. Lakshmibai, *Bases for Quantum Demazure modules,* (The same volume).

[Li] P. Littelmann, *A Littlewood-Richardson rule for symmetrizable Kac-Moody algebras* (preprint).

[LS] V. Lakshmibai and C.S. Seshadri, *Standard Monomial Theory,* "Proceedings of Hyderbad conference on Algebraic Groups", Manoj Prakashan, Madras (1991) 279–323.

Northeastern University
Mathematics Department
Boston, MA 02115, U.S.A.

Moduli Spaces of Abelian Surfaces with Isogeny*

Ch. Birkenhake and H. Lange

To M. S. Narasimhan and C. S. Seshadri on the occasion of their 60th birthdays

Let (X, L) be a polarized abelian surface or type $(1, n)$. An isogeny of type $(1, n)$ is an isogeny of polarized abelian surfaces $\pi: (X, L) \to (Y, P)$ such that P defines a principal polarization on Y. According to [H-W] the coarse moduli space $\mathcal{A}_{(1,n)}$ of such triplets (X, L, π) exists and is analytically isomorphic to the quotient of the Siegel upper half space of degree 2 by the action of $\Gamma = \{M \in \mathrm{Sp}_4(Z) : M = (m_{ij}) \text{ with } n|m_{i4}, i = 1, 2, 3\}$. $\mathcal{A}_{(1,n)}$ is a finite covering of the moduli space of principally polarized abelian surfaces as well as of the moduli space of polarized abelian surfaces of type $(1, n)$. On the other hand, the moduli space of polarized abelian surfaces with level n-structure is a finite covering of $\mathcal{A}_{(1,n)}$. If for example n is a prime, the degrees of these coverings are $(n+1)(n^2+1)$, $(n+1)$ and $n(n-1)$ respectively

The aim of the present paper is to give explicit algebraic descriptions of the moduli spaces $\mathcal{A}_{(1,2)}$ (see Theorem 3.1) and $\mathcal{A}_{(1,3)}$ (see Theorem 6.1). An immediate consequence is that the moduli spaces $\mathcal{A}_{(1,2)}$ and $\mathcal{A}_{(1,3)}$ are rational.

An essential ingredient of the proof is the fact that the moduli space $\mathcal{A}_{(1,n)}$ is canonically isomorphic to the moduli space $\mathcal{C}_2^n$ of cyclic étale n-fold coverings of curves of genus 2. This will be shown in Section 1. The second important tool is the fact that the composition of every $C \to H$ in $\mathcal{C}_2^n$ with the hyperelliptic covering $H \to P_1$ is Galois with the dihedral group D_n as Galois group (see Section 2). Finally we need some results on duality of polarizations on abelian surfaces which we compile in Section 4.

We would like to thank W. Barth and W.-D. Geyer for some valuable discussions.

*Supported by DFG-contract La 318/4 and EC-contract SC1-0398-C(A)

1 Abelian Surfaces with an Isogeny of Type $(1, n)$

In this section we show that there is a canonical isomorphism between the moduli space of polarized abelian surfaces with isogeny of type $(1, n)$ and the moduli space of cyclic étale n-fold coverings of curves of genus two.

Let X be an abelian surface over the field of complex numbers. Any ample line bundle L on X defines a polarization on X. In the notation we do not distinguish between the line bundle L and the corresponding polarization. Denote by $\widehat{X} = \mathrm{Pic}^0(X)$ the dual abelian variety. The polarization L determines an isogeny

$$\phi_L: X \to \widehat{X}\ , \quad x \mapsto t_x^* L \otimes L^{-1}$$

where $t_x: X \to X$ is the translation map $y \mapsto y + x$. The kernel $K(L)$ of φ_L is isomorphic to $(Z/n_1Z \times Z/n_2Z)^2$ for some positive integers n_1, n_2 with $n_1 \mid n_2$. We call (n_1, n_2) the *type of the polarization*. Any polarization of type (n_1, n_2) is the n_1-th power of a unique polarization of type $(1, \frac{n_2}{n_1})$. Hence for moduli problems it suffices to consider polarizations of type $(1, n)$.

From now on let L be a line bundle defining a polarization of type $(1, n)$. An *isogeny of type* $(1, n)$ is by definition an isogeny of polarized abelian varieties $\pi: (X, L) \to (Y, P)$ whose kernel is cyclic of order n. Necessarily P defines a principal polarization on Y and $\ker p$ is contained in $K(L)$. Conversely, according to [L-B] Cor. 6.3.5 any cyclic subgroup of $K(L)$ of order n defines an isogeny of type $(1, n)$ of (X, L). In particular, if n is a prime number, then (X, L) admits exactly $n + 1$ isogenies of type $(1, n)$. According to [L-B] Exercise 8.4 the moduli space $\mathcal{A}_{(1,n)}$ of polarized abelian surfaces with isogeny of type $(1, n)$ exists and is analytically isomorphic to the quotient of the Siegel upper half space h_2 of degree 2 by the group $\{M \in \mathrm{Sp}_4(Z) \mid M = (m_{ij}) \text{ with } n|m_{i4},\ i = 1, 2, 3\}$.

In the sequel a curve of genus two means either a smooth projective curve of genus 2 or a union of two elliptic curves intersecting transversally at the origin. Note that such a union $E_1 + E_2$ is of arithmetic genus 2. Torelli's Theorem implies that the moduli space of principally polarized abelian surfaces can be considered as a moduli space for curves of genus two in this sense.

Let $f: C \to H$ be a cyclic étale covering of degree n of a curve H of genus 2. According to Hurwitz' formula C has arithmetic genus $n+1$. Every line bundle $l \in \mathrm{Pic}^0(H)$ of order n determines such a cyclic étale covering $f: C \to H$ (for an explicit description of the covering see Section 2). Two such line bundles lead to the same covering, if they generate the same group in $\mathrm{Pic}^0(H)$. This implies that the (coarse) moduli space $\mathcal{C}_2^n$ of cyclic étale n-fold coverings of curves of genus two is a finite covering of the moduli space $\mathcal{M}_2$ of curves of genus two. In particular $\mathcal{C}_2^n$ is an algebraic variety of dimension 3. The moduli spaces $\mathcal{A}_{(1,n)}$ and $\mathcal{C}_2^n$ are related as follows.

1.1 Proposition. *There is a canonical biholomorphic map* $\mathcal{A}_{(1,n)} \to \mathcal{C}_2^n$

There seems to be no explicit construction of the moduli space $\mathcal{C}_2^n$ in the literature. One could also interprete Proposition 1.1 as a construction of $\mathcal{C}_2^n$. However it is not difficult to show its existence in a different way and thus the proposition makes sense as stated.

Proof. Step I: The map $\mathcal{A}_{(1,n)} \to \mathcal{C}_2^n$. Let $\pi: (X, L) \twoheadrightarrow (Y, P)$ be an isogeny of type $(1, n)$. We may assume that $\pi^* P \simeq L$ as line bundles. Since (Y, P) is a principally polarized abelian surface there is a curve H of genus 2 (in the above sense) such that $Y = J(H)$, the Jacobian of H, and $P \simeq \mathcal{O}_Y(H)$. Note that for $H = E_1 + E_2$ with elliptic curves E_1 and E_2, $J(H) = \mathrm{Pic}^0(H) \simeq E_1 \times E_2$. By assumption $C := \pi^{-1} H \in |L|$. The étale covering $\pi: X \to Y$ is given by a line bundle $l \in \mathrm{Pic}^0(Y)$ of order n and the covering $\pi|C: C \to H$ corresponds to $l|H$. Since the restriction map $\mathrm{Pic}^0(Y) \xrightarrow{\sim} \mathrm{Pic}^0(H)$ is an isomorphism, the line bundle $l|H$ is of order n and thus $\pi|C: C \to H$ is an element of $\mathcal{C}_2^n$.

Step II: The inverse map $\mathcal{A}_{(1,n)} \to \mathcal{C}_2^n$. Let $f: C \to H$ be a cyclic étale covering in $\mathcal{C}_2^n$ associated to the line bundle $l_H \in \mathrm{Pic}^0(H)$. Via the isomorphism $\mathrm{Pic}^0(J(H)) \xrightarrow{\sim} \mathrm{Pic}^0(H)$ the line bundle l_H extends to a line bundle $l \in \mathrm{Pic}^0(J(H))$ of order n. Let $\pi: X \to Y = J(H)$ denote the cyclic étale n-fold covering associated to l. Then $L = \pi^* \mathcal{O}_Y(H)$ defines a polarization of type $(1, n)$, since $K(L)$ is a finite group of order n^2 (by Riemann-Roch) and contains the cyclic group $\ker \pi$ of order n. Hence $\pi: (X, L) \to (Y, \mathcal{O}_Y(H))$ is an element of $\mathcal{A}_{(1,n)}$.

Obviously the maps $\mathcal{A}_{(1,n)} \to \mathcal{C}_2^n$ and $\mathcal{C}_2^n \to \mathcal{A}_{(1,n)}$ are inverse to each other. Finally, extending the above construction to families of morphisms of curves and abelian varieties one easily sees that the maps are holomorphic. □

2 Cyclic Étale Coverings of Hyperelliptic Curves

Any curve H of genus 2 (in the sense of section 1) admits a natural involution ι with quotient H/ι of arithmetic genus 0. The aim of this section is to show that for any finite cyclic étale covering $f: C \to H$ the composition $C \to H \to H/\iota$ is Galois and to compute its Galois group. We prove the result in greater generality than actually needed, since this makes no difference for the proof.

In this section a *hyperelliptic curve* means a complete, reduced, connected curve admitting an involution whose quotient is of arithmetic genus zero. Let H denote a hyperelliptic curve of arithmetic genus g over k with hyperelliptic covering $H \to P$. Suppose $f: C \to H$ is a cyclic étale covering of degree $n \geq 2$. We first show that the composed map $C \to P$ is a Galois covering with the dihedral group D_n of order $2n$ as Galois group.

Let $\iota: H \to H$ denote the hyperelliptic involution and $\tau: C \to C$ an automorphism generating the group $\mathrm{Gal}(C|H)$. There is a line bundle $L \in \mathrm{Pic}^0(H)$ with $L^n \simeq \mathcal{O}_H$ such that $C = \mathrm{Spec}(A)$ with $A := \mathcal{O}_H \oplus L \oplus \cdots \oplus L^{n-1}$ and where the $\mathcal{O}_H$-algebra structure of A is given by an isomorphism $\sigma: \mathcal{O}_H \xrightarrow{\sim} L^n$. Consider the pull back diagram

$$\begin{array}{ccc} \mathbf{Spec}(\subset^* A) = \subset^* C & \xrightarrow{j} & C = \mathbf{Spec}\, A \\ \downarrow & & \downarrow f \\ H & \xrightarrow{\subset} & H \end{array}$$

Since $\iota^* L = L^{-1}$, the isomorphism σ induces isomorphisms

$$\sigma_\nu = (\sigma \otimes 1_{L^{-\nu}}) \circ \iota^*: L^\nu \to L^{n-\nu}$$

for $\nu = 0, \ldots, n-1$ which yields an $\mathcal{O}_H$-algebra isomorphism $A \to \iota^* A$. Hence we may identify $\iota^* C = C$ and $j: C \to C$ is an automorphism.

2.1 Proposition. *The covering $C \to P$ is Galois with* $\mathrm{Gal}(C|P_1) = D_n$

Proof. It suffices to show $j\tau j = \tau^{-1}$. According to [EGAI, Th. 9.1.4] the automorphism τ of $C = \mathrm{Spec} A$ corresponds to an $\mathcal{O}_H$-algebra automorphism $\widetilde{\tau}: A \to A$, namely

$$\widetilde{\tau}(a_0, a_1, \ldots, a_{n-1}) = (a_0, \xi a_1, \ldots, \xi^{n-1} a_{n-1}) \;\; \textit{with } \xi = \exp(\frac{2\pi i}{n}).$$

Similarly using [EGAI, Cor. 9.1.9] the automorphism j of C corresponds to the algebra automorphism $\tilde{\jmath}: A \to A$ over ι^* defined by

$$\tilde{\jmath}(a_0, a_1, \ldots, a_{n-1}) = (\sigma_0(a_0), \sigma_{n-1}(a_{n-1}), \ldots, \sigma_1(a_1)) \; .$$

We have to show that $\tilde{\jmath}\widetilde{\tau}, \tilde{\jmath} = \widetilde{\tau}^{-1}$. But $\sigma_{n-\nu}\xi^{n-\nu}\sigma_\nu(a_\nu) = \xi^{-\nu} a_\nu$. This implies the assertion. □

The dihedral group D_n contains the involutions $j\tau^\nu$ for $\nu = 0,\dots,n-1$ and for even n also $\tau^{\frac{n}{2}}$. These involutions correspond to double coverings $C \to C_\nu = C/j\tau^\nu$ for $\nu = 0,\dots,n-1$ and $C \to C' = C/\tau^{\frac{n}{2}}$ for even n. If C is smooth and irreducible we have for the genera g_{C_ν} and $g_{C'}$ of C_ν and C'

2.2 Proposition.

a) For n odd: $g_{C_\nu} = \frac{1}{2}(g-1)(n-1)$ *for* $\nu = 0,\dots,n-1$.

b) For n even: $(\frac{n}{2}-1)(g-1) \le g_\nu \le \frac{1}{2}n(g-1)$ *for all* $0 \le \nu \le n-1$ *and* $g_{C'} = \frac{n}{2}+1$.

The proof is an application of the formula of Chevalley-Weil (see [C-W]). We omit the details. The genus of C' can be computed by Hurwitz' formula, since $C \to C'$ is étale.

2.3 Remark. Let $H = E_1 + E_2$ be a reducible curve of genus two as in Section 1. The curve H is hyperelliptic with hyperelliptic involution ι the multiplication by -1 on each curve E_i. The quotient $P = H/\iota$ consists of two copies of P_1 intersecting in one point. In this situation Proposition 2.1 can be seen also in the following way.

If for example the covering $C \to H$ is nontrivial on each component E_i, then C consists of two elliptic curves F_1 and F_2 intersecting in n points. We choose one of these points to be the origin of F_1 and F_2, the remaining intersection points are $x,\dots,(n-1)x$ for some n-division point x on F_1 and F_2. The automorphism $\tau\colon C \to C$ defined as the translation t_x by x on each F_i generates the group of covering transformations of $C \to H$. The involution ι on H lifts to an involution j on C, the multiplication by (-1) on each F_i. Obviously $j\tau j = \tau^{-1}$, so $C \to H/\iota$ is a Galois covering with Galois group $D_n = < j,\tau >$. As in the irreducible case we consider the double coverings $C \to C_\nu = C/j\tau^\nu$ for $\nu = 0,\dots,n-1$ and $C \to C' = C/\tau^{\frac{n}{2}}$ for even n. Also here the result of Proposition 2.2 is valid: for example, if n is odd and the covering $C \to H$ is nontrivial on each component, then C_ν consists of two copies of P_1 intersecting in $\frac{n+1}{2}$ points, the images of kx for $k = 0,\dots,\frac{n-1}{2}$. In particular C_ν has arithmetical genus $\frac{n-1}{2}$. The other cases can be worked out in a similar way.

3 The Moduli Space $\mathcal{A}^0_{(1,2)}$

Denote by $\mathcal{A}^0_{(1,n)}$ the open set in $\mathcal{A}_{(1,n)}$ corresponding to abelian surfaces of type $(1,n)$ with an isogeny onto a Jacobian of a smooth curve of genus 2.

The aim of this section is to give a description of the moduli space $\mathcal{A}^0_{(1,2)}$. From this it is easy to see that $\mathcal{A}_{(1,2)}$ is rational.

Let $\widetilde{\mathcal{A}}_1$ denote the moduli space of elliptic curves E together with a set of four points of E of the form $\{\pm p_1, \pm p_2\}$. Necessarily such a set does not contain any 2-division point of E. We write the elements of $\widetilde{\mathcal{A}}_1$ as pairs $(E, \{\pm p_1, \pm p_2\})$. The main result of this section is

3.1 Theorem. *The moduli space $\mathcal{A}^0_{(1,2)}$ of polarized abelian surfaces with an isogeny of type* $(1,2)$ *onto a Jacobian of a smooth curve of genus* 2 *is canonically isomorphic to $\widetilde{\mathcal{A}}_1$.*

3.2 Corollary. *The moduli space $\mathcal{A}_{(1,2)}$ is rational.*

Proof of Corollary 3.2. Via the j-invariant, $U := C - \{0, 1728\}$ is the moduli space of elliptic curves without nontrivial automorphisms. U admits a universal elliptic curve $p: \mathcal{E} \to U$. Consider the quotient $\bar{p}: \mathcal{E}/(-1) \to U$ by the action of (-1) on every fibre. An open set of $\widetilde{\mathcal{A}}_1$ can be identified with an open set of the relative symmetric product $S^2_{\bar{p}}(\mathcal{E}/(-1))$ over U. Every fibre of $S^2_{\bar{p}}(\mathcal{E}/(-1)) \to U$ is isomorphic to P_2, so $S^2_{\bar{p}}(\mathcal{E}/(-1))$ is a P_2-bundle over U. According to [G] Corollaire 1.2 the Brauer group of U is zero. Hence $S^2_{\bar{p}}(\mathcal{E}/(-1))$ is the projectivization of a vector bundle on U and thus it is rational. □

For the proof of Theorem 3.1 we first describe the map $\mathcal{A}^0_{(1,2)} \to \widetilde{\mathcal{A}}_1$. Let (X, L, π) be an element in $\mathcal{A}^0_{(1,2)}$ and $f: C \to H$ the corresponding étale double covering of a curve H of genus 2 according to Proposition 1.1. As we saw in the last section the automorphism group of C contains the group D_2. As above denote by $\tau \in D_2$ the involution corresponding to the covering $C \to H$ and by $j \in D_2$ a lifting of the hyperelliptic involution on H. Either from the proof of Proposition 2.2 or by considering the ramification points of the 4-fold covering $C \to P_1$ one easily sees that the genera of the curves C/j and $C/j\tau$ are 0 and 1. By eventually interchanging the roles of j and $j\tau$ we may assume that $E = C/j$ is an elliptic curve and $C/j\tau = P_1$. In particular the curve C is hyperelliptic and we have a commutative diagram

$$\begin{array}{ccc} C & \longrightarrow & E \\ {\scriptstyle f}\downarrow & & \downarrow \\ H & \longrightarrow & P_1 \end{array}$$

We can choose the origin in E in such a way that (-1) is the involution on E corresponding to the covering $E \to P_1$, so that the ramification points of $E \to P_1$ are the 2-division points of E. From the commutative diagram we see that the 4 ramification points $p_1, \ldots, p_4 \in E$ of the covering $C \to E$ are different from the 2-division points of E. Since the involutions j and τ commute and τ is a lifting of (-1) on E, the involution (-1) acts on the set $\{p_1, \ldots, p_4\}$. Hence we may assume that $p_3 = -p_1$ and $p_4 = -p_2$. Now define the map

$$\psi: \mathcal{A}^0_{(1,2)} \to \widetilde{\mathcal{A}}_1\,, \quad (X, L, \pi) \mapsto (E, \{\pm p_1, \pm p_2\})\,.$$

Since E and the set $\{\pm p_1, \pm p_2\}$ can be given via algebraic equations out of the covering $C \to H$, the map ψ is holomorphic and it remains to show that it admits an inverse.

Let $(E, \{\pm p_1, \pm p_2\}) \in \widetilde{\mathcal{A}}_1$. Note that E admits exactly four double coverings ramified in $\pm p_1$ and $\pm p_2$, since the line bundle $\mathcal{O}_E(p_1+p_2+(-p_1)+(-p_2))$ admits exactly 4 square roots in $\mathrm{Pic}^2(E)$. They can be given as follows: Let E be given by the equation $y^2 = x(x-1)(x-a)$ and choose as usual the origin to be the flex at infinity. Then the nontrivial 2-division points of E are $(x, y) = (0,0), (1,0), (a,0)$. Write $p_i = (x_i, y_i)$ for $i = 1, 2$ and consider the double coverings $D_i \to P_1$, $i = 0, \ldots, 3$, defined by the equations

$$\begin{aligned} y_0^2 &= x(x-1)(x-a)(x-x_1)(x-x_2) \\ y_1^2 &= x(x-x_1)(x-x_2) \\ y_2^2 &= (x-1)(x-x_1)(x-x_2) \\ y_3^2 &= (x-a)(x-x_1)(x-x_2)\,. \end{aligned}$$

Finally denote by C_i the curve corresponding to the compositum of the function fields of E and D_i. Then we have the following commutative diagram

$$\begin{array}{ccc} C_i & \longrightarrow & D_i \\ \downarrow & & \downarrow \\ E & \longrightarrow & P_1 \end{array}$$

According to Abhyankar's lemma $C_i \to E$ is not ramified over the 2-division points of E, hence $C_0, \ldots, C_3$ are exactly the four double coverings of E ramified in $\pm p_1 = (x_1, \pm y_1)$ and $\pm p_2 = (x_2, \pm y_2)$. Moreover D_0 is of genus 2 and D_1, D_2, D_3 are of genus 1 and $C_i|P_1$ is Galois with $\mathrm{Gal}(C_i|P_1) = D_2 =< j_i, \tau_i >$ where j_i and τ_i are the involutions corresponding to $C_i \to E$ and $C_i \to D_i$ respectively. The third involution in $\mathrm{Gal}(C_i|P_1)$ is $j_i\tau_i$. The corresponding curves $D'_i := C_i/j_i\tau_i$ are given by the equations respectively

$$\begin{aligned} z_0^2 &= (x-x_1)(x-x_2) \\ z_1^2 &= (x-1)(x-a)(x-x_1)(x-x_2) \\ z_2^2 &= x(x-a)(x-x_1)(x-x_2) \\ z_3^2 &= x(x-1)(x-x_1)(x-x_2) \end{aligned}$$

Hence C_0 is the only covering of E ramified in $\pm p_1$ and $\pm p_2$ admitting an étale double covering of a curve of genus 2 in this way. So the data $(E, \{\pm p_1, \pm p_2\})$ determine uniquely an element of $\mathcal{C}_2^2$, namely $C_0 \to D_0$. Let (X, L, π) denote the corresponding element of $\mathcal{A}^0_{(1,2)}$ and define a map

$$\varphi\colon \widetilde{\mathcal{A}}_1 \to \mathcal{A}^0_{(1,2)}\,, \quad (E, \{\pm p_1, \pm p_2\}) \mapsto (X, L, \pi)\,.$$

Obviously φ is holomorphic and inverse to ψ. This completes the proof of Theorem 3.1. □

The above proof easily gives another description of the moduli space $\mathcal{A}^0_{(1,2)}$. Let $\mathcal{H}_3(D_2)$ denote the moduli space of isomorphism classes of curves of genus three given by the following equation

$$y^2 = (x^2-1)(x^2-\alpha)(x^2-\beta)(x^2-\gamma) \tag{1}$$

with pairwise different $\alpha, \beta, \gamma \in C^* - \{1\}$. Every curve in $\mathcal{H}_3(D_2)$ is hyperelliptic and its automorphism group contains $D_2 = \{x \mapsto \pm x, y \mapsto \pm y\}$ which explains the notation.

3.3 Proposition. *There is a canonical isomorphism $\mathcal{A}^0_{(1,2)} \simeq \mathcal{H}_3(D_2)$.*

Proof. Let $(X, L, \pi) \in \mathcal{A}^0_{(1,2)}$ and $C \to H$ be the associated étale double covering in $\mathcal{C}_2^2$. As we saw in the proof above the curve C is hyperelliptic. Moreover $\mathrm{Aut}(C)$ contains D_2 according to Proposition 2.1. It is well known (see e. g. [I]) that every hyperelliptic curve C of genus three with $D_2 \subset \mathrm{Aut}(C)$ admits an equation of the form (1). Hence the assignment $(X, L, p) \mapsto C$ gives a holomorphic map $\mathcal{A}^0_{(1,2)} \to \mathcal{H}_3(D_2)$.

For the inverse map suppose $C \in \mathcal{H}_3(D_2)$ is given by an equation (1). The involution $(x, y) \mapsto (-x, -y)$ induces the double covering $C \to H$, where H is given by the equation $v^2 = u(u-1)(u-\alpha)(u-\beta)(u-\gamma)$. It is easy to see that $C \to H$ is an element of $\mathcal{C}_2^2$ and the assignment $C \mapsto \{C \to H\}$ defines a holomorphic map $\mathcal{H}_3(D_2) \to \mathcal{C}_2^2 \simeq \mathcal{A}^0_{(1,2)}$ which is inverse to the map $\mathcal{A}^0_{(1,2)} \to \mathcal{H}_3(D_2)$ given above. □

3.4 Remark. Let $U = \{(\alpha, \beta, \gamma) \in (C^* - \{1\})^3 : \alpha \neq \beta \neq \gamma \neq \alpha\}$. The moduli space $\mathcal{H}_3(D_2)$ is birational to the quotient of U by a (nonlinear) action of the group $Z_2 \times \mathbf{S}_4$. As a consequence of Corollary 3.2 the quotient $U/Z_2 \times \mathbf{S}_4$ is rational, which seems not to be known from Invariant Theory. □

4 Remarks on Duality on Polarized Abelian Surfaces

In this section we introduce the dual of a polarization of an abelian surface and compile some of its properties needed in the next section. The results easily generalize to abelian varieties of arbitrary dimension.

Let (X, L) be a polarized abelian surface of type $(1, d)$. Recall that the polarization L induces an isogeny from X onto its dual $\varphi_L: X \to \widehat{X}$, $x \mapsto t_x^* L \otimes L^{-1}$. Its kernel $K(L)$ is isomorphic to the group $Z/dZ \times Z/dZ$.

4.1 Proposition. *There is a unique polarization $\widehat{L}$ on $\widehat{X}$ characterized by the following two equivalent properties:*

$$i)\ \varphi_L^* \widehat{L} \equiv L^d \quad \text{and} \quad ii)\ \varphi_{\widehat{L}} \varphi_L = d \cdot 1_X$$

The polarization $\widehat{L}$ is also of type $(1, d)$.

Proof. The equivalence i) $\iff$ ii) follows from the equation $\varphi_{\varphi_L^* \widehat{L}} = \widehat{\varphi}_L \varphi_{\widehat{L}} \varphi_L$, since the polarization L and the isogeny φ_L determine each other and $\widehat{\varphi}_L = \varphi_L$ (see [L-B] Section 2.4). The uniqueness of $\widehat{L}$ follows from ii) and again, since $\widehat{L}$ and $\varphi_{\widehat{L}}$ determine each other.

For the existence of $\widehat{L}$ note that φ_L^{-1} exists in $\mathrm{Hom}(\widehat{X}, X) \otimes Q$ since φ_L is an isogeny. By [L-B] Proposition 1.2.6 $\psi = d\varphi_L^{-1}: \widehat{X} \to X$ is an isogeny. We have

$$\varphi_{\psi^* L} = \widehat{\psi} \varphi_L \psi = \widehat{\psi} d = d\psi \ . \tag{1}$$

According to [L-B] Lemma 2.5.6 there exists a polarization $\widehat{L} \in \mathrm{Pic}(\widehat{X})$ such that $\widehat{L}^d \equiv \psi^* L$ and hence

$$\varphi_{\psi^* L} = \varphi_{\widehat{L}^d} = d\varphi_{\widehat{L}} \ .$$

Together with (1) this implies $\psi = \varphi_{\widehat{L}}$ and thus $\varphi_{\widehat{L}} \varphi_L = d \cdot 1_X$. Moreover ii) implies that $\widehat{L}$ is of type $(1, d)$. □

In the next section we need the following example of a pair of dual polarizations.

4.2 Example. Let E be an elliptic curve and Ξ the polarization on $E \times E$ defined by the divisor $E \times \{0\} + \{0\} \times E + \Delta$, where Δ denotes the diagonal in $E \times E$. If we identify as usual $E = \widehat{E}$ via $\varphi_{\mathcal{O}_E(0)}$, then we have for the dual polarization $\widehat{\Xi}$ on $E \times E$

$$\widehat{\Xi} = \mathcal{O}_{E\times E}(E \times \{0\} + \{0\} \times E + A),$$

where A denotes the antidiagonal in $E \times E$. To see this note that Ξ can be written as $\Xi = p_1^*\mathcal{O}_E(0) + p_2^*\mathcal{O}_E(0) + \alpha^*\mathcal{O}_E(0)$ where $p_i: E \times E \to E$ are the projections and $\alpha: E \times E \to E$ is the difference map $\alpha(x, y) = x - y$. Hence we have for $\varphi_\Xi: E \times E \to E \times E$

$$\begin{aligned}\varphi_\Xi &= \widehat{p_1}\varphi_{\mathcal{O}_E(0)}p_1 + \widehat{p_2}\varphi_{\mathcal{O}_E(0)}p_2 + \widehat{\alpha}\varphi_{\mathcal{O}_E(0)}\alpha \\ &= \begin{pmatrix}1 & 0\\ 0 & 0\end{pmatrix} + \begin{pmatrix}0 & 0\\ 0 & 1\end{pmatrix} + \begin{pmatrix}1 & -1\\ -1 & 1\end{pmatrix} = \begin{pmatrix}2 & -1\\ -1 & 2\end{pmatrix}\end{aligned}$$

Similarly, if Ψ denotes the polarization defined by the divisor $E \times \{0\} + \{0\} \times E + A$, then $\varphi_\Psi = \begin{pmatrix}2 & 1\\ 1 & 2\end{pmatrix}$. This implies

$$\varphi_\Psi\varphi_\Xi = \begin{pmatrix}2 & 1\\ 1 & 2\end{pmatrix}\begin{pmatrix}2 & -1\\ -1 & 2\end{pmatrix} = 3 \cdot 1_{E\times E}.$$

Since both polarizations are of type $(1, 3)$, Proposition 4.1 gives $\Psi = \widehat{\Xi}$.

Let C be a smooth projective curve and (J, Θ) its canonically principally polarized Jacobian variety.

4.3 Proposition. *For a morphism $\varphi: C \to X$ the following statements are equivalent*

i) $(\varphi^*)^*\Theta \equiv \widehat{L}$

ii) $\varphi_*[C] = [L]$ *in* $H^2(X, Z)$.

Both conditions imply that φ is birational onto its image.

Here $[C]$ denotes the fundamental class of C in $H^2(C, Z)$. Similarly $[L]$ denotes the first Chern class of L in $H^2(X, Z)$.

Proof. Identify $J = \widehat{J}$ via φ_Θ. Condition i) is equivalent to

$$\varphi_{\widehat{L}} = \varphi_{(\varphi^*)^*\Theta} = \widehat{\varphi^*}\varphi^*$$

By the Universal Property of the Jacobian φ extends to a homomorphism from $J(C)$ to X also denoted by φ. According to [L-B] Corollary 11.4.2 the homomorphisms $\varphi^*\colon \widehat{X} \to J(C)$ and $\varphi\colon J(C) \to X$ are related by $\widehat{\varphi} = -\varphi^*$. Hence $\varphi_{\widehat{L}} = \varphi\widehat{\varphi}$.

Let $\delta(\varphi(C), L)$ and $\delta(L, L)$ denote the endomorphisms of X associated to the pairs $(\varphi(C), L)$ and (L, L) induced by the intersection product (see [L-B] Section 5.4). Applying [L-B] Propositions 11.6.1 and 5.4.7 condition i) is equivalent to

$$\delta(\varphi(C), L) = -\varphi\widehat{\varphi}\varphi_L = -\varphi_{\widehat{L}}\varphi_L = -d \cdot 1_X = -\frac{(L^2)}{2} 1_X = \delta(L, L) .$$

According to [L-B] Theorem 11.6.4 this is equivalent to $[\varphi(C)] = [L]$. Since L is of type $(1, d)$ and hence primitive, i) as well as ii) imply that φ is birational onto its image. Hence $\varphi_*[C] = [\varphi(C)]$. □

5 Abelian Surfaces of Type $(1, 3)$

Recall that $\mathcal{A}^0_{(1,3)} \subset \mathcal{A}_{(1,3)}$ is the open subset corresponding to abelian surfaces X of type $(1, 3)$ with an isogeny onto a Jacobian of a smooth curve of genus 2. In this section we derive some properties of the elements of $\mathcal{A}^0_{(1,3)}$.

Let $\pi\colon (X, L) \to (Y, P) \simeq (J(H), \mathcal{O}(H))$ be an element of $\mathcal{A}^0_{(1,3)}$ corresponding to the cyclic étale 3-fold covering $f\colon C \to H$ of a smooth curve H of genus 2 (see Proposition 1.1). According to Proposition 2.1 the Galois group of the composed covering $C \to H \to P_1$ is the dihedral group D_3 generated by an involution $j\colon C \to C$ over the hyperelliptic involution ι of H and a covering transformation τ of $f\colon C \to H$. According to Proposition 2.2 the involutions j, $j\tau$, $j\tau^2$ are elliptic. Denote by $f_\nu\colon C \to E_\nu := C/j\tau^\nu$ the corresponding coverings. The automorphisms j and τ of C extend to automorphisms of the Jacobian $J(C)$ which we also denote by j and τ. For any point $c \in C$ we have an embedding

$$\alpha_c\colon C \to J(C) , \quad p \mapsto \mathcal{O}_C(p - c) .$$

Since the double coverings $f_\nu\colon C \to E_\nu$ are ramified, the pull back homomorphism $E_\nu = \mathrm{Pic}^0(E_\nu) \to \mathrm{Pic}^0(C) = J(C)$ is an embedding (see [L-B] Proposition 11.4.3). We always consider the elliptic curves E_ν as abelian subvarieties of $J(C)$. Then the following diagram commutes

$$\begin{array}{ccc} C & \overset{\alpha_c}{\hookrightarrow} & J(C) \\ & {\scriptstyle f_\nu}\searrow \quad \swarrow {\scriptstyle (1 + j\tau^\nu)} & \\ & E_\nu & \end{array} \qquad (1)$$

for $\nu = 0,1,2$ and any $c \in C$. Since $\tau(1 + j\tau^\nu) = (1 + j\tau^{\nu+1})\tau$, the automorphism τ of $J(C)$ restricts to isomorphisms

$$\tau: E_\nu \to E_{\nu+1}$$

for $\nu \in Z/3Z$.

The curve C is contained in the abelian surface X and generates X as a group, since $L = \mathcal{O}_X(C)$ is ample. So the Universal Property of the Jacobian yields a surjective homomorphism $J(C) \to X$, the kernel of which is described by the following

5.1 Proposition. $0 \longrightarrow E_\nu \times E_\nu \overset{p_1+\tau p_2}{\longrightarrow} J(C) \longrightarrow X \longrightarrow 0$ *is an exact sequence of abelian varieties for* $\nu \in Z/3Z$.

Here $p_i: E_\nu \times E_\nu \to E_\nu$ denotes the i-th projection for $i = 1,2$. For the proof of the proposition we need the following

5.2 Lemma. *For a general* $c \in C$ *either* $h^0(\mathcal{O}_C(2c + jc + j\tau c)) = 1$ *or* $h^0(\mathcal{O}_C(2c + jc + j\tau^2 c)) = 1$.

Proof. According to Castelnuovo's inequality (see [ACGH] Exercise C-1 p. 366) C is not hyperelliptic. We identify C with its image in P_3 under the canonical embedding. Assume $h^0(\mathcal{O}_C(2c + jc + j\tau c)) = h^0(\mathcal{O}_C(2c + jc + j\tau^2 c)) = 2$ for all $c \in C$. Denote $P_c = \operatorname{span}(2c, jc, j\tau c)$ and $P'_c = \operatorname{span}(2c, jc, j\tau^2 c)$ in P_3. According to the Geometric Riemann-Roch Theorem (see [ACGH] p. 12) the assumption is equivalent to

$$\dim P_c = \dim P'_c = 2 .$$

For any $p \in C$ denote by T_pC the tangent of C at p in P_3. Applying the Geometric Riemann-Roch Theorem again, we obtain

$$4 - \dim \operatorname{span}(T_cC, T_{jc}C) = h^0(\mathcal{O}_C(2c + 2jc)) \geq h^0(\mathcal{O}_{E_0}(2\pi_0(c))) = 2 .$$

Since $h^0(\mathcal{O}_C(2c+2jc)) \leq 2$ by Clifford's Theorem, $\dim \operatorname{span}(T_cC, T_{jc}C) = 2$. On the other hand, since a general $c \in C$ is not a ramification point of a trigonal pencil, we have $h^0(\mathcal{O}_C(2c+jc)) = 1$ and thus $\dim \operatorname{span}(T_cC, jc) = 2$. Hence

$$P_c = \operatorname{span}(c, jc, T_cC) = \operatorname{span}(c, jc, T_{jc}C) = P_{jc} .$$

Similarly we obtain $P'_c = P'_{jc}$. Since $\deg C = 6$, this implies

$$P_c \cap C = \{2c, 2jc, j\tau c, \tau^2 c\} \quad \text{and} \quad P'_c \cap C = \{2c, 2jc, j\tau^2 c, \tau c\} .$$

In particular $P_c = \operatorname{span}(T_cC, T_{jc}C) = P'_c$. But then the plane P_c contains more than 6 points of C, a contradiction. □

Proof of the Proposition. It suffices to prove the proposition for $\nu = 0$.

Step I: The map $p_1 + \tau p_2$ is injective. According to Lemma 5.2 we may assume $h^0(2c + jc + j\tau c) = 1$ (if $h^0(2c + jc + j\tau^2 c) = 1$, then we work with $\nu = 2$ instead of $\nu = 0$). We have to determine the points $p, q \in C$ satisfying the equation

$$(1 + j)\alpha_c(p) + \tau(1 + j)\alpha_{j\tau^2 c}(q) = 0 .$$

Here we use the fact that $E_0 = (1 + j)\alpha_c(C) = (1 + j)\alpha_{j\tau^2 c}(C)$ according to diagram (1). Since $h^0(2c + jc + j\tau c) = 1$, the above equation is equivalent to the following identity of divisors on C

$$p + jp + \tau q + \tau jq = 2c + jc + j\tau c .$$

But the only solutions are $(p, q) \in \{(c, \tau^2 c), (c, j\tau^2 c), (jc, \tau^2 c), (jc, j\tau^2 c)\}$, all of which represent the point $(0, 0) \in E_0 \times E_0 = (1+j)\alpha_c(C) \times (1+j)\alpha_{j\tau^2 c}(C)$. Hence $p_1 + \tau p_2$ is an injective homomorphism of abelian varieties.

Step II: The sequence is exact at $J(C)$. The homomorphism $J(C) \to X$ fits into the following commutative diagram

$$\begin{array}{ccc} J(C) & & \\ {\scriptstyle N_f}\downarrow & \searrow & \\ & & X \\ & \nearrow{\scriptstyle g} & \\ J(H) & & \end{array} \qquad (2)$$

where N_f is the divisor norm map associated to $f: C \to H$ and g is an isogeny of degree 3. Since the kernel of N_f consists of 3 connected components, the kernel of $J(C) \to X$ is an abelian surface. Hence it suffices to show that $N_f(p_1 + \tau p_2)(E_\nu \times E_\nu) = 0$. But

$$\begin{aligned} N_f(1 + j\tau^\nu) &= (1 + \tau + \tau^2)(1 + j\tau^\nu) \\ &= (1 + \tau + \tau^2 + j + j\tau + j\tau^2) \end{aligned}$$

is the divisor norm map of the covering $C \to P_1$ and hence is the zero map. □

The Proposition implies that the images of $E_\nu \times E_\nu$ in $J(C)$ coincide for $\nu = 0, 1, 2$. Therefore it suffices to consider the case $\nu = 0$.

The automorphism τ of $J(C)$ is of order 3 and induces the identity on $J(H)$. So by diagram (2) it induces the identity on X. Hence there is an automorphism T of $E_0 \times E_0$ of order 3 fitting into the following commutative diagram

$$\begin{array}{ccccccccc} 0 & \longrightarrow & E_0 \times E_0 & \xrightarrow{p_1 + \tau p_2} & J(C) & \longrightarrow & X & \longrightarrow & 0 \\ & & {\scriptstyle T}\downarrow & & {\scriptstyle \tau}\downarrow & & \| & & \\ 0 & \longrightarrow & E_0 \times E_0 & \xrightarrow{p_1 + \tau p_2} & J(C) & \longrightarrow & X & \longrightarrow & 0 \end{array} .$$

5.3 Lemma. $T = \begin{pmatrix} 0 & -1 \\ 1 & -1 \end{pmatrix}$

Proof. Since τ is a covering transformation of the 3-fold covering $f: C \to H$, it satisfies the equation $\tau^2+\tau+1=0$ on $\mathrm{im}\,(E_0 \times E_0) \subset \ker N_f \subset J(C)$. So in terms of matrices we have $p_1 + \tau p_2 = (1, \tau) = (1, -1-\tau^2)$. An immediate computation shows that $T = \begin{pmatrix} 0 & -1 \\ 1 & -1 \end{pmatrix}$ is the only solution of the equation $(1,\tau)T = \tau(1, -1-\tau^2)$. □

The automorphism T restricts to isomorphisms

$$E_0 \times \{0\} \xrightarrow{T} \{0\} \times E_0 \xrightarrow{T} \Delta \xrightarrow{T} E_0 \times \{0\}$$

where Δ denotes the diagonal in $E_0 \times E_0$. Hence $p_1 + \tau p_2$ maps the curves $E_0 \times \{0\}$, $\{0\} \times E_0$ and Δ onto E_0, E_1 and E_2 respectively.

5.4 Lemma. *The canonical principal polarization Θ on $J(C)$ induces a polarization Ξ of type $(1,3)$ on $E_0 \times E_0$ which is invariant with respect to the action of τ. Moreover $\Xi = [E_0 \times \{0\} + \{0\} \times E_0 + \Delta]$ in $H^2(E_0 \times E_0, Z)$.*

Proof. The dual of the divisor norm map N_f is the pull back map f^* (see [L-B] 11.4(2)). So dualizing diagram (2) above we get

$$\begin{array}{ccc} J(C) & \overset{\iota}{\hookleftarrow} & \\ {\scriptstyle f^*}\big\uparrow & & \widehat{X} \\ J(H) & \nearrow_{\widehat{g}} & \end{array}$$

Denote by P the canonical principal polarization of $J(H)$. Since $(f^*)^*\Theta \equiv 3P$ and $\widehat{g}$ is an isogeny of degree 3, the induced polarization $\iota^*\Theta$ on $\widehat{X}$ is of type $(1,3)$. According to Proposition 3.1 and [L-B] Proposition 12.1.3 $(E_0 \times E_0, \widehat{X})$ is a pair of complementary abelian subvarieties of $J(C)$. Hence by [L-B] Corollary 12.1.5 the induced polarization $\Xi := (p_1 + \tau p_2)^*\Theta$ on $E_0 \times E_0$ is also of type $(1,3)$. Moreover Ξ is invariant under T, since the polarization Θ is invariant under τ. It remains to prove the last assertion.

It suffices to prove the equation in the Néron-Severi group. $\mathrm{NS}(E_0 \times E_0)$ is a free abelian group generated by $[E_0 \times \{0\}]$, $[\{0\} \times E_0]$, $[\Delta]$, and, if E_0 admits complex multiplication, also the class $[\Gamma]$ of the graph Γ of an endomorphism γ of E_0. Since Ξ is invariant under T and T permutes the curves $E_0 \times \{0\}$, $\{0\} \times E_0$ and Δ, we have

$$\Xi = a([E_0 \times \{0\}] + [\{0\} \times E_0] + [\Delta]) + b[\Gamma]$$

for some integers a, b.

Assume $b \neq 0$. Then necessarily $[\Gamma]$ is invariant with respect to the action of T. Since $(\{0\} \times E_0 \cdot \Gamma) = 1$, this implies that also

$$1 = (E_0 \times \{0\} \cdot \Gamma) = (\Delta \cdot \Gamma) .$$

On the other hand $(E_0 \times \{0\} \cdot \Gamma) = \deg \gamma$ and $(\Delta \cdot \Gamma) =$ number of fixed points of γ. But on an elliptic curve there is no automorphism with exactly one fix point, a contradiction. So $b = 0$.

Since the polarization Ξ is of type $(1,3)$ and thus

$$6 = (\Xi^2) = a^2(E_0 \times \{0\} + \{0\} \times E_0 + \Delta)^2 = 6a^2 ,$$

this implies the assertion. □

If we identify $J(C)$ and $E_0 \times E_0$ with their dual abelian varieties, the map $(p_1 + \tau p_2)^\wedge$ is a surjective homomorphism $J(C) \to E_0 \times E_0$. The composed map

$$a_c \colon C \xrightarrow{\alpha_c} J(C) \xrightarrow{(p_1+\tau p_2)^\wedge} E_0 \times E_0$$

is called the *Abel-Prym map* of the abelian subvariety $E_0 \times E_0$ of $(J(C), \Theta)$. Recall from Example 4.2 that $\widehat{\Xi} = [E_0 \times \{0\} + \{0\} \times E_0 + A]$ is the dual polarization of Ξ.

5.5 Lemma. *$a_c \colon C \to E_0 \times E_0$ is an embedding and its image $a_c(C)$ defines the polarization $\widehat{\Xi}$.*

Proof. According to [L-B] Corollary 11.4.2 we have $a_c^* = \alpha_c^*(p_1 + \tau p_2) = -(p_1 + \tau p_2) \colon E \times E \to J(C)$. Hence $(a_c^*)^*\Theta = (p_1 + \tau p_2)^*\Theta = \Xi$. So Proposition 4.3 implies that α_c is birational onto its image and $a_{c*}[C] = \widehat{\Xi}$. It remains to show that $a_c(C)$ is smooth. But by the adjunction formula $p_a(a_c(C)) = \frac{(\widehat{\Xi}^2)}{2} + 1 = 4 = p_g(C)$. □

Recall that $K(\widehat{\Xi})$ is the kernel of the isogeny $\varphi_{\widehat{\Xi}} \colon E_0 \times E_0 \to E_0 \times E_0$. According to Example 4.2 $\varphi_{\widehat{\Xi}} = \begin{pmatrix} 2 & 1 \\ 1 & 2 \end{pmatrix}$ and hence

$$K(\widehat{\Xi}) = \{(x, x) \in E_0 \times E_0 : 3x = 0\} .$$

Consider the dual $\widehat{T}$ of the automorphism T of $E \times E$ as an automorphism of $E \times E$. From Lemma 5.3 we deduce that

$$\widehat{T} = \begin{pmatrix} 0 & 1 \\ -1 & -1 \end{pmatrix} .$$

Moreover the polarization $\widehat{\Xi}$ is invariant under $\widehat{T}$, i.e. $\widehat{T}^*\widehat{\Xi} = \widehat{\Xi}$. This follows for example from

$$\varphi_{\widehat{\Xi}} = 3\varphi_{\Xi}^{-1} = 3\varphi_{T^{2*}\Xi}^{-1} = 3T\varphi_{\Xi}^{-1}\widehat{T} = T\varphi_{\widehat{\Xi}}\widehat{T} = \varphi_{\widehat{T}}^{*}\widehat{\Xi} .$$

Denote by $\mathrm{Fix}\widehat{T}$ the set of fixed points of $\widehat{T}$. We obviously have

$$\mathrm{Fix}\widehat{T} = K(\widehat{\Xi}) = \{(x,x) \in E_0 \times E_0 : 3x = 0\} .$$

This shows that $K(\widehat{\Xi})$ is the set of those points y of $E_0 \times E_0$, for which the translation map t_y commutes with $\widehat{T}$. This is the essential argument in the proof of the following

5.6 Lemma. *$|E_0 \times \{0\} + \{0\} \times E_0 + A|$ is the unique linear system defining the polarization $\widehat{\Xi}$ on which $\widehat{T}$ induces the identity, i.e. $\widehat{T}$ restricts to an automorphism of every divisor in $|E_0 \times \{0\} + \{0\} \times E_0 + A|$.*

Proof. Let D be any divisor defining the polarization $\widehat{\Xi}$. We first claim that if $\widehat{T}^*D = D$, and the eigenvalue of the corresponding action on the sections defining D is 1, then $\widehat{T}$ induces the identity on the linear system $|D|$. For any $y \in K(\widehat{\Xi}) = \mathrm{Fix}(\widehat{T})$ we have $\widehat{T}^*t_y^*D = t_y^*\widehat{T}^*D = t_y^*D$. The Stone-von Neumann Theorem implies that translating D by elements of $K(\widehat{\Xi})$ leads to a system of generators of the projective space $|D|$. So $\widehat{T}$ acts as the identity on the linear system $|D|$. This proves the claim.

In order to show that $|E_0 \times \{0\} + \{0\} \times E_0 + A|$ is the only linear system defining $\widehat{\Xi}$ on which $\widehat{T}$ acts as the identity, note first that $\widehat{T}^*(E_0 \times \{0\} + \{0\} \times E_0 + A) = E_0 \times \{0\} + \{0\} \times E_0 + A$. Moreover the eigenvalue of the corresponding action the section defining $E_0 \times \{0\} + \{0\} \times E_0 + A$ is 1. Any linear system on $E_0 \times E_0$ defining $\widehat{\Xi}$ contains a divisor of the form $t_z^*(E_0 \times \{0\} + \{0\} \times E_0 + A)$ for some $z \in E_0 \times E_0$. Since no group of translations acts on the divisor $E_0 \times \{0\} + \{0\} \times E_0 + A$ itself, the divisor $t_z^*(E_0 \times \{0\} + \{0\} \times E_0 + A)$ is invariant under $\widehat{T}$ if and only if $z \in \mathrm{Fix}\widehat{T} = K(\widehat{\Xi})$. This completes the proof of the lemma. □

5.7 Lemma.

a) For any $c \in C$ there exists a point $y = y(c) \in E_0 \times E_0$ uniquely determined modulo $K(\widehat{\Xi})$ such that $t_y^*a_c(C) \in |E_0 \times \{0\} + \{0\} \times E_0 + A|$.

b) The set $\{t_{y+x}^*a_c(C) \mid x \in K(\widehat{\Xi})\}$ does not depend on the choice of c.

Proof.

a) According to Lemma 5.5 the divisor $a_c(C)$ is algebraically equivalent to $E_0 \times \{0\} + \{0\} \times E_0 + A$. Hence there is a $y \in E_0 \times E_0$ such that the divisors $t_y^* a_c(C)$ and $E_0 \times \{0\} + \{0\} \times E_0 + A$ are linearly equivalent. The uniqueness of y modulo $K(\widehat{\Xi})$ follows from the definition of $K(\widehat{\Xi})$.

b) Varying $c \in C$, the subset $\{t_y^* a_c(C) \mid c \in C\}$ of $|E_0 \times \{0\} + \{0\} \times E_0 + A|$ depends continuously on c. On the other hand, the curves in $\{t_y^* a_c(C) \mid c \in C\}$ differ only by translations by elements of the finite group $K(\widehat{\Xi})$. Hence the set $\{t_{y+x}^* a_c(C) \mid x \in K(\widehat{\Xi})\}$ is independent of the point c.

□

6 The Moduli Space $\mathcal{A}^0_{(1,3)}$

In this section we use the results of §5 in order to give an explicit description of the moduli space $\mathcal{A}^0_{(1,3)}$.

Let (X, L, π) be an element of $\mathcal{A}^0_{(1,3)}$ with corresponding étale 3-fold covering $\{C \to H\} \in \mathcal{C}_2^3$. The hyperelliptic covering $H \to P_1$ lifts to an elliptic covering $C \to E$. The elliptic curve E is uniquely determined by $C \to H$. Let $\widehat{\Xi}$ denote the polarization on $E \times E$ defined by the divisor $E \times \{0\} + \{0\} \times E + A$. According to Lemmas 5.5 and 5.7 there is an embedding $C \hookrightarrow E \times E$ uniquely determined modulo translations by elements of $K(\widehat{\Xi})$ whose image is contained in the linear system $|E \times \{0\} + \{0\} \times E + A|$. On the other hand, consider the automorphism $\widehat{T} = \begin{pmatrix} 0 & 1 \\ -1 & -1 \end{pmatrix}$ of $E \times E$. According to Lemma 5.6 it restricts to an automorphism of C which coincides with the covering transformation of $C \to H$. Since $\mathrm{Fix}\widehat{T} = K(\widehat{\Xi})$, this implies $C \cap K(\widehat{\Xi}) = \emptyset$.

Let $\mathcal{M}$ denote the moduli space of pairs (E, C) with E an elliptic curve and C a smooth curve in the linear system $|E \times \{0\} + \{0\} \times E + A|$ modulo translations by elements of $K(\widehat{\Xi})$ such that $C \cap K(\widehat{\Xi}) = \emptyset$. Using level structures it is easy to see that $\mathcal{M}$ exists as a coarse moduli space for this moduli problem.

Summing up we constructed a holomorphic map $\psi \colon \mathcal{A}^0_{(1,3)} \to \mathcal{M}$.

6.1 Theorem. *$\psi \colon \mathcal{A}^0_{(1,3)} \to \mathcal{M}$ is an isomorphism of algebraic varieties.*

Proof. It remains to construct an inverse map. Let $(E, C) \in \mathcal{M}$. According to Lemma 5.6 the automorphism $\widehat{T}$ of $E \times E$ acts on every curve of the linear system. In particular $\widehat{T}$ restricts to an automorphism τ of C which is of order 3, since C generates $E \times E$ as a group. Moreover τ is fixed point free, since $C \cap \mathrm{Fix}\widehat{T} = \emptyset$. So τ induces an étale 3-fold covering $C \to H$ corresponding to an element $(X, L, \pi) \in \mathcal{A}^0_{(1,3)}$. It is easy to see that the map $\mathcal{M} \to \mathcal{A}^0_{(1,3)}$, $(E, C) \mapsto (X, L, \pi)$ is holomorphic and inverse to ψ. □

6.2 Corollary. *$\mathcal{A}_{(1,3)}$ is a rational variety.*

Proof. It suffices to show that the open set $\mathcal{M}^0 = \{(E, C) \in \mathcal{M} : E$ admits no nontrivial automorphisms $\}$ is rational.

The open set $U = C - \{0, 1728\}$ parametrizing elliptic curves without nontrivial automorphisms admit a universal family $\mathcal{E} \to U$. Consider the line bundle $\mathcal{O}_{\mathcal{E}\times_U\mathcal{E}}(\mathcal{E} \times_U \{0\} + \{0\} \times_U \mathcal{E} + \mathcal{A})$ on the fibre product $p \colon \mathcal{E} \times_U \mathcal{E} \to U$ where $\mathcal{A}$ denotes the relative antidiagonal. According to Grauert's Theorem $p_*\mathcal{O}_{\mathcal{E}\times_U\mathcal{E}}(\mathcal{E} \times_U \{0\} + \{0\} \times_U \mathcal{E} + \mathcal{A})$ is a vector bundle of rank 3 over U. The corresponding projective bundle $P_U := P(p_*\mathcal{O}_{\mathcal{E}\times_U\mathcal{E}}(\mathcal{E} \times_U \{0\} + \{0\} \times_U \mathcal{E} + \mathcal{A}))$ parametrizes the linear systems $|E \times \{0\} + \{0\} \times E + A|$. By construction $\mathcal{M}^0$ is an open subset of the quotient $P_U/K(\Xi)$, where $K(\Xi)$ acts as usual on the fibres of $P_U \to U$. Since every vector bundle on U is trivial, $P_U \simeq P^2 \times U$ and $P_U/K(\Xi) \simeq P^2/(Z/3Z \times Z/3Z) \times U$, which is rational by Lüroth's theorem (see [G-H] p. 541). □

References

[ACGA] Arbarello E., Cornalba M., Griffiths P.A., Harris J., *Geometry of algebraic curves I*, Springer Grundlehren **267** (1985).

[C-W] Chevalley C., Weil A., *Über das Verhalten der Integrale erster Gattung bei Automorphismen des Funktionenkörpers*, Hamb. Abh. **10** (1934) 358–361.

[EGAI] Grothendieck A., Dieudonné J., *Eléments de Géometrie Algébrique I*, Springer Grundlehren **166** (1971).

[G] Grothendieck A., *Le Groupe de Brauer III. In: Dix Exposés sur la Cohomologie des Schémas*, North Holland, Amsterdam, (1968) 88–188.

[G-H] Griffiths P., Harris J., *Principles of Algebraic Geometry*, John Wiley & Sons, New York (1978).

[H-W] Hulek K., Weintraub S.H., *Bielliptic Abelian Surfaces*, Math. Ann. **283** (1989) 411–429.

[I] Ingrisch W., *Automorphismengruppen und Moduln hyperelliptischer Kurven*, Dissertation Erlangen (1985).

[L-B] Lange H., Birkenhake Ch., *Complex Abelian Varieties*, Springer Grundlehren **302** (1992).

Ch. Birkenhake
H. Lange
Mathematisches Institut
Bismarckstraße $1\frac{1}{2}$
D-8520 Erlangen

Instantons and Parabolic Sheaves

M. Maruyama

Introduction

S. Donaldson [1] found a beautiful bijection between the set of marked $SU(r)$-instantons and the set of couples of a rank-r vector bundle on $\mathbf{P}^2_{\mathbf{C}}$ and a trivialization on a fixed line. Then, based on Hulek's result in [3], he concluded that the moduli space of marked $SU(r)$-instantons with fixed instanton number is connected. Hulek's result is, however, insufficient to deduce the connectedness. In fact, a vector bundle E on $\mathbf{P}^2_{\mathbf{C}}$ is said to be s-stable in Hulek's sense if $H^0(\mathbf{P}^2_{\mathbf{C}}, E) = 0$ and $H^0(\mathbf{P}^2_{\mathbf{C}}, E^\vee) = 0$. Hulek [3] proved that the set of s-stable vector bundles on $\mathbf{P}^2_{\mathbf{C}}$ with $r(E) = r$, $c_1(E) = 0$ and $c_2(E) = n$ is parametrized by an irreducible algebraic set. There are vector bundles on $\mathbf{P}^2_{\mathbf{C}}$ that correspond to marked $SU(r)$-instantons but are not s-stable. For example, if $c_2(E) < r(E)$, E cannot be s-stable and we have, on the other hand, $SU(r)$-instantons with instanton number $n < r$.

In this article we shall show that we can regard the couple (E, h) of a vector bundle E on $\mathbf{P}^2_{\mathbf{C}}$ and a trivialization h of E on a fixed line as a parabolic stable vector bundle. Then, the connectedness of the moduli space of marked $SU(r)$-instantons reduces to that of the moduli space of parabolic stable sheaves. It is rather complicated but not hard to prove the connectedness of the moduli space of parabolic stable sheaves. The author hopes that he could prove in this way the connectedness of the moduli space of marked $SU(r)$-instantons.

Notation. For a field k and integers m, n, $M(m, n, k)$ denotes the set of $(m \times n)$-matrices over k and $M(n, k)$ does the full matrix ring $M(n, n, k)$. If $f : X \to S$ is a morphism of schemes, E is a coherent sheaf on X and if s is a point (or, geometric point) of S, then $E(s)$ denotes the sheaf $E \otimes_{\mathcal{O}_S} k(s)$. For a coherent sheaf F on a variety Y, we denote the rank of F by $r(F)$. Assuming Y to be smooth and quasi-projective, we can define the i-th Chern class $c_i(F)$ of F.

1 A result of Donaldson

We shall here reproduce briefly the main part of Donaldson's work [1]. Fix a line ℓ in $\mathbf{P}^2_{\mathbf{C}}$. Let E be a vector bundle of rank r on $\mathbf{P}^2_{\mathbf{C}}$ with the following properties:

(1.1.1) $E|_\ell \cong \mathcal{O}_\ell^{\oplus r}$,

(1.1.2) $c_1(E) = 0$ and $c_2(E) = n$.

(1.1.1) implies that E is μ-semi-stable and hence $c_2(E) = n \geq 0$. Since the μ-semi-stability of E implies $H^0(\mathbf{P}^2_{\mathbf{C}}, E(-1)) = 0$, we see that E is the cohomology sheaf of a monad

$$H \otimes_{\mathbf{C}} \mathcal{O}_{\mathbf{P}^2}(-1) \xrightarrow{s} K \otimes_{\mathbf{C}} \mathcal{O}_{\mathbf{P}^2} \xrightarrow{t} L \otimes_{\mathbf{C}} \mathcal{O}_{\mathbf{P}^2}(1),$$

where H, K and L are $\mathbf{C}$-vector spaces of dimension n, $2n+r$ and n, respectively. s is represented by a $(2n+r) \times n$ matrix A whose entries are linear forms on $\mathbf{P}^2_{\mathbf{C}}$. Fixing a system of homogeneous coordinates $(x : y : z)$ of $\mathbf{P}^2_{\mathbf{C}}$, we may write

$$A = A_x x + A_y y + A_z z,$$

where $A_x, A_y, A_z \in M(2n+r, n, \mathbf{C})$. Similarly t is represented by

$$B = B_x x + B_y y + B_z z$$

with $B_x, B_y, B_z \in M(n, 2n+r, \mathbf{C})$. The condition $ts = 0$ is equivalent to $B_x A_x = B_y A_y = B_z A_z = 0$, $B_x A_y = -B_y A_x$, $B_y A_z = -B_z A_y$ and $B_z A_x = -B_x A_z$. E is represented by such a monad uniquely up to the action of $GL(H) \times GL(K) \times GL(L)$.

We may assume that ℓ is defined by $z = 0$. Then $E|_\ell$ is trivial if and only if $\det B_x A_y \neq 0$. We can find bases of H, K and L so that $B_x A_y = I_n$, where I_n is the identitiy matrix of degree n. Changing the basis of L, we have

$$A_x = \begin{pmatrix} I_n \\ 0 \\ 0 \end{pmatrix} \begin{matrix} n \\ n \\ r \end{matrix}, \quad A_y = \begin{pmatrix} 0 \\ I_n \\ 0 \end{pmatrix} \begin{matrix} n \\ n \\ r \end{matrix}$$

and

$$B_x = \overset{\quad n \quad\; n \quad\; r}{(\ 0,\ \ I_n,\ \ 0\ )}, \quad B_y = \overset{\quad n \quad\;\; n \quad r}{(\ -I_n,\ \ 0,\ \ 0\ )}.$$

Then, setting

$$A_z = \begin{pmatrix} \alpha_1 \\ \alpha_2 \\ a \end{pmatrix}$$

with $\alpha_1, \alpha_2 \in M(n, \mathbf{C})$ and $a \in M(r, n, \mathbf{C})$, the equations $B_y A_z = -B_z A_y$ and $B_z A_x = -B_x A_z$ imply $B_z = (-\alpha_2, \alpha_1, b)$ with $b \in M(n, r, \mathbf{C})$. The last equation $B_z A_z = 0$ means

(1.2) $[\alpha_1, \alpha_2] + ba = 0$.

On ℓ we have an exact sequence

$$0 \longrightarrow \mathcal{O}_\ell(-1) \xrightarrow{u} \mathcal{O}_\ell^{\oplus 2} \xrightarrow{v} \mathcal{O}_\ell(1) \longrightarrow 0.$$

The restriction of our monad to ℓ is

$$\mathcal{O}_\ell(-1)^{\oplus n} \xrightarrow{\oplus^n u \oplus 0} (\oplus^n \mathcal{O}_\ell^{\oplus 2}) \oplus \mathcal{O}_\ell^{\oplus r} \xrightarrow{\oplus^n v \oplus 0} \mathcal{O}_\ell(1)^{\oplus n}$$

and the trivialization of $E|_\ell$ comes from the last term of the middle. The equivalence defined by the action of $GL(H) \times GL(K) \times GL(L)$ induces an action of the group

$$\left\{ \left(p, \begin{pmatrix} p^{-1} & 0 & 0 \\ 0 & p^{-1} & 0 \\ 0 & 0 & q \end{pmatrix}, p\right) \;\middle|\; p \in GL(n, \mathbf{C}), q \in GL(r, \mathbf{C}) \right\}$$

on the above normalized matrices. The action of q is nothing but changing the trivializations of $E|_\ell$.

The condition that the above normalized couple (A, B) gives rise to a vector bundle is that for every $(\lambda, \mu, \nu) \in \mathbf{P}^2_{\mathbf{C}}$,

$$\lambda A_x + \mu A_y + \nu A_z = \begin{pmatrix} \lambda I_n + \nu\alpha_1 \\ \mu I_n + \nu\alpha_2 \\ \nu a \end{pmatrix}$$

is an injection of H to K and

$$\lambda B_x + \mu B_y + \nu B_z = (-\mu I_n - \nu\alpha_2, \lambda I_n + \nu\alpha_1, \nu b)$$

is a surjection of K to L. If $\nu = 0$, then these conditions are trivially satisfied. Thus we have the following.

1,3 Proposition. *The set* $\{(E, h) \mid E$ has the properties (1.1.1) and (1.1.2). h *is a trivialization of* $E|_\ell\}/\cong$ *is in bijective correspondence with the set of quadruples* $(\alpha_1, \alpha_2, a, b)$ *of matrices with the following properties* (1.3.1), (1.3.2) *and* (1.3.3) *modulo an action of* $GL(n, \mathbf{C})$:

(1.3.1) $\alpha_1, \alpha_2 \in M(n, \mathbf{C})$, $a \in M(r, n, \mathbf{C})$ *and* $b \in M(n, r, \mathbf{C})$,

(1.3.2) $[\alpha_1, \alpha_2] + ba = 0$,

(1.3.3) *for all* $(\lambda, \mu) \in \mathbf{C}^2$,

$$\begin{pmatrix} \lambda I_n + \alpha_1 \\ \mu I_n + \alpha_2 \\ a \end{pmatrix}$$

is injective and $(-\mu I_n - \alpha_2, \lambda I_n + \alpha_1, b)$ *is surjective.*

Here $p \in GL(n, \mathbf{C})$ *sends* $(\alpha_1, \alpha_2, a, b)$ *to* $(p\alpha_1 p^{-1}, p\alpha_2 p^{-1}, ap^{-1}, pb)$.

Let us embed $\mathbf{P}^2_{\mathbf{C}}$ into $\mathbf{P}^3_{\mathbf{C}}$ as the plane defined by $w = 0$, where $(x : y : z : w)$ is a system of homogeneous coordinates of $\mathbf{P}^3_{\mathbf{C}}$. Now we look at the vector bundle defined by a monad

$$H \otimes_{\mathbf{C}} \mathcal{O}_{P^3}(-1) \xrightarrow{c} K \otimes_{\mathbf{C}} \mathcal{O}_{P^3} \xrightarrow{d} L \otimes_{\mathbf{C}} \mathcal{O}_{P^3}(1),$$

and trivialized on $\ell = \{z = w = 0\}$. c and d are represented by

$$\begin{aligned} A &= A_x x + A_y y + A_z z + A_w w \\ B &= B_x x + B_y y + B_z z + B_w w \end{aligned}$$

with

$$A_x = \begin{pmatrix} I_n \\ 0 \\ 0 \end{pmatrix}, \quad A_y = \begin{pmatrix} 0 \\ I_n \\ 0 \end{pmatrix}, \quad A_z = \begin{pmatrix} \alpha_1 \\ \alpha_2 \\ a \end{pmatrix}, \quad A_w = \begin{pmatrix} \hat{\alpha}_1 \\ \hat{\alpha}_2 \\ \hat{a} \end{pmatrix}$$

and

$$B_x = (0, I_n, 0), \quad B_y = (-I_n, 0, 0), \quad B_z = (-\alpha_2, \alpha_1, b), \quad B_w = (-\hat{\alpha}_2, \hat{\alpha}_1, \hat{b}).$$

The condition $dc = 0$ of the monad means

(1.4.1) $[\alpha_1, \alpha_2] + ba = 0$,

(1.4.2) $[\hat{\alpha}_1, \hat{\alpha}_2] + \hat{b}\hat{a} = 0$,

(1.4.3) $[\alpha_1, \hat{\alpha}_2] + [\hat{\alpha}_1, \alpha_2] + b\hat{a} + \hat{b}a = 0.$

Let $\mathbf{H} = \mathbf{R} + \mathbf{R}i + \mathbf{R}j + \mathbf{R}k$ be the algebra of quaternions over $\mathbf{R}$. Regarding $\mathbf{C}^4$ as $\mathbf{H}^2$, $\mathbf{H}$ acts on $\mathbf{C}^4$ from left. Hence we have a j-invariant real analytic map

$$\pi : \mathbf{P}^3_{\mathbf{C}} = \{\mathbf{C}^4 \setminus \{0\}\}/\mathbf{C}^* \longrightarrow S^4 \cong \mathbf{P}^2_{\mathbf{H}}.$$

A vector bundle E of rank r on $\mathbf{P}^3_{\mathbf{C}}$ is called an instanton bundle if it comes from an $SU(r)$-instanton on S^4, or equivalently if the following conditions are satisfied:

(1.5.1) there is an isomorphism λ of E to $\overline{j^*(E^\vee)}$ such that the composition $\overline{j^*({}^t\lambda^{-1})} \cdot \lambda$ is equal to id_E when we identify $j^*(\overline{j^*(E^\vee)}^\vee)$ with E,

(1.5.2) $E|_{\pi^{-1}(x)}$ is trivial for all $x \in S^4$.

An instanton bundle E is the cohomology of a monad which appeared in the above. The instanton structure of E provides the space K with a Hermitian structure and an isomorphism of L to $\overline{H}^\vee$.

For $\lambda = (\lambda_1, \lambda_2, \lambda_3, \lambda_4) \in \mathbf{C}^4$, we set

$$\begin{aligned} A(\lambda) &= A_x\lambda_1 + A_y\lambda_2 + A_z\lambda_3 + A_w\lambda_4 \\ B(\lambda) &= B_x\lambda_1 + B_y\lambda_2 + B_z\lambda_3 + B_w\lambda_4. \end{aligned}$$

Since $A(\lambda)^* = {}^t\overline{A(\lambda)}$ defines a linear map of $K \cong \overline{K}^\vee$ to $\overline{H}^\vee \cong L$, we have a linear map

$$\mathcal{A}(\lambda) = A(\lambda)^* \oplus B(\lambda) : K \longrightarrow L \oplus L.$$

The quaternion algebra $\mathbf{H}$ acts on $L \oplus L$ by

$$i\begin{pmatrix} u \\ v \end{pmatrix} = \begin{pmatrix} -\sqrt{-1}u \\ \sqrt{-1}v \end{pmatrix}, \quad j\begin{pmatrix} u \\ v \end{pmatrix} = \begin{pmatrix} v \\ -v \end{pmatrix}.$$

Then the condition that

$$\mathcal{A}(q\lambda)(v) = q\mathcal{A}(\lambda)(v) \quad \text{for all } q \in \mathbf{H}, v \in K$$

is equivalent to the condition that (A, B) gives rise to an $SU(r)$-instanton bundle.

Now the above condition can be written down in the form

$$\begin{aligned} \mathcal{A}(j(0,0,z,w)) &= \begin{pmatrix} -{}^t\bar{A}_z w + {}^t\bar{A}_w z \\ -B_z\bar{w} + B_w\bar{z} \end{pmatrix} \\ &= j\begin{pmatrix} {}^t\bar{A}_z\bar{z} + {}^t\bar{A}_w\bar{w} \\ B_z z + B_w w \end{pmatrix} = \begin{pmatrix} B_z z + B_w w \\ -{}^t\bar{A}_z\bar{z} - {}^t\bar{A}_w\bar{w} \end{pmatrix} \end{aligned}$$

that is,

$${}^t\bar{A}_w = B_z \text{ and } {}^t\bar{A}_z = -B_w.$$

Thus we have $\hat{\alpha}_1 = -\alpha_2^*$, $\hat{\alpha}_2 = \alpha_1^*$, $\hat{a} = b^*$ and $\hat{b} = -a^*$. On the other hand, $A(\lambda)$ is injective if and only if $A(\lambda)^*$ is surjective. Thus our monad defines a vector bundle if and only if $\mathcal{A}(\lambda)$ is surjective for all $\lambda \in \mathbf{C}^4 \setminus \{0\}$. (1.3.3) implies that $\mathcal{A}(\lambda)$ is surjective for $\lambda \in \mathbf{C}^3 \setminus \{0\}$ which gives our plane $w = 0$ and then so is $\mathcal{A}(q\lambda)$. Since $\{q\lambda \mid q \in \mathbf{H}, \lambda \in \mathbf{C}^3 \setminus \{0\}\}$ sweeps $\mathbf{C}^4 \setminus \{0\}$, (1.3.3) is equivalent to the condition that $\mathcal{A}(\lambda)$ is surjective for all $\lambda \in \mathbf{C}^4 \setminus \{0\}$

Now the condition (1.4.2) is the adjoint of (1.4.1) or (1.3.2) and the condition (1.4.3) becomes

$$[\alpha_1, \alpha_1^*] + [\alpha_2, \alpha_2^*] + bb^* - a^*a = 0.$$

Let $M(SU(r), n)$ be the set of isomorphism classes of marked $SU(r)$-instantons with instanton number n. $M(SU(r), n)$ is the set of the isomorphism classes of the couples (∇, g) of an $SU(r)$-instanton ∇ and an element g of the fiber over northpole of the $SU(r)$-principal fiber bundle where the instanton is defined. What we have seen in the above is stated as follows.

1.6 Proposition. *$M(SU(r), n)$ is in bijective correspondence with the $U(n)$-quotient of the set of quadruples $\{(\alpha_1, \alpha_2, a, b)\}$ of matrices with the properties* (1.3.1), (1.3.2), (1.3.3) *and*

(1.6.1) $[\alpha_1, \alpha_1^*] + [\alpha_2, \alpha_2^*] + bb^* - a^*a = 0.$

Assume that $G = GL(n, \mathbf{C})$ acts on $\mathbf{C}^N$ with a fixed norm structure such that $U(n)$ does isometrically. Take a G-invariant subscheme W of $\mathbf{C}^N$ whose points are all stable. Let W_0 be the set of points that are nearest to the origin in its G-orbit.

1.7 Proposition. ([4]). *W/G is isomorphic to $W_0/U(n)$.*

Let us look at the $\mathbf{C}$-linear space $V = M(n, \mathbf{C}) \times M(n, \mathbf{C}) \times M(r, n, \mathbf{C}) \times M(n, r, \mathbf{C})$ where $GL(n, \mathbf{C})$ acts as in Proposition 1.3. Then $U(n)$ acts on V isometrically with respect to the obvious norm of V. Let W be the subscheme of V defined by (1.3.2) and (1.3.3). Then we have the following key results.

1.8 Lemma. [1, p. 458]. *A point $(\alpha_1, \alpha_2, a, b)$ in W is contained in W_0 if and only if it has the property* (1.6.1).

1.9 Lemma. [1, Lemma]. *W is contained in the set of stable points of V with respect to the action of $GL(n, \mathbf{C})$.*

Therefore, we come to the main result of [1].

1.10 Theorem. *The set* $\{(\alpha_1, \alpha_2, a, b) \mid (1.3.1), (1.3.2)$ and $(1.3.3)$ are satisfied $\}/GL(n, \mathbf{C})$ *is in bijective correspondence with the set* $\{(\alpha_1, \alpha_2, a, b) \mid (1.3.1), (1.3.2), (1.3.3)$ and $(1.6.1)$are satisfied$\}/U(n)$. *Hence* $\{(E, h) \mid E$ has the properties (1.1.1) and (1.1.2). h is a trivialization of $E|_\ell\}/\cong$ *is isomorphic to the space* $M(SU(r), n)$.

2 Parabolic sheaves

Let $(X, \mathcal{O}_X(1), D)$ be a triple of a non-singular projective variety X over an algebraically closed field k, an ample line bundle $\mathcal{O}_X(1)$ on X and an effective Cartier divisor D on X. A coherent torsion free sheaf E is said to be a parabolic sheaf if the following data are assigned to it:

(2.1.1) a filtration $0 = F_{t+1} \subset F_t \subset \cdots \subset F_1 = E \otimes_{\mathcal{O}_X} \mathcal{O}_D$ by coherent subsheaves,

(2.1.2) a system of weights $0 \le \alpha_1 < \alpha_2 < \cdots < \alpha_t < 1$.

We denote the parabolic sheaf by (E, F_*, α_*).

For a parabolic sheaf (E, F_*, α_*), we define

$$par\text{-}\chi(E(m)) = \chi(E(-D)(m)) + \sum_{i=1}^{t} \alpha_i \chi(F_i/F_{i+1}(m)).$$

If E' is a coherent subsheaf of E with E/E' torsion free, then we have an induced parabolic structure. In fact, since $E' \otimes_{\mathcal{O}_X} \mathcal{O}_D$ can be regarded as a subsheaf of $E \otimes_{\mathcal{O}_X} \mathcal{O}_D$, we have a filtration $0 = F'_{s+1} \subset F'_s \subset \cdots \subset F'_1 = E' \otimes_{\mathcal{O}_X} \mathcal{O}_D$ such that $F'_j = E' \otimes_{\mathcal{O}_X} \mathcal{O}_D \cap F_i$ for some i. The weight α'_j of F'_j is defined to be α_i with $i = \max\{k \mid F'_j = E' \otimes_{\mathcal{O}_X} \mathcal{O}_D \cap F_k\}$.

2.2 Definition. A parabolic sheaf (E, F_*, α_*) is said to be stable if for every coherent subsheaf E' with E/E' torsion free and with $1 \le r(E') < r(E)$ and for all sufficiently large integers m, we have

$$par\text{-}\chi(E'(m))/r(E') < par\text{-}\chi(E(m))/r(E),$$

where the parabolic structure of E' is the induced from that of E.

Let S be a scheme of finite type over a universally Japanese ring and now let $(X, \mathcal{O}_X(1), D)$ be a triple of a smooth, projective, geometrically integral scheme X over S, an S-ample line bundle $\mathcal{O}_X(1)$ on X and an effective relative Cartier divisor D on X over S.

2.3 Lemma. *If T is a locally noetherian S-scheme and if E is a T-flat coherent sheaf on $X \times_S T$ such that for every geometric point t of T, $E(t)$ is torsion free, then $E|_D = E \otimes_{\mathcal{O}_X} \mathcal{O}_D$ is flat over T.*

Proof. For every point y of T, $E(y)$ is torsion free and D_y is a Cartier divisor on X_y. Thus the natural homomorphism $E(-D)(y) \to E(y)$ is injective. Then, since E is T-flat, $E|_D = E/E(-D)$ is T flat. Q.E.D.

Let E be a coherent sheaf on $X \times_S T$ which satisfies the condition in the above lemma. E is said to be a T-family of parabolic sheaves if the following data are assigned to it:

(2.4.1) a filtration $0 = F_{t+1} \subset F_t \subset \cdots \subset F_1 = E|_D$ by coherent subsheaves such that for $1 < i < t$, $E|_D/F_i$ is flat over T,

(2.4.2) a system of weights $0 \leq \alpha_1 < \alpha_2 < \cdots < \alpha_t < 1$.

As in the absolute case we denote by (E, F_*, α_*) the family of parabolic sheaves. For T-families of parabolic sheaves (E, F_*, α_*) and (E', F'_*, α'_*), they are said to be equivalent and we denote $(E, F_*, \alpha_*) \sim (E', F'_*, \alpha'_*)$ if there is an invertible sheaf L on T such that E is isomorphic to $E' \otimes_{\mathcal{O}_T} L$, the filtration F_* is equal to $F'_* \otimes_{\mathcal{O}_T} L$ under this isomorphism and if the systems of weights are the same.

Fixing polynomials $H(x), H_1(x), \ldots, H_t(x)$ and a system of weights $0 \leq \alpha_1 < \cdots < \alpha_t < 1$, we set

$$par\text{-}\Sigma(H, H_*, \alpha_*)(T) = \left\{ (E, F_*, \alpha_*) \;\middle|\; \begin{array}{l} (E, F_*, \alpha_*) \text{ is a } T\text{-family} \\ \text{of parabolic sheaves with} \\ \text{the properties (2.5.1) and} \\ \text{(2.5.2)} \end{array} \right\} \Big/ \sim$$

(2.5.1) for every geometric point y of T and $1 \leq i \leq t$, $\chi(E(y)(m)) = H(m)$ and $\chi((E(y)/F_{i+1}(y))(m)) = H_i(m)$,

(2.5.2) for every geometric point y of T, $(E(y), F_*(y), \alpha_*)$ is stable.

Obviously $par\text{-}\Sigma(H, H_*, \alpha_*)$ defines a contravariant functor of the category (Sch/S) of locally noetherian S-schemes to that of sets. Note that for every geometric point s of S and every member

$$(E, F_*, \alpha_*) \in par\text{-}\Sigma(H, H_*, \alpha_*)(s),$$

we have

$$par\text{-}\chi(E(m)) = H(m) - \sum_{i=1}^{t} \varepsilon_i H_i(m),$$

where $\varepsilon_i = \alpha_{i+1} - \alpha_i$ with $\alpha_{t+1} = 1$.

One of main results on parabolic stable sheaves is stated as follows.

2.6 Theorem. [7]. *Assume that all the weights $\alpha_1, \ldots, \alpha_t$ are rational numbers. Then the functor par-$\Sigma(H, H_*, \alpha_*)$ has a coarse moduli scheme $M_{X/S}(H, H_*, \alpha_*)$ of locally of finite type over S. If S is a scheme over a field of characteristic zero, then the coarse moduli scheme is quasi-projective over S.*

Now let us go back to the situation of the preceding section. We have a fixed line ℓ in $\mathbf{P}^2_{\mathbf{C}}$, a torsion free, coherent sheaf E of rank r on $\mathbf{P}^2_{\mathbf{C}}$ and a trivialization of $E|_\ell$. There is a bijective correspondence between the set of trivialzations of $E|_\ell$ and the set

$$\mathcal{T}_E = \{\varphi : E|_\ell \to \mathcal{O}_\ell(r-1) \mid H^0(\varphi) : H^0(E|_\ell) \xrightarrow{\sim} H^0(\mathcal{O}_\ell(r-1))\}/\cong$$

where $\cong$ means isomorphisms as quotient sheaves.

2.7 Lemma. *Let E be a torsion free, coherent sheaf of rank r on $\mathbf{P}^2_{\mathbf{C}}$ such that $E|_\ell$ is a trivial vector bundle. For every element $(\varphi : E|_\ell \to \mathcal{O}_\ell(r-1))$ of $\mathcal{T}_E$,* $\ker(\varphi)$ *is isomorphic to $\mathcal{O}_\ell(-1)^{\oplus r-1}$*

Proof. Since $\ker(\varphi)$ is a vector bundle of rank $r-1$ on the line ℓ, it is isomorphic to a direct sum $\mathcal{O}_\ell(a_1) \oplus \cdots \oplus \mathcal{O}_\ell(a_{r-1})$ of line bundles. Our condition on φ means that $H^0(\ell, \ker(\varphi)) = 0$. Combining this and the fact that $\deg(\ker(\varphi)) = 1 - r$, we see that $a_1 = \cdots = a_{r-1} = -1$. Q.E.D.

Fixing a system of weights $\alpha_1 = 1/3$, $\alpha_2 = 1/2$, every element φ of $\mathcal{T}_E$ gives rise to a parabolic structure of E:

$$0 = F_3 \subset F_2 = \ker(\varphi) \cong \mathcal{O}_\ell(-1)^{\oplus r-1} \subset F_1 = E|_\ell.$$

2.8 Proposition. *If E is as in Lemma 2.7. Assume E has the properties (1.1.1) and (1.1.2), then the above parabolic sheaf is stable.*

Proof. Set $r = r(E)$. We know that

$$\begin{aligned} par\text{-}\chi(E(m))/r(E) &= \frac{(m-1)^2}{2} + \frac{3(m-1)}{2} + 1 - \frac{n}{r} \\ &\quad + \frac{(r-1)m}{2r} + \frac{(r+m)}{3r} \\ &= \frac{m^2}{2} + \left(1 - \frac{1}{6r}\right) m + \frac{1}{3} - \frac{n}{r}. \end{aligned}$$

Pick a coherent subsheaf E' of E with E/E' torsion free and write

$$par\text{-}\chi(E'(m))/r(E') = \frac{m^2}{2} + a_1 m + a_0.$$

Then we see

$$a_1 = \mu(E') + \frac{1}{2} + \frac{1}{r(E')}\left(\frac{s}{2} + \frac{t}{3}\right),$$

where $s + t = r(E')$ and $0 \leq t \leq 1$. Thus if $\mu(E') < 0$, then we have

$$a_1 \leq \frac{-1}{r(E')} + 1 < \frac{-1}{r} + 1 < 1 - \frac{1}{6r}$$

and hence we obtain the desired inequality. We may assume therefore that $c_1(E') = 0$. Since E/E' is torsion free, E' is locally free in a neighborhood of ℓ and $E'|_\ell$ is a subsheaf of $E|_\ell$. Thus the triviality of $E|_\ell$ implies that $E'|_\ell \cong \mathcal{O}_\ell^{\oplus r(E')}$ and hence $E'|_\ell$ is not contained in F_2. Since $E'|_\ell/(F_2 \cap E'|_\ell)$ is a subsheaf of $F_1/F_2 \cong \mathcal{O}_\ell(r-1)$, it is of rank 1. We have two cases.

Case 1. $F_2 \cap E'|_\ell = 0$ and $r(E') = 1$. Then the length of the filtration of $E'|_\ell$ is 1 and the weight is 1/3. Thus we see that

$$a_1 = \frac{1}{2} + \frac{1}{3} = 1 - \frac{1}{6} < 1 - \frac{1}{6r}.$$

Case 2. $F_2 \cap E'|_\ell \neq 0$. In this case $F_2' = F_2 \cap E'|_\ell$ is a subsheaf of $\mathcal{O}_\ell(-1)^{\oplus(r-1)}$ and hence for $m \geq 0$,

$$\chi(F_2'(m)) \leq (r(E') - 1)m.$$

On the other hand, since $E'|_\ell/F_2'$ is a subsheaf of $\mathcal{O}_\ell(r-1)$, we have that for $m \geq 0$,

$$\chi(E'|_\ell/F_2'(m)) \leq m + r.$$

Combining these, we get

$$a_1 \leq \frac{1}{2} + \frac{1}{2} - \frac{1}{2r(E')} + \frac{1}{3r(E')} = 1 - \frac{1}{6r(E')} < 1 - \frac{1}{6r}.$$

This completes our proof.

3 Connectedness of the moduli of instantons

Let us set

$$H(x) = \frac{rx^2}{2} + \frac{3rx}{2} + r - n$$

$$H_1(x) = x + r$$

$$H_2(x) = rx + r$$

$$\alpha_1 = \frac{1}{3} \text{ and } \alpha_2 = \frac{1}{2}.$$

For these invariants, we have the moduli space $\tilde{M}(r,n) = M(H, H_*, \alpha_*)$ of parabolic stable sheaves on $(\mathbf{P}^2_{\mathbf{C}}, \mathcal{O}_{\mathbf{P}^2}(1), \ell)$. There is an open subscheme $M(r,n)$ of $\tilde{M}(r,n)$ consisting of (E, F_*, α_*) with the properties

(3.1.1) $E|_\ell \cong \mathcal{O}_\ell^{\oplus r}$ and $E|_\ell/F_2 \cong \mathcal{O}_\ell(r-1)$,

(3.1.2) for the surjection $\varphi : E|_\ell \to E|_\ell/F_2$, $H^0(\varphi)$ is isomorphic.

$M(r,n)$ contains a slightly smaller open subscheme $M(r,n)_0$ consisting of locally free sheaves. What we have seen in the above is $M(SU(r),n) \cong M(r,n)_0$.

3.2 Proposition. *$M(r,n)$ is smooth and of pure dimension $2rn$.*

To prove this proposition we shall follow the way we used in [5]. Let us start with the general setting in Theorem 2.6. Let Σ be the family of the classes of parabolic stable sheaves on the fibers of X over S with fixed invariants. For simplicity we assume that Σ is bounded (Proposition 3.2 is the case). If m is a sufficiently large integer and if (E, F_*, α_*) is a representative of a member of Σ over a geometric point s of S, then we have that

(3.3.1) both $E(-D_s)(m)$ and $E(m)$ are generated by their global sections and for all $i > 0$, $H^i(E(-D_s)(m)) = H^i(E(m)) = 0$,

(3.3.2) for $1 \leq j \leq t$ and $i > 0$, $H^i(F_j/F_{j+1}(m)) = 0$ and $F_j(m)$ is generated by its global sections.

Replacing every member

$$(E, F_*, \alpha_*) \in par\text{-}\Sigma(H, H_*, \alpha_*)(T)$$

by $(\mathcal{O}_X(m) \otimes_{\mathcal{O}_S} E, \mathcal{O}_X(m) \otimes_{\mathcal{O}_S} F_*, \alpha_*)$, we may assume $m = 0$. Setting $N = \dim H^0(X(s), E)$ for a member (E, F_*, α_*) of Σ on a fiber X_s over s, fixing a free $\mathcal{O}_S$-module V of rank N and putting $V_X = V \otimes_{\mathcal{O}_S} \mathcal{O}_X$, there is an open subscheme R of $Q = \mathrm{Quot}^H_{V_X/X/S}$ such that for every algebraically closed field k, a k-valued point z of Q is in $R(k)$ if and only if the following conditions are satisfied:

(3.3.3) For the universal quotient $\varphi : V_X \otimes_{\mathcal{O}_S} \mathcal{O}_Q \to \tilde{E}$, the induced map $H^0(\varphi(z)) : V \otimes_{\mathcal{O}_S} k(s) \to H^0(X_s, \tilde{E}(z))$ is an isomorphism, where s is the image of z in $S(k)$.

(3.3.4) For every $i > 0$, $H^i(X(s), \tilde{E}(z)) = 0$.

(3.3.5) $\tilde{E}(z)$ is torsion free.

We denote the restriction of $\tilde{E}$ to $X_R = X \times_S R$ by the same $\tilde{E}$. Then, by Lemma 2.3 we see that $\tilde{E}|_{D_R}$ is flat over R. Let R_t be the R-scheme $\mathrm{Quot}^{H_t}_{\tilde{E}|_{D_R}/X_R/R}$ and let

$$\tilde{E}|_{D_R} \otimes_{\mathcal{O}_R} \mathcal{O}_{R_t} \longrightarrow \tilde{E}_t$$

be the universal quotent. Assume that we have a sequence

$$R_j \to R_{j+1} \to \cdots \to R_t \to R$$

of schemes R_j and a sequence of sheaves $\tilde{E}_j, \tilde{E}_{j+1} \ldots, \tilde{E}_t, \tilde{E}|_{D_R}$ such that $\tilde{E}_i$ is R_i-flat coherent sheaf on X_{R_i} and that there is a surjection

$$\tilde{E}_i \otimes_{\mathcal{O}_{R_i}} \mathcal{O}_{R_{i-1}} \longrightarrow \tilde{E}_{i-1}.$$

Set R_{j-1} to be $\mathrm{Quot}^{H_{j-1}}_{\tilde{E}_j/X_{R_j}/R_j}$ and take the universal quotient

$$\tilde{E}_j \otimes_{\mathcal{O}_{R_j}} \mathcal{O}_{R_{j-1}} \longrightarrow \tilde{E}_{j-1}.$$

on R_{j-1}. By induction on j we come to R_1 and we have a sequence of surjections

$$\tilde{E}|_{D_R} \otimes_{\mathcal{O}_R} \mathcal{O}_{R_1} \to \tilde{E}_t \otimes_{\mathcal{O}_{R_t}} \mathcal{O}_{R_1} \to \cdots \to \tilde{E}_2 \otimes_{\mathcal{O}_{R_2}} \mathcal{O}_{R_1} \to \tilde{E}_1.$$

Setting

$$\begin{aligned} E &= \tilde{E} \otimes_{\mathcal{O}_R} \mathcal{O}_{R_1} \\ F_i &= \ker(\tilde{E}|_{D_R} \otimes_{\mathcal{O}_R} \mathcal{O}_{R_1} \to \tilde{E}_i \otimes_{\mathcal{O}_{R_i}} \mathcal{O}_{R_1}) \end{aligned}$$

we get an R_1-family of parabolic sheaves. There is an open subscheme U of R_1 such that for every algebraically closed field k, a k-valued point z of R_1 is in $U(k)$ if and only if $(E(z), F_*(z), \alpha_*)$ is stable. We shall denote the restriction of (E, F_*, α_*) to U by the same (E, F_*, α_*).

The S-group scheme $G = GL(V)$ naturally acts on Q and R is G-invariant. There is a canonical G-linearization on $\tilde{E}$ and hence G so acts on R_t that the natural morphism of R_t to R is G-invaraiant. Then $\tilde{E}_t$ carries a natural G-linearization. Tracing these procedure to R_1, we come eventually to a G-action on U and a G-linearization of the family (E, F_*, α_*). Obviously the center $\mathbf{G}_{m,S}$ of G acts trivially on U and we have an action of $\overline{G} = G/\mathbf{G}_{m,S}$. We can show that there exists a geomeric quotient of U by $\overline{G}$. Then we see

(3.4) The quotient $U/\overline{G}$ is the moduli scheme in Theorem 2.6.

For a T-family of parabolic sheaves (E, F_*, α_*), we put

$$K_i = \ker(E \to E|_{D_T}/F_i)$$

and then we get a sequence of T-flat coherent subsheaves $K_{t+1} = E(-D_T) \subset K_t \subset \cdots \subset K_2 \subset K_1 = E$. For a real number α, there is an integer i such that $1 \leq i \leq t+1$ and $\alpha_{i-1} < \alpha - [\alpha] \leq \alpha_i$, where $\alpha_0 = \alpha_t - 1$ and $\alpha_{t+1} = 1$. Then we set

$$E_\alpha = K_i(-[\alpha]D_T).$$

Thus we obtain a filtration $\{E_\alpha\}_{\alpha \in \mathbf{R}}$ of E parametrized by real numbers that has the following properties

(3.5.1) E/E_α is T-flat and if $\alpha \leq \beta$, then $E_\alpha \supseteq E_\beta$.

(3.5.2) If ε is a sufficiently small positive real number, then $E_{\alpha-\varepsilon} = E_\alpha$.

(3.5.3) For every real number α, we have $E_{\alpha+1} = E_\alpha(-D_T)$.

(3.5.4) $E_0 = E$.

(3.5.5) The length of the filtration for $0 \leq \alpha \leq 1$ is finite.

Convesely, if E is a T-flat coherent sheaf on X_T such that for every geometric point y of T, $E(y)$ is torsion free and that E has a filtration parametrized by $\mathbf{R}$ with the above properties, then we have a T-family of parabolic sheaves. Thus we may use the notation E_* for a T-family of parabolic sheaves (E, F_*, α_*).

3.6 Definition. Let E_* and E'_* be T-familes of parabolic sheaves. A homomorphism $f : E \to E'$ of the undelying coherent sheaves is said to be a homomorphism of parabolic sheaves if for all $0 \leq \alpha < 1$, we have

$$f(E_\alpha) \subset E'_\alpha.$$

$\mathrm{Hom}^{\mathrm{par}}(E_*, E'_*)$ denotes the set of all homomorphisms of parabolic sheaves of E_* to E'_*.

If one notes that for a stable parabolic sheaf E_* on a projective varaiety, a homomorphism of E_* to itself is the multiplication by an element of the ground field k or $\mathrm{Hom}^{\mathrm{par}}(E_*, E_*) \cong k$, then one can prove the following lemma by the same argument as the proof of Lemma 6.1 in [5].

3.7 Lemma. *Let A be an artinian local ring with residue field k and let E_* be a* Spec(A)*-family of parabolic sheaves. Assume that the restriction $\overline{E} = E_* \otimes_A k$ to the closed fiber is stable. Then the natural homomorphism $A \to \mathrm{Hom}^{\mathrm{par}}(E_*, E_*)$ is an isomorphism.*

Let us go back to the situation of (3.4). Replacing the role of Lemma 6.1 in the argument of Lemma 6.3 of [5] by the above lemma, we get a basic result on the action of $\overline{G}$ on U.

3.8 Lemma. *The action of $\overline{G}$ on U is free.*

It is well-known that this lemma implies the following (see [5, Proposition 6.4].

3.9 Proposition. *The natural morphism $\pi : U \to W = U/\overline{G}$ is a principal fiber bundle with group $\overline{G}$*

Now we come to our proof of Proposition 3.2.

Proof of Proposition 3.2. There is a universal space U whose quotient by the group $\overline{G}$ is $\tilde{M}(r,n) = M(H, H_*, \alpha_*)$ under the notation before Proposition 3.2. Since our moduli space $M(r,n)$ is open in $\tilde{M}(r,n)$, we have a $\overline{G}$-invariant open subscheme P of U which mapped onto $M(r,n)$. Thanks to Proposition 3.9 the natural quotient morphism $\pi : P \to M(r,n)$ is a principal fiber bundle with group $\overline{G}$ and hence we have only to show that P is smooth and has the right dimension. Put $X = \mathbf{P}^2_{\mathbf{C}}$. Fix an integer m which satisfies the conditions (3.3.1) and (3.3.2) for our $M(r,n)$. We set $H[m](x) = H(x+m)$. Take a point E of $M(r,n)(\mathbf{C})$ and a $\mathbf{C}$-vector space V of dimension $N = H^0(X, E(m))$. We have an surjection

$$\theta : V_X = V \otimes_{\mathbf{C}} \mathcal{O}_X \longrightarrow E(m).$$

Since the kernel K of θ is locally free, $\mathrm{Hom}_{\mathcal{O}_X}(K, E(m))$ is the tangent space of $Q = \mathrm{Quot}^{H[m]}_{V_X/X/\mathbf{C}}$ at the point q that is given by the above sequence and an obstraction of the smoothness of Q at q is in $H^1(X, K^\vee \otimes_{\mathcal{O}_X} E(m))$. Since $\mathrm{Ext}^2_{\mathcal{O}_X}(E, E)$ is a dual space of $\mathrm{Hom}_{\mathcal{O}_X}(E, E(-3))$, it vanishes for stable E.

We can apply the same argument as in the proofs of Propositions 6.7 and 6.9 in [5] to our situation and we see that Q is smooth and of dimension

$$2rn - r^2 + N^2$$

at the point q. P is a fiber space over an open subscheme of Q whose fibers are an open subscheme of $\mathrm{Quot}_{\mathcal{O}_\ell^{\oplus r}/\ell/\mathbf{C}}$ consisting of surjections

$$\mathcal{O}_\ell^{\oplus r} \longrightarrow \mathcal{O}_\ell(r-1)$$

such that the induced map of global sections is bijective. By Lemma 2.7 the space of obstructions for the smoothness of P over Q is $H^1(\ell, \mathcal{O}_\ell(r)^{\oplus r-1}) = 0$. Thus P is smooth over Q and hence so is over $\mathrm{Spec}(\mathbf{C})$. Moreover, the dimension of the fiber is equal to $\dim H^0(\ell, \mathcal{O}_\ell(r)^{\oplus r-1}) = r^2 - 1$. Combining this and the above result on the dimension of Q, we see that $\dim P = 2rn + N^2 - 1$. Since $\dim \overline{G} = N^2 - 1$, $M(r,n)$ is of dimension $2rn$ at every point.

Q.E.D.

Based on Proposition 3.2 and Hulek's result stated in Introduction, we can prove the following.

3.10 Theorem. *$M(r,n)_0$ is connected.*

Proof. Our proof is divided into several steps. We set as before X to be $\mathbf{P}^2_{\mathbf{C}}$.

(I) If $r = 1$, then $M(r,n)$ is the moduli space of ideals with colength n which define 0-dimensional closed subschemes in $X \setminus \ell$. It is well known that this is irreducible [2].

(II) Assume that $r \geq 2$. By Hulek's result we see that

$$U(0) = \{E_* \in M(r,n)_0 \mid H^0(X,E) = H^0(X,E^\vee) = 0\}$$

is irreducible. In fact, we have a surjective morphism of a $PGL(r)$-bundle over Hulek's parameter space of s-stable bundles to $U(0)$. Let us set

$$\begin{aligned} U(a) &= \{E_* \in M(r,n)_0 \mid \dim H^0(X,E) = a\}, \\ U(a,b) &= \{E_* \in U(a) \mid E \cong \mathcal{O}_X^{\oplus b} \oplus E_1, E_1 \not\cong \mathcal{O}_X \oplus E_1'\}. \end{aligned}$$

Then $U(a)$ is locally closed and $U(a,b)$ is constructible in $M(r,n)_0$.

(III) We shall compute the dimension of $U(a,b)$. For an $E_* \in U(a,b)$, there is an extension

$$0 \longrightarrow \mathcal{O}_X^{\oplus a} \longrightarrow E \longrightarrow J \longrightarrow 0$$

because E is μ-semi-stable and $c_1(E) = 0$ [6, Lemma 1.1]. We have moreover that $J|_\ell \cong \mathcal{O}_\ell^{\oplus r-a}$. We may assume that the first b direct factors of $\mathcal{O}_X^{\oplus a}$ is a direct summand of E and put $E = \mathcal{O}_X^{\oplus b} \oplus E_1$. Consider the exact sequence

$$0 \longrightarrow \mathcal{O}_X^{\oplus a-b} \longrightarrow E_1 \longrightarrow J \longrightarrow 0.$$

For the double dual J' of J, we set $T = J'/J$ and $c = \dim H^0(X,T)$. Then we see

$$\mathcal{E}xt^2_{\mathcal{O}_X}(T, \mathcal{O}_X) \cong \mathcal{E}xt^1_{\mathcal{O}_X}(J, \mathcal{O}_X)$$

and $\dim H^0(X, \mathcal{E}xt^2_{\mathcal{O}_X}(T, \mathcal{O}_X)) = c$. On the other hand, since $c_2(J^\vee) = n - c$, $J^\vee$ is locally free and since $H^2(X, J^\vee) = 0$, we have

$$\dim H^1(X, J^\vee) = n - c - (r-a) + d,$$

where $d = \dim H^0(X, J^\vee)$. Thus we see

$$\begin{aligned} \dim \operatorname{Ext}^1_{\mathcal{O}_X}(J, \mathcal{O}_X^{\oplus(a-b)}) &= \dim H^1(X, (J^\vee)^{\oplus a-b}) + \\ &\quad \dim H^0(X, \mathcal{E}xt^1_{\mathcal{O}_X}(J, \mathcal{O}_X)^{\oplus a-b}) \\ &= (a-b)\{n - (r-a) + d\}. \end{aligned}$$

If we change free bases of $\mathcal{O}_X^{\oplus a-b}$, then we obtain the same sheaf. Hence if $\dim \operatorname{Ext}^1_{\mathcal{O}_X}(J, \mathcal{O}_X) < a-b$, then every extension of J by $\mathcal{O}_X^{\oplus a-b}$ contains $\mathcal{O}_X$ as a direct factor, which is not the cace. Therefore, we get

$$\begin{aligned} a - b &\leq n - (r-a) + d \qquad \text{or} \\ n - r + b + d &\geq 0 \end{aligned}$$

The extensions of J by $\mathcal{O}_X^{\oplus a-b}$ are now parametrized by a space of dimensin $(a-b)\{n-(r-a)+d\}$ and each point of the space is contained in a subspace of dimension $(a-b)^2 + \dim \operatorname{End}_{\mathcal{O}_X}(J) - 1$ whose points parmetrize the same extension.

IV) Let us fix a system of homogeneous coordinates $(x_0 : x_1)$ of ℓ and identify $J|_\ell$ with the free sheaf $\oplus_{i=1}^{r-a} \mathcal{O}_\ell e_i$. If we have a surjection

$$\varphi : J|_\ell = \oplus_{i=1}^{r-a} \mathcal{O}_\ell e_i \longrightarrow \mathcal{O}_\ell(r-a-1)$$

and if $f_1(x_0, x_1) = \varphi(e_1), \ldots, f_{r-a}(x_0, x_1) = \varphi(e_{r-a})$ are linearly independent, then for general homogeneous forms $g_1, \ldots, g_{r-a}$ of degree $a-1$ and $g_{r-a+1}, \ldots, g_r$ of degree $r-1$, we define a map $\tilde{\varphi}$ of $\oplus_{i=1}^r \mathcal{O}_\ell e_i$ to $\mathcal{O}_\ell(r-1)$ by

$$\begin{aligned} \tilde{\varphi}(e_i) &= g_i x_0^{r-a} + x_1^a f_i \qquad 1 \le i \le r-a \\ \tilde{\varphi}(e_j) &= g_j \qquad\qquad\qquad r-a+1 \le j \le r. \end{aligned}$$

Choosing g_j suitably, we obtain a surjective $\tilde{\varphi} : \oplus_{i=1}^r \mathcal{O}_\ell e_i \to \mathcal{O}_\ell(r-1)$ which induces a bijection between the spaces of global sections. Conversely, if we have a homomorphism ψ of $\oplus_{i=1}^r \mathcal{O}_\ell e_i$ to $\mathcal{O}_\ell(r-1)$ such that $h_1 = \psi(e_1), \ldots, h_r = \psi(e_r)$ are linearly independent. Then we can write h_i uniquely in the following way:

$$h_i = h_i' x_0^{r-a} + x_1^a h_i''$$

where h_i' (or, h_i'') is of degree $a-1$ (or, $r-a-1$, resp.). There is a permutation σ of $\{1, \ldots, r\}$ such that $h''_{\sigma(1)}, \ldots, h''_{\sigma(r-a)}$ are linearly independent. Thus, after a permutation of indices, the above procedure produces ψ from a homomorphism $J|_\ell \to \mathcal{O}_\ell(r-a-1)$ which induces a bijection between the spaces of global sections. We see therefore that the parabolic structures on E are parametrized by a fiber space over the space of parabolic structures of J whose fibers are a finite union of open subschemes of an affine space of dimension $2ra - a^2$.

(V) Every element of $\mathrm{End}_{\mathcal{O}_X}(E)$ induces an endomorhism of the space $H^0(X, E)$ and hence gives rise to an element of $\mathrm{End}_{\mathcal{O}_X}(J)$. Let $\mathrm{End}^J(E)$ denote the subspace of $\mathrm{End}_{\mathcal{O}_X}(E)$ consisting of the elements which induce the identity on J. Then we have

$$\dim \mathrm{End}^J(E) \ge \dim \mathrm{End}_{\mathcal{O}_X}(E) - \dim \mathrm{End}_{\mathcal{O}_X}(J).$$

On the other hand, we see

$$\begin{aligned} \mathrm{End}_{\mathcal{O}_X}(E) &= \mathrm{End}_{\mathcal{O}_X}(\mathcal{O}_X^{\oplus b}) \oplus \mathrm{Hom}_{\mathcal{O}_X}(\mathcal{O}_X^{\oplus b}, E_1) \\ &\quad \oplus \mathrm{Hom}_{\mathcal{O}_X}(E_1, \mathcal{O}_X^{\oplus b}) \oplus \mathrm{End}_{\mathcal{O}_X}(E_1). \end{aligned}$$

Since $\dim H^0(X, J^\vee) = d$, there is an exact sequence

$$0 \longrightarrow J_1 \longrightarrow J \longrightarrow G \longrightarrow 0$$

with $(G^\vee)^\vee \cong \mathcal{O}_X^{\oplus d}$. For a ξ in $\mathrm{Hom}_{\mathcal{O}_X}(G, \mathcal{O}_X^{\oplus a-b}) \cong H^0(X, \mathcal{O}_X^{\oplus d(a-b)})$, we have a member of $\mathrm{End}(E_1)$

$$E_1 \longrightarrow J \longrightarrow G \xrightarrow{\ \xi\ } \mathcal{O}_X^{\oplus a-b} \longrightarrow E_1.$$

Thus $\dim \mathrm{End}_{\mathcal{O}_X}(E_1) \geq d(a-b)+1$. Therefore, we get the following inequality

$$\begin{aligned}\dim \mathrm{End}^J(E) &\geq \dim \mathrm{End}_{\mathcal{O}_X}(E) - \dim \mathrm{End}_{\mathcal{O}_X}(J) \\ &\geq b^2 + b(a-b) + db + d(a-b) + 1 - \dim \mathrm{End}_{\mathcal{O}_X}(J) \\ &= ab + ad + 1 - \dim \mathrm{End}_{\mathcal{O}_X}(J)\end{aligned}$$

(VI) There is a couple $(A, \tilde{J}_*)$ of a scheme A and an A-family $\tilde{J}_*$ of parabolic sheaves which parametrizes all parabolic stable sheaves J_* with rank $r-a$ and $c_2(J) = n$ such that the restriction $J|_\ell$ is a trivial vector bundle and $H^0(X, J) = 0$. We may assume that A is reduced and quasi-finite over $M(r-a, n)$, and hence $\dim A \leq 2n(r-a)$. By Proposition 2.8 the sheaf J defined in the step (III) appears as the underlying sheaf of a prabolic sheaf parametrized by $\tilde{J}_*$. For the underlying sheaf $\tilde{J}$ of the family $\tilde{J}_*$, we have a resolution by a locally free sheaves

$$0 \longrightarrow B_1 \longrightarrow B_0 \longrightarrow \tilde{J} \longrightarrow 0.$$

Using this resolusion and splitting out A into the direct sum of suitable subschemes, we can construct a locally free sheaf C on A such that for every point y of A, we have a natural isomorphism

$$C(y) \cong \mathrm{Ext}^1_{\mathcal{O}_X}(\tilde{J}(y), \mathcal{O}_X).$$

Note that C is not necessarily of constant rank. On $D = \mathbf{V}((C^\vee)^{\oplus a-b})$ we have a universal section ξ of the sheaf $g^*(C^{\oplus a-b})$, where $g : X \times D \to X \times A$ is the natural projection. Let ξ_i be the projection of ξ to the i-th direct factor of $C^{\oplus a-b}$. The subset $D_0 = \{y \in D \mid \xi_1(y), \cdots, \xi_{a-b}(y)$ span a linear subspace of rank $a-b$ in $C(y)\}$ is open in D. $\xi|_{D_0}$ gives rise to the universal family of extensions

$$0 \longrightarrow \mathcal{O}^{\oplus a-b}_{X\times D_0} \longrightarrow \tilde{E}_1 \longrightarrow g^*(\tilde{J}) \longrightarrow 0,$$

where we denote $g|_{D_0}$ by g. Let H be the maximal open subscheme of D_0 where $\tilde{E}_1$ is locally free. Set $\tilde{E} = \tilde{E}_1|_{X\times H} \oplus \mathcal{O}^{\oplus b}_{X\times H}$. (III) tells us

$$\dim D_0 \leq 2n(r-a) + (a-b)\{n - (r-a) + d\}.$$

Note here that d may depend on connected components of D_0. Furthermore, by the result of (III) again, each point of H is contained in a subspace of dimension $(a-b)^2 + \dim \mathrm{End}_{\mathcal{O}_X}(J) - 1$ where $\tilde{E}_1$ parmetrizes the same extensions.

(VII) By breaking up H into the direct sum of suitable subschemes, we may assume that $g^*(\tilde{J})|_{\ell\times H}$ has a constant trivialization and hence so is $\tilde{E}|_{\ell\times H}$. By the result of (IV), the parabolic structure of $g^*(\tilde{J})$ provides us with a fiber space $p : Z \to H$ and a Z-family $\tilde{E}_*$ of parabolic stable vector bundles such that $(Z, \tilde{E}_*)$ parametrizes all the parabolic stable sheaves contained in $U(a, b)$ and that every fiber of p is of dimension $2ra - a^2$. Thus we see that

$$\begin{aligned} \dim Z &\leq 2n(r-a) + (a-b)\{n - (r-a) + d\} + 2ra - a^2 \\ &= 2nr - na - nb + ra + rb - ab + ad - bd \end{aligned}$$

Moreover, the conclusion of (V) shows that on the fiber of p, each point is contained in a closed subscheme of dimension at least $ab+ad+1-\dim \mathrm{End}_{\mathcal{O}_X}(J)$ whose points define the same parabolic sheaf.

(VIII) The family $\tilde{E}_*$ gives rise to a morphism of Z to $M(r, n)_0$ whose image is exactly our $U(a, b)$. Combining the results of (VI) and (VII) we get

$$\begin{aligned} \dim U(a,b) &\leq 2nr - na - nb + ra + rb - ab + ad - bd - (a-b)^2 - \\ &\qquad \dim \mathrm{End}_{\mathcal{O}_X}(J) + 1 - ab - ad - 1 + \dim \mathrm{End}_{\mathcal{O}_X}(J) \\ &= 2nr - (n - r + a)a - (n - r + b + d)b \end{aligned}$$

Since for every member E_* of $U(a, b)$, the underlying sheaf E is μ-semi-stable, Riemann-Roch implies

$$n - r + a = \dim H^1(X, E) \geq 0.$$

This and the inequality we obtained in (III) show that

$$\dim U(a, b) \leq 2nr.$$

Replacing E by $E^\vee$ in the definition of $U(a)$ and $U(a, b)$, we define $U^\vee(a)$ and $U^\vee(a, b)$. Then, by the same argument as above we come to a family (Z', E'_*) of the dual bundles of the members of $U^\vee(a, b)$. By taking the dual basis of the trivial sheaf $E'|_{\ell\times Z'}$, we have an isomorphism $E'^\vee|_{\ell\times Z'} \to E'|_{\ell\times Z'}$. Combining this isomorphism and the parabolic structure of E'_*, we obtain a Z'-family $E'^\vee_*$ of parabolic sheaves which parametrizes all the members of $U^\vee(a, b)$. The dimension of $U^\vee(a, b)$ is the same as $U(a, b)$.

(IX) Assume that $n \geq r$. In this case we see that

$$M(r, n)_0 = U(0) \bigcup \left(\bigcup_{a\geq 1, b\geq 0} U(a, b) \right) \bigcup \left(\bigcup_{a\geq 1, b\geq 0} U^\vee(a, b) \right).$$

By our result in (VIII) we see that if $a \geq 1$, then both $\dim U(a, b)$ and $\dim U^\vee(a, b)$ are less than $2rn$. On the other hand, we know that $U(0)$ is irreducible by [3]. This completes the proof of this case.

(X) Assume that $n < r$. Then, Riemann-Roch implies that for every member E_* of $M(r,n)_0$, we have $\dim H^0(X,E) \geq r-n$. Thus we see that

$$M(r,n)_0 = U(r-n) \bigcup \left(\bigcup_{a>r-n, b\geq 0} U(a,b) \right).$$

As in (IX) if $a > r-n$, then $\dim U(a,b) < 2rn$. This means that it is sufficient to prove that $U(r-n)$ is connected. For a member E_* of $U(r-n)$, there is an exact sequence

$$0 \longrightarrow \mathcal{O}_X^{\oplus r-n} \longrightarrow E \longrightarrow J \longrightarrow 0.$$

According to the type of J, $U(r-n)$ is deivided into three subschemes:

$$\begin{aligned} V_0 &= \{E_* \in U(r-n) \mid J \text{ is locally free}, H^0(X, J^\vee) = 0\} \\ V_1 &= \{E_* \in U(r-n) \mid J \text{ is locally free}, H^0(X, J^\vee) \neq 0\} \\ V_2 &= \{E_* \in U(r-n) \mid J \text{ is not locally free}\}. \end{aligned}$$

For J of $E_* \in V_0$, we have that $H^0(X, J^\vee) = 0$, $c_1(J^\vee) = 0$, $c_2(J^\vee) = n$, $r(J^\vee) = n$ and $J^\vee$ is μ-semi-stable. Hence Riemann-Roch implies that $\mathrm{Ext}^1_{\mathcal{O}_X}(J, \mathcal{O}_X) = H^1(X, J^\vee) = 0$. Then V_0 is contained in $U(r-n, r-n)$ or the undelying sheaf E of a member of V_0 is written in a form $\mathcal{O}_X^{\oplus r-n} \oplus J$ with J s-stble. Since these J's are parametrized by an irreducible variety, so are the members of V_0. This proves the irreducibilty of V_0. Applying the argument before (VIII) to the set $\{J^\vee \mid E_* \in V_1\}$, we find that $\{J_* \mid E_* \in V_1\}$ is of dimension less than $2n^2$. Then the dimension of A in (VI) for V_1 is of dimension less than $2n^2$ and hence $\dim V_1 < 2nr$. A similar argument tells us that our remaining problem is to prove that $L = \{J_* \mid E_* \in V_2\}$ is of dmension less than $2n^2$.

(XI) For the underlying sheaf J of a member J_* of L, we set J' to be the double dual of J. Since J is locally free in a neighborhood of ℓ, the parabolic structure of J_* induces that of J'. J'/J is a torsion sheaf supported by a 0-dimensional subscheme of X. L is the disjoint union of $L_1, \dots, L_n$, where $L_m = \{J_* \in L \mid \dim H^0(X, J'/J) = m\}$. Pick an J_* of L_m and let $\{x_1, \dots, x_k\}$ be the support of J'/J. If the length of the artinian module $(J'/J)_{x_i}$ is a_i, then there is a filtration $0 \subset T^{(i)}_{a_i} \subset \cdots \subset T^{(i)}_1 = (J'/J)_{x_i}$ such that $T^{(i)}_j / T^{(i)}_{j+1} \cong k(x_i)$. Let $J^{(1)}_j$ be the kernel of the surjection $J' \to T^{(1)}_1 / T^{(1)}_{j+1}$ and $J^{(2)}_j$ be the kernel of $J^{(1)}_{a_1} \to T^{(2)}_1 / T^{(2)}_{j+1}$. Thus we can define a filtration $J^{(k)}_{a_k} \subset \cdots \subset J^{(k)}_1 \subset \cdots \subset J^{(1)}_{a_1} \subset \cdots \subset J^{(1)}_1 \subset J'$ such that $J^{(i)}_j / J^{(i)}_{j+1} \cong k(x_i)$

and $J^{(i)}_{a_i}/J^{(i+1)}_1 \cong k(x_{i+1})$ and we see that $J^{(k)}_{a_k} = J$. Since $J^{(i-1)}_{a_{i-1}}(x_i)$ is an n-dimensional vector space, the surjection of $J^{(i-1)}_{a_{i-1}}$ to $k(x_i)$ is parametrized by an $(n-1)$-dimensional projective space. Since $\mathrm{Tor}^{\mathcal{O}}_1(k(x_i), k(x_i))$ is isomorphic to $k(x_i)^{\oplus 2}$, the exact sequence

$$\mathrm{Tor}^{\mathcal{O}}_1(k(x_i), k(x_i)) \longrightarrow J^{(i)}_j(x_i) \longrightarrow J^{(i)}_{j-1}(x_i) \longrightarrow k(x_i) \longrightarrow 0$$

shows us that $\dim J^{(i)}_j(x_i) \leq \dim J^{(i)}_{j-1}(x_i) + 1$. Therefore, surjections of $J^{(i)}_j$ to $k(x_i)$ is parametrized by a projective space of dimension less than or equal to $n + j - 2$, where $\mathcal{O} = \mathcal{O}_{X,x_i}$. Fixing J'_*, $\{J_* \in L_m \mid (J^\vee)^\vee \cong J'\}$ is parametrized by a space of dimension less than or equal to

$$\begin{aligned}
\delta_m &= \sum_{i=1}^{k}\sum_{j=1}^{a_i}(n+j-2) + 2k \\
&= nm + \sum_{i=1}^{k}\frac{a_i(a_i-3)}{2} + 2k \\
&= nm + \sum_{i=1}^{k}\left(\frac{a_i^2}{2} - \frac{3a_i}{2} + 2\right).
\end{aligned}$$

On the other hand, the space $\{J'_* \mid J_* \in L_m\}$ is of dimension $2n(n-m)$. Therefore, L_m is of dimension at most $2n(n-m) + \delta_m$. Now we have

$$\begin{aligned}
2n(n-m) + \delta_m &= 2n^2 - nm + \sum_{i=1}^{k}\left(\frac{a_i^2}{2} - \frac{3a_i}{2} + 2\right) \\
&\leq 2n^2 - m^2 + \sum_{i=1}^{k}\left(\frac{a_i^2}{2} - \frac{3a_i}{2} + 2\right) \\
&= 2n^2 - \sum_{i=1}^{k}\left(\frac{a_i^2}{2} + \frac{3a_i}{2} - 2 + a_i\sum_{j\neq i} a_j\right).
\end{aligned}$$

It is easy to see that

$$\sum_{i=1}^{k}\left(\frac{a_i^2}{2} + \frac{3a_i}{2} - 2 + a_i\sum_{j\neq i} a_j\right)$$

is non-negative and equal to 0 if and only if $n = 1$. If $n = 1$, then both V_0 and V_1 are empty and V_2 is exactly $\{E_* \mid J = I_x$ with I_x the ideal of a point $x \in X \setminus \ell\}$. There is a unique locally free sheaf G_x which is an extension of

I_x by $\mathcal{O}_X$. Finally we see that in this case the set of undelying sheaves of the members of $U(r-n)$ is

$$\{\mathcal{O}_X^{\oplus r-2} \oplus G_x \mid x \in X \setminus \ell\}$$

which is parametrized by the irreducible variety $X \setminus \ell$. Q.E.D.

Now we come to the connectedness of the moduli space of marked $SU(r)$-instantons.

3.11 Corollary. *The moduli space $M(SU(r), n)$ of marked $SU(r)$-instantons with instanton number n is connected.*

References

[1] S. K. Donaldson, *Instantons and geometric invariant theory* , Comm. Math. Phys. **93** (1984) 453–460.

[2] J. Fogarty, *Algebraic families on an algebraic surface*, Amer. J. Math. **90** (1968) 511–521.

[3] K. Hulek, *On the classification of stable rank-r vector bundles on the projective plane,* Proc. Nice conference on vector bundles and differential equations, Birkhäuser (1983) 113–142.

[4] G. Kempf and L. Ness, *Lengths of vectors in representation spaces*, Lect. Notes in Math. **732**, Springer-Verlag, Berlin–Heidelberg–New York–Tokyo, (1978) 233–243.

[5] M. Maruyama, *Moduli of stable sheaves II,* J. Math. Kyoto Univ. **18** (1978) 557–614.

[6] M. Maruyama, *On a compactification of a moduli space of stable vector bundles on a rational surface,* Alg. Geom. and Comm. Alg. in Honor of Masayoshi NAGATA, Kinokuniya, Tokyo (1987) 233–260.

[7] M. Maruyama and K. Yokogawa,*Moduli of parabolic stable sheaves*, to appear in Math. Ann.

[8] C. H. Taubes, *Pathconnected Yang-Mills moduli spaces,* J. Diff. Geom. **19** (1984) 337–392.

Department of Mathematics
Faculty of Sciense
Kyoto University
Sakyo-ku, Kyoto 606-01
Japan

Numerically Effective line bundles which are not ample

V.B. Mehta and S. Subramanian

1 Introduction

In [6], there is a construction of a line bundle on a complex projective nonsingular variety which is ample on every proper subvariety but which is nonample on the ambient variety. The example is obtained as the projective bundle associated to a "general" stable vector bundle of degree zero on a compact Riemann surface of genus $g \geq 2$. Now, we can show by an algebraic argument valid in any characteristic, the existence of a variety of dimension ≤ 3 with a line bundle as above. The details of the proof will appear elsewhere.

2. Let C be a complete nonsingular curve defined over an uncountable algebraically closed field (of any characteristic). Let M_r^s denote the moduli space of stable bundles of rank r and degree zero on C and M_r^{ss} the moduli of semistable bundles of rank r and degree zero. We assume throughout that the curve C is ordinary. We can show

2.1 Proposition. *Let the characteristic of the ground field be positive and F the Frobenius morphism on C. There is a proper closed subset of M_r^s such that for any stable bundle V in the complement of this closed set, F^*V is also stable.*

Proof. We use Artin's theorem on the algebraisation of formal moduli space for proving the above proposition. We have as a corollary to the proof of Proposition (2.1).

2.1.1 Corollary. *Let C be an ordinary curve. Then the rational map $f : M_r^{ss} \to M_r^{ss}$ induced by the Frobenius $F : C \to C$, is etale on an open set, and in particular, dominant.*

2.1.2 Corollary. *Let C be an ordinary curve. For any positive integer k, there is a nonempty open subset of M_r^s such that for V in this open set, $F^{m*}V$ is stable for $1 \leq m \leq k$.*

Proof. We apply Proposition (2.1) and Corollary (2.1.1) successively. QED.

We have

2.2 Proposition. *Given a finite etale morphism $p : C_1 \to C$, there exists a proper closed subset of M_r^s such that any vector bundle in the complement of this closed set remains stable on C_1.*

We also have

2.3 Proposition. *For a fixed positive integer k, there is a nonempty open subset of M_r^s such that for any stable bundle V in this open set, there is no nonzero homomorphism from a line bundle of degree zero to S^kV.*

Using the above results, we can show

2.4. Theorem. *Let C be a nonsingular ordinary curve of genus ≥ 2 over an uncountable algebraically closed field (of any characteristic). There is a dense subset of M_r^s such that for any stable bundle V in this dense set, we have*

1) *$F^{k^*}(V)$ is stable for all $k \geq 1$.*

2) *For any separable finite morphism $\pi : \tilde{C} \to C, \pi^*(V)$ is stable.*

3) *There is no nonzero homomorphism from a line bundle of degree zero to the symmetric power $S^k(V)$ for any $k \geq 1$.*

Remark. If C is a smooth curve defined over a finite field (of characteristic p) then any continuous irreducible representation $\rho : \pi_1^{alg}(C) \to SL(r, \overline{\mathbf{F}}_p)$ of the algebraic fundamental group of C of rank r over the finite field defines a stable vector bundle V on C such that $F^{m*}V \simeq V$ for some $m \geq 1$ (see [4]). Such a bundle V satisfies $F^{k*}(V)$ is stable for all $k \geq 1$. We can construct such representations for any curve C of genus $g \geq 2$ when r is coprime to p, and for an ordinary curve C when p divides r.

3. Let C be a nonsingular ordinary curve of genus ≥ 2, and V a stable vector bundle of rank 3 and degree zero on C satisfying the conditions of Theorem (2.4) above. Let $\pi : P(V) \to C$ be the projective bundle associated to V and $L = \mathcal{O}_{\mathbf{P}(V)}(1)$ the universal line bundle on $\mathbf{P}(V)$. Then we have

3.1 Theorem. *The line bundle L is ample on every proper subvariety of $\mathbf{P}(V)$, but L is not ample on $\mathbf{P}(V)$.*

Proof. We can check that $L.C > 0$ for any integral curve C, and that $L^2.D > 0$ for any irreducibl divisor D. This implies that $L \mid D$ is ample on D. This shows that L is ample on divisors in $\mathbf{P}(V)$ and hence on any proper subvariety of $\mathbf{P}(V)$. Also, L, is not ample on $\mathbf{P}(V)$. QED.

3.2 Remark. The case $r = 2$ is covered by the first part of Theorem (3.1).

References

1. M. Artin, *The implicit function theorem in Algebraic Geometry,* Proceedings International Colloquium on Algebraic Geometry, Bombay 1968, Oxford University Press (1969) 13–34.

2. T. Fujita, *Semipositive line bundles,* Jour. of Fac. Sc. Univ. of Tokyo, Vol. **30** (1983) 353–378.

3. R. Hartshorne, *Ample subvarieties of algebraic varieties*, Lecture Notes No. **156** (springer-Verlag) (1970)

4. H. Lange and U.Stuhler, *Vektor bundel and Kurven und Darstellungen der algebraischen Fundamental gruppe,* Math. Z., **156** (1977) 73–83.

5. V.B. Mehta and A. Ramanathan, *Restriction of stable sheaves and representations of the fundamental group,* Inv. Math. **77** (1984) 163–172.

6. S. Subramanian, *Mumford's example and a general construction,* Proc. Ind. Acad. of Sci. Vol. **99**, December (1989) 97–208.

School of Mathematcs
Tata Institute of Fundamental Research
Homi Bhabha Road, Colaba
Bombay 400 005
INDIA.

Moduli of logarithmic connections

Nitin Nitsure

The talk was based on the paper [N] which will appear elsewhere. What follows is a summary of the results.

Let X be a non-singular projective variety, with $S \subset X$ a divisor with normal crossings. A logarithmic connection $E = (\mathcal{E}, \nabla)$ on X with singularity over S is a torsion free coherent sheaf $\mathcal{E}$ together with a $\mathbf{C}$-linear map $\nabla : \mathcal{E} \to \Omega^1_X[\log S] \otimes \mathcal{E}$ satisfying the Leibniz rule and having curvature zero, where $\Omega^1_X[\log S]$ is the sheaf of 1-forms on X with logarithmic singularities over S. By a theorem of Deligne [D], a connection with curvature zero on a non-nonsingular quasi-projective variety Y is regular if and only if given any Hironaka completion X of Y (so that X is non-singular projective and $S = X - Y$ is a divisor with normal crossings), the connection extends to a logarithmic connection on X with singularity over S.

Carlos Simpson has constructed in [S] a moduli scheme for non-singular connections (with zero curvature) on a projective variety. A simple example (see [N]) shows that a moduli scheme for regular connections on a quasi-projective variety does not in general exist. Therefore, we have to consider the moduli problem for logarithmic connections on a projective variety.

The main difference between non-singular connections and logarithmic connections is that for logarithmic connections, we have to define a notion of (semi-)stability, and restrict ourselves to these. We say that a logarithmic connection is (semi-)stable, if the usual inequality between normalized Hilbert polynomials is satisfied for any ∇-invariant coherent subsheaf. In the case of non-singular connections on a projective variety, the normalized Hilbert polynomial is always the same, so semi-stability is automatically fulfilled. Following Simpson's method, with the extra feature of keeping track of (semi-)stability, we prove the existence of a coarse moduli scheme for (S-equivalence classes of) semistable-logarithmic connections which have a given Hilbert polynomial. We also show that the infinitesimal deformations of a locally free logarithmic connection E are parametrized by the first hypercohomology of the logarithmic de Rham complex associated with $\mathrm{End}(E)$.

A given regular connection on a quasi-projective variety Y has infinitely many extensions as logarithmic connections on a given Hironaka completion X of Y. A canonical choice of such an extension is given by the fundamental construction of Deligne [D], which gives a locally free logarithmic extension. Using our description of the infinitesimal deformations of a locally free logarithmic connections, we show that certain extensions of any given regular connection are *rigid*, that is, they have no infinitesimal deformations which keep the underlying regular connection fixed. The criterion for this is that no two distinct eigenvalues of the residue of the logarithmic connection must differ by an integer. In particular, this shows that Deligne's construction gives a rigid extension.

References

[D] Deligne, P., *Equations differentielles a points singuliers reguliers,* Lect. Notes in Math. **163** Springer (1970)

[N] Nitsure, N. , *Moduli of semi-stable logarithmic connections,* Journal of Amer. Math. Soc. **6** (1993) 597–609.

[S] Simpson, C., *Moduli of representations of the fundamental group of a smooth projective variety,* Preprint (1990) Princeton University.

School of Mathematics
Tata Institute of Fundamental Research
Homi Bhabha Road
Bombay 400 005
India

The Borel-Weil theorem and the Feynman path integral

Kiyosato Okamoto

Introduction

Let p and q be the canonical momentum and coordinate of a particle. In the operator method of quantization, corresponding to p and q there are operators P, Q which in the coordinate representation have the form:

$$Q = q, \quad P = -\sqrt{-1}\frac{d}{dq}.$$

The quantization means the correspondence between the Hamiltonian function $h(p, q)$ and the Hamiltonian operator $\mathbf{H} = h(P, Q)$, where a certain procedure for ordering noncommuting operator arguments P and Q is assumed. The path integral quantization is the method to compute the kernel function of the unitary operator $\exp(-\sqrt{-1}t\mathbf{H})$.

Any mathematically strict definition of the path integral has not yet given. If one tries to compute path integrals in general one may encounter the difficulty of divergence of the path integrals. Many examples, however, show that the path integral is a very powerful tool to compute the kernel function of the operator explicitly.

The purpose of this lecture is to explain what kinds of divergence we have when we try to compute the path integral for complex polarizations of a connected semisimple Lie group which contains a compact Cartan subgroup and to show that we can regularize the path integral by the process of "normal ordering" (cf. Chapter 13 in [10]). The details and proofs of these results are given in the forthcoming paper [7] .

Since a few in the audience do not seem to know about the Feynman path integral I would like to start with explaining it from a point of view of the theory of unitary representations, using the Heisenberg group which is most deeply related with the quantum mechanics.

The Feynman's idea of the path integral can be easily and clearly understood if one computes the path integral on the coadjoint orbits of the Heisenberg group:

$$G = \left\{ \begin{pmatrix} 1 & p & r \\ & 1 & q \\ & & 1 \end{pmatrix} ;\ p, q, r \in \mathbf{R} \right\}.$$

The Lie algebra of G is given by

$$\mathfrak{g} = \left\{ \begin{pmatrix} 0 & a & c \\ & 0 & b \\ & & 0 \end{pmatrix} ; \quad a, b, c \in \mathbf{R} \right\}.$$

The dual space of $\mathfrak{g}$ is identified with

$$\mathfrak{g}^* = \left\{ \begin{pmatrix} 0 & & \\ \xi & 0 & \\ \sigma & \eta & 0 \end{pmatrix} ; \quad \xi, \eta, \sigma \in \mathbf{R} \right\}$$

by the pairing

$$\mathfrak{g} \times \mathfrak{g}^* \ni (X, \lambda) \longmapsto \mathrm{tr}(\lambda X) \in \mathbf{R}.$$

Any nontrivial coadjoint orbit is given by an element

$$\lambda_\sigma = \begin{pmatrix} 0 & & \\ 0 & 0 & \\ \sigma & 0 & 0 \end{pmatrix} \quad \text{for some } \sigma \neq 0.$$

Then the isotropy subgroup at λ_σ is given by

$$G_{\lambda_\sigma} = \left\{ \begin{pmatrix} 1 & 0 & r \\ & 1 & 0 \\ & & 1 \end{pmatrix} ; \quad r \in \mathbf{R} \right\},$$

and the Lie algebra of G_{λ_σ} is

$$\mathfrak{g}_{\lambda_\sigma} = \left\{ \begin{pmatrix} 0 & 0 & c \\ & 0 & 0 \\ & & 0 \end{pmatrix} ; \quad c \in \mathbf{R} \right\}.$$

We consider the real polarization:

$$\mathfrak{p} = \left\{ \begin{pmatrix} 0 & a & c \\ & 0 & 0 \\ & & 0 \end{pmatrix} ; \quad a, c \in \mathbf{R} \right\}.$$

Then the analytic subgroup of G corresponding to $\mathfrak{p}$ is given by

$$P = \left\{ \begin{pmatrix} 1 & p & r \\ & 1 & 0 \\ & & 1 \end{pmatrix}; \quad p, r \in \mathbf{R} \right\}.$$

Clearly the Lie algebra homomorphism

$$\mathfrak{p} \ni \begin{pmatrix} 0 & a & c \\ & 0 & 0 \\ & & 0 \end{pmatrix} \longmapsto -\sqrt{-1}\sigma c \in \sqrt{-1}\mathbf{R}.$$

lifts to the unitary character ξ_{λ_σ} :

$$P \ni \begin{pmatrix} 1 & p & r \\ & 1 & 0 \\ & & 1 \end{pmatrix} \longmapsto e^{-\sqrt{-1}\sigma r} \in U(1).$$

Let $L_{\xi_{\lambda_\sigma}}$ denote the line bundle associated with ξ_{λ_σ} over the homogeneous space G/P. Then the space $C^\infty(L_{\xi_{\lambda_\sigma}})$ of all complex valued C^∞-sections of $L_{\xi_{\lambda_\sigma}}$ can be identified with

$$\left\{ f \in C^\infty(G); f(gp) = \xi_{\lambda_\sigma}(p)^{-1} f(g) \quad (g \in G, p \in P) \right\}.$$

For any $g \in G$ we define an operator $\pi^{\mathfrak{p}}_{\lambda_\sigma}(g)$ on $C^\infty(L_{\xi_{\lambda_\sigma}})$: For $f \in \mathbf{C}^\infty(L_{\xi_{\lambda_\sigma}})$

$$(\pi^{\mathfrak{p}}_{\lambda_\sigma}(g)f)(x) = f(g^{-1}x) \qquad (x \in G).$$

Let $\mathcal{H}^{\mathfrak{p}}_{\lambda_\sigma}$ be the Hilbert space of all square integrable sections of L_{ξ_λ}. Then $\pi^{\mathfrak{p}}_{\lambda_\sigma}$ is a unitary representation of G on $\mathcal{H}^{\mathfrak{p}}_{\lambda_\sigma}$.

We put

$$M = \left\{ \begin{pmatrix} 1 & 0 & 0 \\ & 1 & q \\ & & 1 \end{pmatrix}; \quad q \in \mathbf{R} \right\}.$$

Then as is easily seen the product mapping $M \times P \longrightarrow G$ is a real analytic isomorphism which is surjective. Let $f \in C^\infty(L_{\xi_{\lambda_\sigma}})$. Then, since

$$f(g \begin{pmatrix} 1 & p & r \\ & 1 & 0 \\ & & 1 \end{pmatrix}) = e^{\sqrt{-1}\sigma r} f(g) \quad \text{for } g \in G, \begin{pmatrix} 1 & p & r \\ & 1 & 0 \\ & & 1 \end{pmatrix} \in P,$$

f can be uniquely determined by its values on M. From this we obtain the following onto-isometry:

$$\mathcal{H}^{\mathfrak{p}}_{\lambda_\sigma} \ni f \longmapsto F \in L^2(\mathbf{R})$$

where

$$F(q) = f(\begin{pmatrix} 1 & 0 & 0 \\ & 1 & q \\ & & 1 \end{pmatrix}) \quad (q \in \mathbf{R}).$$

For any $g = \exp \begin{pmatrix} 0 & a & c \\ & 0 & b \\ & & 0 \end{pmatrix} \in G$, we define a unitary operator $U^{\mathfrak{p}}_{\lambda_\sigma}(g)$ on $L^2(\mathbf{R})$ such that the diagram below is commutative:

$$\begin{array}{ccc} \mathcal{H}^{\mathfrak{p}}_{\lambda_\sigma} & \longrightarrow & L^2(\mathbf{R}) \\ \pi^{\mathfrak{p}}_{\lambda_\sigma}(g) \downarrow & & \downarrow U^{\mathfrak{p}}_{\lambda_\sigma}(g) \\ \mathcal{H}^{\mathfrak{p}}_{\lambda_\sigma} & \longrightarrow & L^2(\mathbf{R}). \end{array}$$

Then we have

$$\begin{aligned} (U^{\mathfrak{p}}_{\lambda_\sigma}(g)F)(q) &= f(g^{-1}\begin{pmatrix} 1 & 0 & 0 \\ & 1 & q \\ & & 1 \end{pmatrix}) \\ &= f(\begin{pmatrix} 1 & -a & -c+\frac{ab}{2} \\ & 1 & -b \\ & & 1 \end{pmatrix}\begin{pmatrix} 1 & 0 & 0 \\ & 1 & q \\ & & 1 \end{pmatrix}) \\ &= f(\begin{pmatrix} 1 & 0 & 0 \\ & 1 & q-b \\ & & 1 \end{pmatrix}\begin{pmatrix} 1 & -a & -c+\frac{ab}{2}-aq \\ & 1 & 0 \\ & & 1 \end{pmatrix}) \\ &= e^{\sqrt{-1}\sigma(-c+\frac{ab}{2}-aq)}F(q-b). \end{aligned}$$

Now we show that the above unitary operator is obtained by the path integral.

In the following, for the definition of the connection form θ_{λ_σ}, the hamiltonian H_Y and the action $\int_0^T \gamma^*\alpha - H_Y dt$ in the general case the audience may refer to the introduction of the paper [5].

We use the local coordinates q, p, r of $g \in G$ as follows:

$$G \ni g = \begin{pmatrix} 1 & 0 & 0 \\ & 1 & q \\ & & 1 \end{pmatrix}\begin{pmatrix} 1 & p & r \\ & 1 & 0 \\ & & 1 \end{pmatrix}.$$

Since the canonical 1-form θ is given by $g^{-1}dg$, we have

$$\theta_{\lambda_\sigma} = <\lambda_\sigma, \theta> = \mathrm{tr}(\lambda_\sigma g^{-1}dg) = \sigma(dr - pdq).$$

We choose

$$\alpha_{\mathfrak{p}} = -\sigma p dq.$$

Then

$$\frac{d\alpha_{\mathfrak{p}}}{2\pi} = \frac{-\sigma dp \wedge dq}{2\pi}.$$

For $Y \in \mathfrak{g}$, the hamiltonian H_Y is given by

$$H_Y = \mathrm{tr}(\lambda_\sigma g^{-1} Y g) = \sigma(aq - bp + c)$$

where $Y = \begin{pmatrix} 0 & a & c \\ & 0 & b \\ & & 0 \end{pmatrix}$. The action is given by

$$\int_0^T \gamma^* \alpha - H_Y dt = \int_0^T \{-\sigma p(t)\dot{q}(t) - \sigma(aq - bp + c)\}\, dt.$$

We divide the time interval $[0, T]$ into N-equal small intervals $[\frac{k-1}{N}T, \frac{k}{N}T]$

$$\int_0^T \gamma^* \alpha - H_Y dt = \sum_{k=1}^{N} \int_{\frac{k-1}{N}T}^{\frac{k}{N}T} \gamma^* \alpha - H_Y dt.$$

The physicists' calculation rule asserts that one should take the "ordering":

$$\sum_{k=1}^{N} \left\{ -\sigma p_{k-1}(q_k - q_{k-1}) - \sigma(a\frac{q_k + q_{k-1}}{2} - bp_{k-1} + c)\frac{T}{N} \right\}.$$

This choice of the ordering can be mathematically formulated as follows. We take the paths: for $t \in [\frac{k-1}{N}T, \frac{k}{N}T]$

$$\begin{aligned} q(t) &= q_{k-1} + (t - \frac{k-1}{N}T)\frac{q_k - q_{k-1}}{T/N}, \\ p(t) &= p_{k-1}, \\ q(0) &= q, \text{ and } q(T) = q'. \end{aligned}$$

Then the action for the above path becomes

$$\begin{aligned} &\sum_{k=1}^{N} \int_{\frac{k-1}{N}T}^{\frac{k}{N}T} \{-\sigma p(t)\dot{q}(t) - \sigma(aq - bp + c)\}\, dt \\ &= \sum_{k=1}^{N} \left\{ -\sigma p_{k-1}(q_k - q_{k-1}) - \sigma(a\frac{q_k + q_{k-1}}{2} - bp_{k-1} + c)\frac{T}{N} \right\}. \end{aligned}$$

The Feynman path integral asserts that the transition amplitude between the point $q = q_0$ and the point $q' = q_N$ is given by the kernel function which is computed as follows :

$$K_Y^{\mathfrak{p}}(q', q; T)$$

$$= \lim_{N\to\infty} \int_{-\infty}^{\infty} \cdots \int_{-\infty}^{\infty} \int_{-\infty}^{\infty} \cdots \int_{-\infty}^{\infty} |\sigma| \frac{dp_0}{2\pi} \cdots |\sigma| \frac{dp_{N-1}}{2\pi} dq_1 \cdots dq_{N-1}$$

$$\times \exp\left\{\sqrt{-1}\sigma \sum_{k=1}^{N} \left[-p_{k-1}(q_k - q_{k-1}) - (a\frac{q_k + q_{k-1}}{2} - bp_{k-1} + c)\frac{T}{N}\right]\right\}$$

$$= \lim_{N\to\infty} \int_{-\infty}^{\infty} \cdots \int_{-\infty}^{\infty} dq_1 \cdots dq_{N-1} \prod_{k=1}^{N} \delta(-q_k + q_{k-1} + b\frac{T}{N})$$

$$\times \exp\left\{-\sqrt{-1}\sigma \sum_{k=1}^{N} (\frac{a(q_k + q_{k-1})}{2} + c)\frac{T}{N}\right\}$$

$$= \lim_{N\to\infty} \delta(-q_N + q_0 + bT) \exp\left\{-\sqrt{-1}\sigma(aT(q_0 + \frac{bT}{2}) + cT)\right\}$$

$$= \delta(-q' + q + bT) \exp\left\{-\sqrt{-1}\sigma(aqT + \frac{abT^2}{2} + cT)\right\}.$$

For $F \in C_c^\infty(\mathbf{R})$ we have

$$\int_{-\infty}^{\infty} K_Y^{\mathfrak{p}}(q', q; T) F(q) dq$$

$$= \exp\left\{-\sqrt{-1}\sigma\left(aq'T - \frac{abT^2}{2} + cT\right)\right\} F(q' - bT)$$

$$= (U_{\lambda_\sigma}^{\mathfrak{p}}(\exp TY)F)(q').$$

Thus the path integral gives our unitary operator.

In the above quantization, the hamiltonian function $H_Y(p, q)$ corresponds to

$$\sqrt{-1}\frac{d}{dt} U_{\lambda_\sigma}^{\mathfrak{p}}(\exp tY)|_{t=0} = \sigma(aq + c) - \sqrt{-1}b\frac{d}{dq},$$

which is slightly different from

$$H_Y(-\sqrt{-1}\frac{d}{dq}, q) = \sigma(aq + c + \sqrt{-1}b\frac{d}{dq}).$$

This difference comes from the fact that we chose $\alpha_{\mathfrak{p}} = -\sigma p dq$ whereas physicists usually take pdq.

1 Coherent representation

In this section, we compute the path integral for unitary representations realized by the Borel-Weil theorem for the Heisenberg group. In other words, we compute the path integral for a complex polarization which is called by physicists the path integral for the coherent representation. We shall show that the path integral, also in this case, gives unitary operators of these representations.

The complexification $G^{\mathbf{C}}$ of G and $\mathfrak{g}^{\mathbf{C}}$ of $\mathfrak{g}$ are given by

$$\begin{aligned} G^{\mathbf{C}} &= \left\{ \begin{pmatrix} 1 & p & r \\ & 1 & q \\ & & 1 \end{pmatrix} ; p, q, r \in \mathbf{C} \right\}, \\ \mathfrak{g}^{\mathbf{C}} &= \left\{ \begin{pmatrix} 0 & a & c \\ & 0 & b \\ & & 0 \end{pmatrix} ; a, b, c \in \mathbf{C} \right\}. \end{aligned}$$

We consider the complex polarization defined by

$$\mathfrak{p} = \left\{ \begin{pmatrix} 0 & \sqrt{-1}b & c \\ & 0 & b \\ & & 0 \end{pmatrix} ; b, c \in \mathbf{C} \right\}.$$

We denote by P the complex analytic subgroup of $G^{\mathbf{C}}$ corresponding to $\mathfrak{p}$. We put $W = GP = G^{\mathbf{C}}$. Then it is easy to see that the Lie algebra homomorphism

$$\mathfrak{p} \ni \begin{pmatrix} 0 & \sqrt{-1}b & c \\ & 0 & b \\ & & 0 \end{pmatrix} \longmapsto -\sqrt{-1}\sigma c \in \mathbf{C}$$

lifts uniquely to the holomorphic character ξ_{λ_σ} :

$$P \ni \begin{pmatrix} 1 & \sqrt{-1}b & c + \frac{\sqrt{-1}}{2}b^2 \\ & 1 & b \\ & & 1 \end{pmatrix} \longmapsto e^{-\sqrt{-1}\sigma c} \in \mathbf{C}^*.$$

We denote by $L_{\xi_{\lambda_\sigma}}$ the holomorphic line bundle on $G^{\mathbf{C}}/P$ associated with the character ξ_{λ_σ}.

We denote by $\Gamma(L_{\xi_{\lambda_\sigma}})$ the space of all holomorphic sections of $L_{\xi_{\lambda_\sigma}}$ and by $\Gamma(\mathbf{C})$ the space of all holomorphic functions on $\mathbf{C}$.

We use the coordinates for $g \in G$:

$$g = \exp\begin{pmatrix} 0 & -\frac{\sqrt{-1}}{2}z & 0 \\ & 0 & \frac{1}{2}z \\ & & 0 \end{pmatrix} \exp\begin{pmatrix} 0 & \frac{\sqrt{-1}}{2}\bar{z} & r+\frac{\sqrt{-1}}{4}|z|^2 \\ & 0 & \frac{1}{2}\bar{z} \\ & & 0 \end{pmatrix}$$

$$= \begin{pmatrix} 1 & -\frac{\sqrt{-1}}{2}z & -\frac{\sqrt{-1}}{8}z^2 \\ & 1 & \frac{1}{2}z \\ & & 1 \end{pmatrix}\begin{pmatrix} 1 & \frac{\sqrt{-1}}{2}\bar{z} & r+\frac{\sqrt{-1}}{4}|z|^2+\frac{\sqrt{-1}}{8}\bar{z}^2 \\ & 1 & \frac{1}{2}\bar{z} \\ & & 1 \end{pmatrix}$$

where $z \in \mathbf{C}, r \in \mathbf{R}$.

We have the isomorphism

$$\Gamma(L_{\xi_{\lambda_\sigma}}) \ni f \longmapsto F \in \Gamma(\mathbf{C})$$

where

$$F(z) = f(\begin{pmatrix} 1 & -\frac{\sqrt{-1}}{2}z & -\frac{\sqrt{-1}}{8}z^2 \\ & 1 & \frac{1}{2}z \\ & & 1 \end{pmatrix}).$$

We denote by $\Gamma^2(L_{\xi_{\lambda_\sigma}})$ the Hilbert space of all square integrable holomorphic sections of $L_{\xi_{\lambda_\sigma}}$ and by $\Gamma^2(\mathbf{C}, \frac{|\sigma|}{2\pi}e^{-\frac{\sigma}{2}|z|^2})$ the space of all square integrable holomorphic functions on $\mathbf{C}$ with respect to the Gaussian measure with the density function $\frac{|\sigma|}{2\pi}e^{-\frac{\sigma}{2}|z|^2}$.

For any $g \in G$ we define an operator $\pi^{\mathfrak{p}}_{\lambda_\sigma}(g)$ on $\Gamma^2(L_{\xi_{\lambda_\sigma}})$: For $f \in \Gamma^2(L_{\xi_{\lambda_\sigma}})$

$$(\pi^{\mathfrak{p}}_{\lambda_\sigma}(g)f)(x) = f(g^{-1}x) \qquad (x \in G).$$

Then $\pi^{\mathfrak{p}}_{\lambda_\sigma}$ is a unitary representation of G on $\Gamma^2(L_{\xi_{\lambda_\sigma}})$. Since

$$\begin{aligned} \int_{G/G_{\lambda_\sigma}} |f(g)|^2\,\omega_{\lambda_\sigma} &= \int_{\mathbf{C}} \left|e^{-\frac{\sigma|z|^2}{4}}F(z)\right|^2 |\sigma|\frac{dzd\bar{z}}{2\pi} \\ &= \int_{\mathbf{C}} |F(z)|^2\, e^{-\frac{\sigma|z|^2}{2}} |\sigma|\frac{dzd\bar{z}}{2\pi}, \end{aligned}$$

where we denote by ω_{λ_σ} the canonical symplectic form on the coadjoint orbit $\mathcal{O}_{\lambda_\sigma} = G/G_{\lambda_\sigma}$ and we put

$$dzd\bar{z} = \frac{\sqrt{-1}}{2}dz \wedge d\bar{z}.$$

The above isomorphism gives an isometry of $\Gamma^2(L_{\xi_{\lambda_\sigma}})$ onto $\Gamma^2(\mathbf{C}, \frac{|\sigma|}{2\pi}e^{-\frac{\sigma}{2}|z|^2})$.

As is easily seen $\Gamma^2(\mathbf{C}, \frac{|\sigma|}{2\pi}e^{-\frac{\sigma}{2}|z|^2}) \neq \{0\}$ if and only if $\sigma > 0$. It follows that

$$\Gamma^2(L_{\xi_{\lambda_\sigma}}) \neq \{0\} \text{ if and only if } \sigma > 0.$$

For the rest of the section we assume that $\sigma > 0$.

For any $g = \exp\begin{pmatrix} 0 & a & c \\ & 0 & b \\ & & 0 \end{pmatrix} \in G$, we define a unitary operator $U^{\mathfrak{p}}_{\lambda_\sigma}(g)$ on $\Gamma^2(\mathbf{C}, \frac{\sigma}{2\pi}e^{-\frac{\sigma}{2}|z|^2})$ such that the diagram below is commutative:

$$\begin{array}{ccc} \Gamma^2(L_{\xi_{\lambda_\sigma}}) & \longrightarrow & \Gamma^2(\mathbf{C}, \frac{\sigma}{2\pi}e^{-\frac{\sigma}{2}|z|^2}) \\ \pi^{\mathfrak{p}}_{\lambda_\sigma}(g)\downarrow & & \downarrow U^{\mathfrak{p}}_{\lambda_\sigma}(g) \\ \Gamma^2(L_{\xi_{\lambda_\sigma}}) & \longrightarrow & \Gamma^2(\mathbf{C}, \frac{\sigma}{2\pi}e^{-\frac{\sigma}{2}|z|^2}). \end{array}$$

Then we have

$$\begin{aligned}
&(U^{\mathfrak{p}}_{\lambda_\sigma}(g)F)(z) \\
&= f(g^{-1}\begin{pmatrix} 1 & -\frac{\sqrt{-1}}{2}z & -\frac{\sqrt{-1}}{8}z^2 \\ & 1 & \frac{1}{2}z \\ & & 1 \end{pmatrix}) \\
&= f(\begin{pmatrix} 1 & -a & -c+\frac{ab}{2} \\ & 1 & -b \\ & & 1 \end{pmatrix}\begin{pmatrix} 1 & -\frac{\sqrt{-1}}{2}z & -\frac{\sqrt{-1}}{8}z^2 \\ & 1 & \frac{1}{2}z \\ & & 1 \end{pmatrix}) \\
&= f(\begin{pmatrix} 1 & -\frac{\sqrt{-1}}{2}(z-\gamma) & -\frac{\sqrt{-1}}{8}(z-\gamma)^2 \\ & 1 & \frac{1}{2}(z-\gamma) \\ & & 1 \end{pmatrix} \\
&\qquad \times \begin{pmatrix} 1 & -\frac{\sqrt{-1}\bar{\gamma}}{2} & -c+\frac{\sqrt{-1}}{4}|\gamma|^2 - \frac{\sqrt{-1}}{2}\bar{\gamma}z + \frac{\sqrt{-1}}{8}\bar{\gamma}^2 \\ & 1 & -\frac{\bar{\gamma}}{2} \\ & & 1 \end{pmatrix}) \\
&= e^{\sigma(-\sqrt{-1}c-\frac{1}{4}|\gamma|^2+\frac{1}{2}\bar{\gamma}z)}F(z-\gamma),
\end{aligned}$$

where $\gamma = b + \sqrt{-1}a$.

It is well-known that $U^{\mathfrak{p}}_{\lambda_\sigma}$ is an irreducible unitary representation of G on $\Gamma^2(\mathbf{C}, \frac{\sigma}{2\pi}e^{-\frac{\sigma}{2}|z|^2})$.

Using the parametrization:

$$g = \begin{pmatrix} 1 & -\frac{\sqrt{-1}}{2}z & -\frac{\sqrt{-1}}{8}z^2 \\ & 1 & \frac{1}{2}z \\ & & 1 \end{pmatrix}\begin{pmatrix} 1 & \frac{\sqrt{-1}}{2}\bar{z} & r+\frac{\sqrt{-1}}{4}|z|^2+\frac{\sqrt{-1}}{8}\bar{z}^2 \\ & 1 & \frac{1}{2}\bar{z} \\ & & 1 \end{pmatrix},$$

we have

$$\theta_{\lambda_\sigma} = \mathrm{tr}(\lambda_\sigma g^{-1}dg) = \sigma(dr + \sqrt{-1}\frac{zd\bar{z} - \bar{z}dz}{4}),$$

$$H_Y = \sigma\left(\sqrt{-1}\frac{\bar{\gamma}z - \gamma\bar{z}}{2} + c\right).$$

We choose $\alpha_{\mathfrak{p}} = -\frac{\sqrt{-1}\sigma}{2}\bar{z}dz$. Then we have

$$\frac{1}{2\pi}d\alpha_{\mathfrak{p}} = \frac{\sqrt{-1}\sigma dz \wedge d\bar{z}}{\cdot 4\pi}.$$

For fixed $z, z' \in \mathbf{C}$ we define the paths: For $t \in [\frac{k-1}{N}T, \frac{k}{N}T]$

$$\begin{aligned} \bar{z}(t) &= \bar{z}_{k-1}, \\ z(t) &= z_{k-1} + (t - \frac{k-1}{N}T)\frac{z_k - z_{k-1}}{T/N}, \\ z(0) &= z \text{ and } z(T) = z'. \end{aligned}$$

Then the action becomes

$$\int_0^T \left\{\frac{1}{2}\sigma\bar{z}(t)\dot{z}(t) - \sqrt{-1}\sigma\left(\frac{\sqrt{-1}\bar{\gamma}z(t) - \sqrt{-1}\gamma\bar{z}(t)}{2} + c\right)\right\} dt$$

$$= \sum_{k=1}^{N}\int_{\frac{k-1}{N}T}^{\frac{k}{N}T} \left\{\frac{1}{2}\sigma\bar{z}(t)\dot{z}(t) - \sqrt{-1}\sigma\left(\frac{\sqrt{-1}\bar{\gamma}z(t) - \sqrt{-1}\gamma\bar{z}(t)}{2} + c\right)\right\} dt$$

$$= \sigma\sum_{k=1}^{N}\left[\frac{1}{2}\bar{z}_{k-1}(z_k - z_{k-1}) - \left(\frac{\gamma}{2}\bar{z}_{k-1} - \frac{\bar{\gamma}}{4}(z_k + z_{k-1}) + \sqrt{-1}c\right)\frac{T}{N}\right].$$

The following lemma can be easily proved.

Lemma 1. *We have the following formula for* $c_1, c_2 \in \mathbf{C}$.

$$\int_{\mathbf{C}} \frac{\sigma dz'd\bar{z}'}{2\pi} \exp\sigma\left\{-\frac{1}{2}|z|^2 + z'(\frac{1}{2}\bar{z} + c_1) + \bar{z}'(\frac{1}{2}z'' - c_2)\right\}$$

$$= \exp\sigma\left\{z''(\frac{\bar{z}}{2} + c_1) - 2c_2(\frac{\bar{z}}{2} + c_1)\right\}.$$

Using this lemma, we can compute the path integral explicitly as follows:

$$
\begin{aligned}
&K_Y^{\mathfrak{p}}(z', z; T) \\
&= \lim_{N\to\infty} \int_{\mathbf{C}} \cdots \int_{\mathbf{C}} \frac{\sigma dz_1\, d\bar{z}_1}{2\pi} \cdots \frac{\sigma dz_{N-1} d\bar{z}_{N-1}}{2\pi} \\
&\times \exp\left\{ \sigma \sum_{k=1}^{N} \left[\frac{1}{2}\bar{z}_{k-1}(z_k - z_{k-1}) - \left(\frac{\gamma}{2}\bar{z}_{k-1} - \frac{\bar{\gamma}}{4}(z_k + z_{k-1}) + \sqrt{-1}c \right) \frac{T}{N} \right] \right\} \\
&= \lim_{N\to\infty} \int_{\mathbf{C}} \cdots \int_{\mathbf{C}} \frac{\sigma dz_1\, d\bar{z}_1}{2\pi} \cdots \frac{\sigma dz_{N-1} d\bar{z}_{N-1}}{2\pi} \\
&\qquad \times \exp\left\{ \sigma \textstyle\sum_{k=1}^{N} \left(-\frac{1}{2}\left|z_{k-1}\right|^2 + \bar{z}_{k-1}\left(\frac{z_k}{2} - \frac{\gamma T}{2N}\right) + z_{k-1}\frac{\bar{\gamma} T}{2N} \right) \right. \\
&\qquad\qquad \left. + \sigma(z_N - z_0)\frac{\bar{\gamma}T}{4N} - \sqrt{-1}\sigma c T \right\}
\end{aligned}
$$

$$
\begin{aligned}
&= \lim_{N\to\infty} \int_{\mathbf{C}} \cdots \int_{\mathbf{C}} \frac{\sigma dz_1\, d\bar{z}_1}{2\pi} \cdots \frac{\sigma dz_{N-1} d\bar{z}_{N-1}}{2\pi} \\
&\quad \times \exp\left\{ \sigma \left(-\frac{1}{2}|z_0|^2 - \bar{z}_0 \frac{\gamma T}{2N} + z_0 \frac{\bar{\gamma} T}{2N} \right) \right. \\
&\qquad + \sigma \left(-\frac{1}{2}|z_1|^2 + \bar{z}_1\left(\frac{z_2}{2} - \frac{\gamma T}{2N}\right) + z_1\left(\frac{\bar{z}_0}{2} + \frac{\bar{\gamma}T}{2N}\right) \right) \\
&\qquad + \sigma \left(-\frac{1}{2}|z_2|^2 + \bar{z}_2\left(\frac{z_3}{2} - \frac{\gamma T}{2N}\right) + z_2 \frac{\bar{\gamma}T}{2N} \right) \\
&\qquad + \sigma \sum_{k=4}^{N} \left(-\frac{1}{2}\left|z_{k-1}\right|^2 + \bar{z}_{k-1}\left(\frac{z_k}{2} - \frac{\gamma T}{2N}\right) + z_{k-1}\frac{\bar{\gamma}T}{2N} \right) \\
&\qquad \left. + \sigma(z_N - z_0)\frac{\bar{\gamma}T}{4N} - \sqrt{-1}\sigma c T \right\}
\end{aligned}
$$

$$
\begin{aligned}
&= \left(\lim_{N\to\infty} \int_{\mathbf{C}} \cdots \int_{\mathbf{C}} \frac{\sigma dz_2 d\bar{z}_2}{2\pi} \cdots \frac{\sigma dz_{N-1} d\bar{z}_{N-1}}{2\pi} \right) \\
&\qquad \times \exp\left\{ \sigma \left(-\frac{1}{2}|z_0| - \bar{z}_0 \frac{\gamma T}{2N} + z_0 \frac{\bar{\gamma}T}{2N} \right) \right. \\
&+ \sigma \left(-\frac{1}{2}|z_2|^2 + \bar{z}_2\left(\frac{z_3}{2} - \frac{\gamma T}{2N}\right) + z_2\left(\frac{\bar{z}_0}{2} + 2\frac{\bar{\gamma}T}{2N}\right) - \frac{\gamma T}{N}\left(\frac{\bar{z}_0}{2} + \frac{\bar{\gamma}T}{2N}\right) \right) \\
&+ \sigma \sum_{k=4}^{N} \left(-\frac{1}{2}\left|z_{k-1}\right|^2 + \bar{z}_{k-1}\left(\frac{z_k}{2} - \frac{\gamma T}{2N}\right) + z_{k-1}\frac{\bar{\gamma}T}{2N} \right)
\end{aligned}
$$

$$+ \quad \sigma(z_N - z_0)\frac{\bar{\gamma}T}{4N} - \sqrt{-1}\sigma cT\Big\}$$

repeating the above procedure,

$$\begin{aligned} &= \lim_{N\to\infty} \exp\left\{\sigma\left(-\frac{1}{2}|z_0|^2 + z_N(\frac{\bar{z}_0}{2} + \frac{\bar{\gamma}T}{2}) - \gamma T(\frac{\bar{z}_0}{2} + \frac{1}{2}\frac{\bar{\gamma}T}{2}) - \sqrt{-1}cT\right)\right. \\ &\qquad\qquad \left. +\sigma\left(-\bar{z}_0\frac{\gamma T}{2N} + (z_0 - z_N)\frac{\bar{\gamma}T}{4N} + \frac{\gamma T}{2N}(\bar{z}_0 + \bar{\gamma}T)\right)\right\} \\ &= \exp\left\{\sigma\left(-\frac{1}{2}|z|^2 + z'(\frac{\bar{z}}{2} + \frac{\bar{\gamma}T}{2}) - \gamma T(\frac{\bar{z}}{2} + \frac{\bar{\gamma}T}{4}) - \sqrt{-1}cT\right)\right\} \end{aligned}$$

Thus for any $Y = \begin{pmatrix} 0 & a & c \\ & 0 & b \\ & & 0 \end{pmatrix} \in \mathfrak{g}$, we have

$$\begin{aligned} &\int_{\mathbf{C}} \frac{\sigma dz d\bar{z}}{2\pi} K_Y^{\mathfrak{p}}(z', z; T) F(z) \\ &= \int_{\mathbf{C}} \frac{\sigma dz d\bar{z}}{2\pi} \exp\left\{\sigma\left(-\frac{1}{2}|z|^2 + \frac{1}{2}\bar{z}(z' - \gamma T) + \frac{1}{2}z'\bar{\gamma}T\right.\right. \\ &\qquad\qquad\qquad \left.\left. -\frac{1}{4}|\gamma|^2 T^2 - \sqrt{-1}cT\right)\right\} F(z) \\ &= \exp\left\{\sigma\left(\frac{1}{2}z'\bar{\gamma}T - \frac{1}{4}|\gamma|^2 T^2 - \sqrt{-1}cT\right)\right\} F(z' - \gamma T) \\ &= (U_{\lambda_\sigma}^{\mathfrak{p}}(\exp TY)F)(z'). \end{aligned}$$

2 Borel-Weil theorem

In this section, we consider unitary representations realized by the Borel-Weil theorem for semisimple Lie groups.

Let G be a connected semisimple Lie group such that there exists a complexification $G^{\mathbf{C}}$ with $\pi_1(G^{\mathbf{C}}) = \{1\}$ and such that rank G=dim T, T a maximal torus of G. Let K be a maximal compact subgroup of G which contains T, and $\mathfrak{k}$ the Lie algebra of K. Note that G can be realized as a matrix group. We denote the conjugation of $G^{\mathbf{C}}$ with respect to G, and that of $\mathfrak{g}^{\mathbf{C}}$ with respect to $\mathfrak{g}$, both by $^-$ Let $\mathfrak{g}$ and $\mathfrak{h}$ be the Lie algebras of G

and T. We denote complexifications of $\mathfrak{g}$ and $\mathfrak{h}$ by $\mathfrak{g}^{\mathbf{C}}$ and $\mathfrak{h}^{\mathbf{C}}$, respectively. Then $\mathfrak{h}^{\mathbf{C}}$ is a Cartan subalgebra of $\mathfrak{g}^{\mathbf{C}}$.

Let Δ denote the set of all nonzero roots and Δ^+ the set of all positive roots. Then we have the root space decomposition

$$\mathfrak{g}^{\mathbf{C}} = \mathfrak{h}^{\mathbf{C}} + \sum_{\alpha\in\Delta} \mathfrak{g}^{\alpha}.$$

Define

$$\mathfrak{n}^{\pm} = \sum_{\alpha\in\Delta^+} \mathfrak{g}^{\pm\alpha}, \qquad \mathfrak{b} = \mathfrak{h}^{\mathbf{C}} + \mathfrak{n}^{-}.$$

Let N, N^-, B and $T^{\mathbf{C}}$ be the analytic subgroups corresponding to $\mathfrak{n}^+, \mathfrak{n}^-, \mathfrak{b}$ and $\mathfrak{h}^{\mathbf{C}}$, respectively.

We fix an integral form Λ on $\mathfrak{h}^{\mathbf{C}}$. Then

$$\xi_\Lambda : T \longrightarrow U(1), \qquad \exp H \longmapsto e^{\Lambda(H)}$$

defines a unitary character of T. And ξ_Λ extends uniquely to a holomorphic one-dimensional representation of B:

$$\xi_\Lambda : B = T^{\mathbf{C}} N^- \longrightarrow \mathbf{C}^\times, \qquad \exp H \cdot n^- \longmapsto e^{\Lambda(H)}.$$

Let $\tilde{L}_\Lambda$ be the holomorphic line bundle over $G^{\mathbf{C}}/B$ associated to the holomorphic one-dimensional representation ξ_Λ of B. We denote by L_Λ the restriction of $\tilde{L}_\Lambda$ to the open submanifold G/T of $G^{\mathbf{C}}/B$:

$$\begin{array}{ccccc} G & \hookrightarrow & GB & \hookrightarrow & G^{\mathbf{C}} \\ \downarrow & & \downarrow & & \downarrow \\ G/T & = & GB/B & \hookrightarrow & G^{\mathbf{C}}/B \end{array}$$

and

$$\begin{array}{ccc} L_\Lambda & \hookrightarrow & \tilde{L}_\Lambda \\ \downarrow & & \downarrow \\ G/T & \hookrightarrow & G^{\mathbf{C}}/B. \end{array}$$

Then we can identify the space of all holomorphic sections of L_Λ with

$$\Gamma(L_\Lambda) = \left\{ f : GB \overset{\text{hol.}}{\longrightarrow} \mathbf{C};\ f(xb) = \xi_\Lambda(b)^{-1} f(x), x \in GB, b \in B \right\}.$$

Let π_Λ be a representation of G on $\Gamma(L_\Lambda)$ defined by

$$\pi_\Lambda(g)f(x) = f(g^{-1}x) \quad \text{for } g \in G, x \in GB \text{ and } f \in \Gamma(L_\Lambda).$$

For any $f \in \Gamma(L_\Lambda)$ we define

$$\|f\|^2 = \int_G |f(g)|^2 dg,$$

where dg is the Haar measure on G. We put

$$\Gamma_2(L_\Lambda) = \{f \in \Gamma(L_\Lambda) ; \|f\| < +\infty\}.$$

Then the Borel-Weil theorem asserts that $(\pi_\Lambda, \Gamma_2(L_\Lambda))$ is an irreducible unitary representation of G (Bott[1], Kostant[8] and Harish-Chandra [2][3][4]).

For the moment we assume that G is noncompact.

We fix a Cartan decomposition of $\mathfrak{g}$:

$$\mathfrak{g} = \mathfrak{k} + \mathfrak{p}.$$

We denote complexifications of $\mathfrak{k}$ and $\mathfrak{p}$ by $\mathfrak{k}^{\mathbf{C}}$ and $\mathfrak{p}^{\mathbf{C}}$, respectively. Let Δ_c and Δ_n denote the set of all compact roots and noncompact roots, respectively.

Now we assume that $\Gamma_2(L_\Lambda) \neq 0$. Then there exists an ordering in the dual space of $\mathfrak{h}_{\mathbf{R}} = i\mathfrak{h}$ so that every positive noncompact root is larger than every compact positive root. The ordering determines sets of compact positive roots Δ_c^+ and noncompact positive roots Δ_n^+. Furthermore Λ satisfies the following two conditions:

$$\begin{aligned} \langle \Lambda, \alpha \rangle &\geq 0 \quad \text{for } \alpha \in \Delta_c^+, \\ \langle \Lambda + \rho, \alpha \rangle &< 0 \quad \text{for } \alpha \in \Delta_n^+, \end{aligned}$$

where $\rho = \frac{1}{2}\sum_{\alpha\in\Delta^+}\alpha$. Then there exists a unique element ψ_Λ in $\Gamma(L_\Lambda)$ which satisfies the following conditions:

$$\begin{aligned} \pi_\Lambda(h)\psi_\Lambda &= \xi_\Lambda(h)\psi_\Lambda \qquad \text{for } h \in T, \\ d\pi_\Lambda(X)\psi_\Lambda &= 0 \qquad \text{for } X \in \mathfrak{n}^+, \\ \psi_\Lambda(e) &= 1, \end{aligned}$$

where $d\pi_\Lambda$ is the complexification of the differential representation of π_Λ. One can show that ψ_Λ is an element of $\Gamma_2(L_\Lambda)$. We normalize dg so that $\int_G |\psi_\Lambda(g)|^2 dg = 1$.
Define D to be an open subset of $\mathfrak{n}^+$ which satisfies $\exp D \cdot B = GB \cap NB$, where exp is the exponential map of $\mathfrak{n}^+$ onto N:

$$\begin{array}{cccc} \exp: & \mathfrak{n}^+ & \xrightarrow{\sim} & N \\ & \cup & & \cup \\ & D & \longrightarrow & \exp D. \end{array}$$

For each $\alpha \in \Delta$, we choose an E_α of $\mathfrak{g}^\alpha$ such that

$$B(E_\alpha, E_{-\alpha}) = 1$$

and

$$E_\alpha - E_{-\alpha}, \quad \sqrt{-1}(E_\alpha + E_{-\alpha}) \in \mathfrak{g}_u,$$

where $B(\cdot,\cdot)$ is the Killing form of $\mathfrak{g}^{\mathbf{C}}$ and $\mathfrak{g}_u = \mathfrak{k} + \sqrt{-1}\mathfrak{p}$, the compact real form of $\mathfrak{g}^{\mathbf{C}}$. Note that

$$\overline{E}_\alpha = \begin{cases} -E_{-\alpha} & \text{for } \alpha \in \Delta_c, \\ E_{-\alpha} & \text{for } \alpha \in \Delta_n. \end{cases}$$

We put m=dim $\mathfrak{n}^+$ and introduce holomorphic coordinates on $\mathfrak{n}^+$ and $\mathfrak{n}^-$ by

$$\mathbf{C}^m \longrightarrow \mathfrak{n}^+, \quad (z_\alpha)_{\alpha\in\Delta+} \longmapsto z = \sum_{\alpha\in\Delta+} z_\alpha E_\alpha,$$

$$\mathbf{C}^m \longrightarrow \mathfrak{n}^-, \quad (w_\alpha)_{\alpha\in\Delta+} \longmapsto w = \sum_{\alpha\in\Delta+} w_\alpha E_{-\alpha}.$$

We put

$$\begin{aligned} n_z &= \exp \sum_{\alpha\in\Delta+} z_\alpha E_\alpha \in N, \\ n_w^- &= \exp \sum_{\alpha\in\Delta+} w_\alpha E_{-\alpha} \in N^-. \end{aligned}$$

Let $\Gamma(D)$ be the space of all holomorphic functions on D. The following correspondence gives an isomorphism of $\Gamma(L_\Lambda)$ into $\Gamma(D)$:

$$\Phi : \Gamma(L_\Lambda) \longrightarrow \Gamma(D), \quad f \longmapsto F,$$

where

$$F(z) = f(n_z) \quad \text{for } z \in D.$$

We put $\mathcal{H}_\Lambda = \Phi(\Gamma_2(L_\Lambda))$. Let us denote by $U_\Lambda(g)$ the representation of G on $\mathcal{H}_\Lambda$ such that the diagram

$$\begin{array}{ccc} \Gamma_2(L_\Lambda) & \longrightarrow & \mathcal{H}_\Lambda \\ \pi_\Lambda(g) \downarrow & & \downarrow U_\Lambda(g) \\ \Gamma_2(L_\Lambda) & \longrightarrow & \mathcal{H}_\Lambda \end{array}$$

is commutative for all $g \in G$.

We normalize the invariant measure μ on G/T such that

$$\int_G f(g)dg = \int_{G/T} \left(\int_T f(gh)dh \right) d\mu(gT) \quad \text{for any } f \in C_c^\infty(G),$$

where dh is the Haar measure on T such that $\int_T dh = 1$.

We denote the measure on D also by μ which is induced by the complex analytic isomorphism:

$$\phi : D \hookrightarrow G/T.$$

By the definition of D, $\phi(D)$ is open dense in G/T. For any $x \in NT^{\mathbf{C}}N^-$ we denote the N-, $T^{\mathbf{C}}$- and N^--component by $n(x), h(x)$ and $n^-(x)$, respectively. Then, for any $f \in \Gamma(L_\Lambda)$, $g \in G$ and $h \in T$ we have

$$|f(gh)| = |f(g)| \quad \text{and} \quad |\xi_\Lambda(h(gh))| = |\xi_\Lambda(h(g))|.$$

This shows that $|f(g)|$ and $|\xi_\Lambda(h(g))|$ can be regarded as functions on G/T.

We put

$$J_\Lambda(z) = |\xi_\Lambda(h(\phi(z)))|^{-2}.$$

Then we have

$$\begin{aligned} \int_G |f(g)|^2 dg &= \int_{G/T} |f(g)|^2 d\mu(gT) \\ &= \int_D |F(z)|^2 J_\Lambda(z) d\mu(z). \end{aligned}$$

We define

$$\Gamma_2(D) = \{F \in \Gamma(D)\,;\, \|F\| < +\infty\},$$

where

$$\|F\|^2 = \int_D |F(z)|^2 J_\Lambda(z) d\mu(z).$$

In case that G is compact, we remark in the above that

$$G = K,\ GB = G^{\mathbf{C}},\ D = \mathfrak{n}^+,\ \mathfrak{g} = \mathfrak{k},\ \mathfrak{p} = 0,$$

$$\Delta_c = \Delta,\ \Delta_n = \emptyset\,,\ \Gamma_2(L_\Lambda) = \Gamma(L_\Lambda),$$

and $\Gamma(L_\Lambda) \neq \{0\}$ if and only if Λ is dominant.

For the rest of this paper we assume that $\Gamma_2(L_\Lambda) \neq \{0\}$.

Suppose that G is noncompact. We put

$$\mathfrak{p}_\pm = \sum_{\alpha \in \Delta_n^\pm} \mathfrak{g}^\alpha.$$

We denote by $K^{\mathbf{C}}$, P_+ and P_- be the analytic subgroups of $G^{\mathbf{C}}$ corresponding to $\mathfrak{k}^{\mathbf{C}}$, $\mathfrak{p}_+$ and $\mathfrak{p}_-$, respectively. Then there is a unique open subset Ω of $\mathfrak{p}_+$ such that $GB = GK^{\mathbf{C}}P_- = \exp \Omega K^{\mathbf{C}}P_-$. We put $W = P_+K^{\mathbf{C}}P_-$. Then ψ_Λ is uniquely extended to a holomorphic function on W

Henceforth, throughout the paper, the discussions are valid for the compact case as well as for the noncompact case.

Define a scalar function $\mathcal{K}_\Lambda$ on $GB \times \overline{GB}$ by

$$\mathcal{K}_\Lambda(g_1, \bar{g}_2) = \psi_\Lambda(g_2^* g_1).$$

Then $\mathcal{K}_\Lambda(\cdot, \bar{g}_2)$, with g_2 fixed, can be regarded as an element of $\Gamma_2(L_\Lambda)$.

We define a scalar function K_Λ on $D \times \overline{D}$ by

$$K_\Lambda(z', \bar{z}'') = \mathcal{K}_\Lambda(n_{z'}, \bar{n}_{z''}).$$

Note that $K_\Lambda(z', \bar{z}'')$ is holomorphic in the first variable and anti-holomorphic in the second and that it can be regarded, with $n_{z''}$ fixed, as an element of $\mathcal{H}_\Lambda$.

Now we define operators $\mathcal{K}_\Lambda$ and K_Λ on $\Gamma_2(L_\Lambda)$ and $\mathcal{H}_\Lambda$ by

$$(\mathcal{K}_\Lambda f)(g'') = \int_G \mathcal{K}_\Lambda(g'', \bar{g}')f(g')dg' \quad \text{for} \quad f \in \Gamma_2(L_\Lambda)$$

and

$$(K_\Lambda F)(z'') = \int_D K_\Lambda(z'', \bar{z}')F(z')J_\Lambda(z')d\mu(z') \quad \text{for} \quad F \in \mathcal{H}_\Lambda,$$

where dg' is the Haar measure on G. Then we have the following commutative diagram:

$$\begin{array}{ccc} \Gamma_2(L_\Lambda) & \longrightarrow & \mathcal{H}_\Lambda \\ \mathcal{K}_\Lambda \downarrow & & \downarrow K_\Lambda \\ \Gamma_2(L_\Lambda) & \longrightarrow & \mathcal{H}_\Lambda. \end{array}$$

The important fact which we use in the next section is that K_Λ is the identity operator.

3 Path Integrals

We keep the notation of the previous section.

In [5], we tried to compute path integrals for the complex polarizations of $SU(1,1)$ and $SU(2)$ and we encountered the crucial difficulty of divergence of the path integrals. In [6], by taking the operator ordering into account and then regularizing the path integrals by use of the explicit form of the integrand, we computed path integrals and proved that the path integral gives the kernel function of the irreducible unitary representation of $SU(1,1)$ and

$SU(2)$. In [7], we gave an idea how to regularize the path integrals for complex polarizations of any connected semisimple Lie group G which contains a compact Cartan subgroup T and showed, along this idea, that the path integral gives the kernel function of the irreducible unitary representation of G realized by Borel-Weil theory.

Our idea is, in a sense, nothing but to regularize the path integral using "normal ordering" (cf. Chapter 13 in [10]) and can be explained as follows.

Put $\lambda = \sqrt{-1}\Lambda$. We extend λ to an element of the dual space of $\mathfrak{g}$ which vanishes on the orthogonal complement of $\mathfrak{h}$ in $\mathfrak{g}$ with respect to the Killing form. Then for any element Y of the Lie algebra of G, the Hamiltonian on the flag manifold G/T is defined by

$$\begin{aligned} H_Y(g) &= \langle \mathrm{Ad}^*(g)\lambda, Y\rangle \\ &= \sqrt{-1}\Lambda(\mathrm{Ad}(g^{-1})Y). \end{aligned}$$

Since the path integral of this Hamiltonian is divergent we regularize it by replacing

$$\begin{aligned} e^{-\sqrt{-1}H_Y(g)} &= e^{\Lambda(H(\mathrm{Ad}(g^{-1})Y))} \\ &= \xi_\Lambda(\exp(H(\mathrm{Ad}(g^{-1})Y))) \end{aligned}$$

by

$$\xi_\Lambda(h(\exp(\mathrm{Ad}(g^{-1})Y))),$$

where H and h denote the projection operators:

$$H : \mathfrak{n}^+ + \mathfrak{h}^{\mathbf{C}} + \mathfrak{n}^- \longrightarrow \mathfrak{h}^{\mathbf{C}},$$

$$h : \exp\mathfrak{n}^+ \exp\mathfrak{h}^{\mathbf{C}} \exp\mathfrak{n}^- \longrightarrow \exp\mathfrak{h}^{\mathbf{C}} = T^{\mathbf{C}}.$$

Remark 1. For simplicity we assume that G is realized by a linear group. For any $X \in \mathfrak{g}^{\mathbf{C}}$, we decompose $X = X_+ + X_0 + X_-$, where $X_+ \in \mathfrak{n}^+$, $X_0 \in \mathfrak{h}^{\mathbf{C}}$ *and* $X_- \in \mathfrak{n}^-$. We define the "normal ordering" $:\ :$ by the rule such that the elements in $\mathfrak{n}^+$ appear in the left, the elements in $\mathfrak{h}^{\mathbf{C}}$ in the middle and the elements in $\mathfrak{n}^-$ in the right. Then we have

$$:\exp(X_+ + X_0 + X_-): = \exp X_- \exp X_0 \exp X_-.$$

For any $X \in \mathfrak{g}^{\mathbf{C}}$ we define $\xi_\Lambda(\exp X) = e^{\Lambda(X)}$. Then the above regularization means that we replace

$$\xi_\Lambda(\exp(X_+ + X_0 + X_-))$$

by

$$\xi_\Lambda(\exp X_-)\xi_\Lambda(\exp X_0)\xi_\Lambda(\exp X_-).$$

Before we start computing path integrals on the flag manifold G/T we prepare several lemmas.

Lemma 2. *For any $g \in NB$ we decompose it as*

$$g = n_z n_w^- t \quad \textit{where} \quad n_z \in N, n_w^- \in N^-, t \in T^{\mathbf{C}}.$$

Suppose that $g \in G \cap NB$. Then we can express w in terms of z and $\bar{z}$ which we denote w by $w(z,\bar{z})$. And we have

$$K_\Lambda(z,\bar{z}) = \xi_\Lambda(t^*t) \textit{ and } J_\Lambda(z) = K_\Lambda(z,\bar{z})^{-1}.$$

For any $z \in D$, we put $g(z,\bar{z}) = n_z n_{w(z,\bar{z})}^-$.

Let d denote the exterior derivative on D. We decompose it as $d = \partial + \bar{\partial}$, where ∂ and $\bar{\partial}$ are holomorphic part and anti-holomorphic part of d, respectively.

Define

$$\theta = \lambda(g^{-1}dg) = \lambda(n_w^{-\,-1} n_z^{-1} \partial n_z n_w^-) + \lambda(t^{-1}dt),$$

where $g = n_z n_w^- t \in G$. And we choose

$$\alpha = \lambda(n_w^{-\,-1} n_z^{-1} \partial n_z n_w^-).$$

For any $Y \in \mathfrak{g}$, the Hamiltonian function is given by

$$H_Y(g) = \langle \mathrm{Ad}^*(g)\lambda, Y\rangle = \langle \mathrm{Ad}^*(g(z,\bar{z}))\lambda, Y\rangle,$$

where $g = g(z,\bar{z}) \in G$.

If we decompose $g = n_z n_w^- t \in G \cap NB$ as in Lemma 2, then we can show that

$$\begin{aligned}\Lambda(n_w^{-\,-1} n_z^{-1} \partial n_z n_w^-) &= -\Lambda((tt^*)^{-1}\partial(tt^*)) \\ &= -\partial \log K_\Lambda(z,\bar{z}).\end{aligned}$$

It follows that $\alpha = -\sqrt{-1}\partial \log K_\Lambda(z,\bar{z})$.

Now we consider the Hamiltonian part of the action. Let

$$K_{\bar{w}}(z) = K_\Lambda(z,\bar{w})$$

and regard it as an element of $\mathcal{H}_\Lambda$.

For any $X \in \mathfrak{g}^{\mathbf{C}}$, we decompose it as $X = X_+ + H + X_-$ with $X_\pm \in \mathfrak{n}^\pm$ and $H \in \mathfrak{h}^{\mathbf{C}}$. Then we put $H(X) = H$.

Lemma 3. *For any $X \in \mathfrak{g}^{\mathbf{C}}$ and $g = n_z n_w^- t \in G \cap NB$, using the above notation, we have*

$$\xi_\Lambda(h(\exp \varepsilon X)) = \xi_\Lambda(\exp \varepsilon H(X)) + O(\varepsilon^2)$$

and

$$(U_\Lambda(\exp \varepsilon X)K_{\bar{z}})(z) = K_{\bar{z}}(z)\xi_\Lambda(h(g^{-1} \exp \varepsilon X g)),$$

for sufficiently small ε. We put $z_0 = z$ and $z_N = z'$. First we compute the path integrals without Hamiltonians. Taking the same paths as in [5], we generalize Propositions 6.1 and 6.2 in [5] as follows:

$$\begin{aligned}
&\int \mathcal{D}(z,\bar{z}) \exp\left(\sqrt{-1}\int_0^T \gamma^*\alpha\right) \\
&= \lim_{N\to\infty} \int_D \cdots \int_D \prod_{i=1}^{N-1} d\mu(z_i) \exp\left(\sum_{k=1}^{N} \int_{\frac{k-1}{N}T}^{\frac{k}{N}T} \partial \log K_\Lambda(z(t), \bar{z}_{k-1})\right) \\
&= \lim_{N\to\infty} \int_D \cdots \int_D \prod_{i=1}^{N-1} d\mu(z_i) \exp\left(\sum_{k=1}^{N} \log \frac{K_\Lambda(z_k, \bar{z}_{k-1})}{K_\Lambda(z_{k-1}, \bar{z}_{k-1})}\right) \\
&= \lim_{N\to\infty} \int_D \cdots \int_D \prod_{i=1}^{N-1} d\mu(z_i) \prod_{k=1}^{N} \frac{K_\Lambda(z_k, \bar{z}_{k-1})}{K_\Lambda(z_{k-1}, \bar{z}_{k-1})} \\
&= \lim_{N\to\infty} \int_D \cdots \int_D J_\Lambda(z_0) \prod_{i=1}^{N-1} J_\Lambda(z_i) d\mu(z_i) \prod_{k=1}^{N} K_\Lambda(z_k, \bar{z}_{k-1}) \\
&= J_\Lambda(z) K_\Lambda(z', \bar{z}),
\end{aligned}$$

where we used Lemma 2 and the fact that K_Λ is the identity operator.

Next, for any $Y \in \mathfrak{g}$, we quantize the Hamiltonian H_Y by choosing the following ordering:

$$\textbf{(C1)} \qquad z \to z_k, \qquad \bar{z} \to \bar{z}_{k-1}.$$

In [5] we proposed to compute the path integral in the following way:

$$\begin{aligned}
&\int \mathcal{D}(z,\bar{z}) \exp\left(\sqrt{-1}\int_0^T \gamma^*\alpha - H_Y(g(z,\bar{z}))dt\right) \\
&= \lim_{N\to\infty} \int_D \cdots \int_D \prod_{i=1}^{N-1} d\mu(z_i) \exp\left(\sum_{k=1}^{N} \log \frac{K_\Lambda(z_k, \bar{z}_{k-1})}{K_\Lambda(z_{k-1}, \bar{z}_{k-1})}\right)
\end{aligned}$$

$$\times \exp\left(\sum_{k=1}^{N} \Lambda(\mathrm{Ad}(g(z_k, \bar{z}_{k-1}))^{-1} Y)\frac{T}{N}\right).$$

However this integral diverges. Therefore we replace

$$e^{\Lambda(\mathrm{Ad}(g(z_k,\bar{z}_{k-1}))^{-1}Y)\frac{T}{N}} = \xi_\Lambda(\exp H(\frac{T}{N}\mathrm{Ad}(g(z_k, \bar{z}_{k-1}))^{-1}Y))$$

by

(C2) $$\xi_\Lambda(h(\exp(\frac{T}{N}\mathrm{Ad}(g(z_k, \bar{z}_{k-1}))^{-1}Y))).$$

Then our path integral, which generalizes the path integral given in [6], becomes

$$\lim_{N\to\infty} \int_D \cdots \int_D J_\Lambda(z_0) \prod_{i=1}^{N-1} J_\Lambda(z_i) d\mu(z_i) \prod_{k=1}^{N} K_\Lambda(z_k, \bar{z}_{k-1}) \times \xi_\Lambda(h(\exp(\frac{T}{N}\mathrm{Ad}(g(z_k, \bar{z}_{k-1})^{-1})Y))).$$

By Lemma 3, we see that

$$K_\Lambda(z', \bar{z}'')\xi_\Lambda(h(g(z', \bar{z}'')^{-1} \exp \frac{T}{N} Y g(z', \bar{z}'')))$$

is extended to the function

$$\psi_\Lambda(n_{z''}^* \exp(-\frac{T}{N}Y)n_{z'})$$

defined on $D \times \overline{D}$ which is holomorphic in z' and anti-holomorphic in z''.

To proceed further, we need the following

Lemma 4. *For any $X \in \mathfrak{g}$ and $g', g'' \in GB$,*

$$\int_D \psi_\Lambda(n_z^* g'')\psi_\Lambda(g'^* \exp X n_z) J_\Lambda(z) d\mu(z) = \int_D \psi_\Lambda(n_z^* \exp X g'')\psi_\Lambda(g' n_z) J_\Lambda(z) d\mu(z).$$

Applying this lemma to the path integral by taking X, g'' and g' in Lemma 4 as $-\frac{T}{N}Y$, $\exp(\frac{T}{N}Y)n_{z_k}$ and $n_{z_{k-1}}$ for each k, respectively, we see that the path integral equals

$$J_\Lambda(z)\psi_\Lambda(n_z^* \exp(-TY)n_{z'}) = J_\Lambda(z)\mathcal{K}_\Lambda(\exp(-TY)n_{z'}, \bar{n}_z).$$

Furthermore, we have

$$\int_D J_\Lambda(z)\mathcal{K}_\Lambda(\exp(-TY)n_{z'}, n_{\bar{z}})F(z)d\mu(z) = (U_\Lambda(\exp(TY))F)(z')$$

for any $F \in \mathcal{H}_\Lambda$.

Thus we have obtained the following theorem.

Theorem. *For any $Y \in \mathfrak{g}$, choosing the ordering* (C1) *and taking the regularization* (C2)*, the path integral of the Hamiltonian H_Y gives the kernel function of the operator* $U_\Lambda(\exp(TY))$.

Remark 2. In case that G is compact, the theorem is valid for any $Y \in \mathfrak{g}^{\mathbf{C}}$, because Lemma 4 holds for any $X \in \mathfrak{g}^{\mathbf{C}}$ in this case.

References

[1] R.Bott, *Homogeneous vector bundles*, Ann. of Math. **66** (1957) 203–248.

[2] Harish-Chandra, *Representations of semisimple Lie groups IV*, Amer. Math. J. **77** (1955) 743–777.

[3] Harish-Chandra, *Representations of semisimple Lie groups V*, Amer. Math. J. **78** (1956) 1–41.

[4] Harish-Chandra, *Representations of semisimple Lie groups VI*, Amer. Math. J. **78** (1956) 564–628.

[5] T.Hashimoto, K.Ogura, K.Okamoto, R.Sawae and H.Yasunaga, *Kirillov-Kostant theory and Feynman path integrals on coadjoint orbits I*, Hokkaido Math. J. **20** (1991) 353–405.

[6] T.Hashimoto, K.Ogura, K.Okamoto, R.Sawae, *Kirillov-Kostant theory and Feynman path integrals on coadjoint orbits of $SU(2)$ and $SU(1,1)$*, to appear in Proceedings of the RIMS Research Project 91 on Infinite Analysis.

[7] T.Hashimoto, K.Ogura, K.Okamoto, R.Sawae, *Borel-Weil theory and Feynman path integrals on flag manifolds*, Hiroshima Math. J. (to appear).

[8] B.Kostant, *Lie algebra cohomology and the generalized Borel-Weil theorem*, Ann. of Math. **74** (1961) 329–387.

[9] M.S.Narasimhan and K.Okamoto, *An analogue of the Borel-Weil-Bott theorem for hermitian symmetric pairs of non-compact type*, Ann. of Math. **91** (1970) 486–511.

[10] A.Pressley and G. Segal, Loop Groups, Clarendon Press, Oxford, 1986.

[11] H.Yasunaga, *Kirillov-Kostant theory and Feynman path integrals on coadjoint orbits II*, Hiroshima Math. J. (to appear).

Department of Mathematics
Faculty of Science
Hiroshima University

Geometric Super-rigidity

Yum-Tong Siu *

Introduction

A more descriptive title for this talk should be: "The superrigidity of Margulis as a consequence of the nonlinear Matsushima vanishing theorem". What is presented in this talk is the culmination of an investigation in the theory of geometric superrigidity which Sai-Kee Yeung and I started about two years ago.

We first used the method of averaging and invariants to obtain Bochner type formulas which yield geometric superrigidity for the Grassmannians and some other cases. Finally we obtained a general Bochner type formula which includes the usual formulas of Bochner, Kodaira, Matsushima, and Corlette as well as those obtained by averaging so that all cases of geometric superrigidity in its most general form can be derived from such a general Bochner type formula. I would like to point out that, for the difficult cases such as those with a Grassmannian of rank at least two as domain and a Riemannian manifold with nonpositive sectional curvature as target, the formula from the Matsushima vanishing theorem does not yield geometric superrigidity. For those difficult cases one needs the cases of the general Bochner type formula motivated by the method of averaging and invariants. Even with the other simpler cases for which the formula from the Matsushima vanishing

*Partially supported by a grant from the National Science Foundation

theorem yields geometric superrigidity, to get the result with only the assumption of nonnegative sectional curvature for the target manifold instead of the stronger assumption of nonnegative curvature operator condition, one needs the use of an averaging argument.

Geometric superrigidity means the Archimedian case of Margulis's superrigidity [Mar] formulated geometrically by assuming the target manifold to be only a Riemannian manifold with nonpositive curvature condition instead of locally symmetric. The complex case of Mostow's strong rigidity theorem [Mos] is a consequence of the nonlinear version of Kodaira's vanishing theorem which yields a stronger result requiring only the target manifold to be suitably nonpositively curved rather than locally symmetric [Si]. It turns out that in the same way the Archimedian case of Margulis's superrigidity is a consequence of the nonlinear version of Matsushima's vanishing theorem for the first Betti number [Mat]. Again the result is stronger in that the target manifold is required only to be suitably nonpositively curved instead of locally symmetric. Moreover, this approach provides a common platform for Margulis's superrigidity for the case of rank at least two and the recent superrigidity result of Corlette [Co] for the hyperbolic spaces of the quaternions and the Cayley numbers. The reason for such vanishing theorems is the holonomy group which explains why superrigidity works for rank at least two as well as the hyperbolic spaces of quaternions and the Cayley numbers. The curvature $R(X,Y)$ as an element of the Lie algebra of $End(T_M)$ generates the Lie algebra of the holonomy group. The minimum condition is that the holonomy group is $O(n)$ which simply says that $R(X,Y)$ is skew-symmetric. To get a useful vanishing theorem one needs an additional condition to remove a term involving only the curvature of the domain manifold. The Kähler case is the same as the holonomy group being $U(n)$. Then $R(X,Y)$ is $\mathbf{C}$-linear as an element of $End(T_M)$. This additional condition enables one to obtain the Kodaira vanishing theorem for negative line bundles. Other holonomy groups help yield vanishing theorems for geometric superrigidity. One can also get vanishing theorems for some of the special holonomy groups.

The approach to geometric superrigidity as the nonlinear version of Matsushima's vanishing theorem is motivated by a remark which E. Calabi made privately to me during the Arbeitstagung of 1981 when I delivered a lecture on the newly discovered approach to the complex case of Mostow's strong rigidity as the nonlinear version of Kodaira's vanishing theorem. Calabi remarked that there is another vanishing theorem, namely Matsushima's, which one should look at. He also remarked that Kodaira's theorem involves linearly the curvature tensor whereas Matsushima's vanishing theorem involves the curvature tensor quadratically [Ca]. Actually the early rigidity result of A. Weil [W] already depends on Calabi's idea of integrating the square of the curvature [Mat, p. 316] and this early rigidity result launched the theory of strong rigidity and superrigidity. ¿From this point of view it is not surprising that superrigidity can be approached from Matsushima's vanishing theorem. We state first here the final result we obtained.

Theorem 1. *Let M be a compact locally symmetric irreducible Riemannian manifold of nonpositive curvature whose universal cover is not the real or complex hyperbolic space. Let N be a Riemannian manifold whose complexified sectional curvature is nonpositive. If f is a nonconstant harmonic map from M to N, then the map from the universal cover of M to that of N induced by f is a totally geodesic isometric embedding.*

Here nonpositive complexified sectional curvature means that

$$R^N(V, W; \overline{V}, \overline{W}) \leq 0$$

for any complexified tangent vectors V, W at any $x \in N$, where R^N is the curvature tensor of N. In this talk we follow the convention in Matsushima's paper [Mat] that R_{ijij} is negative for a negative curvature tensor [Mat, p. 314, line 6].

Theorem 2. *In Theorem 1 when the rank of M is at least two, one can replace the curvature condition of N by the weaker condition that the Riemananian sectional curvature of N is nonpositive.*

When the universal cover of M is a bounded symmetric domain of rank at least two, Theorem 1 was proved by Mok [Mo]. When the universal cover of M is the hyperbolic space of the quaternions and the Cayley numbers, Corlette's result differs from Theorem 1 only in that Corlette's result requires the stronger curvature condition that the quadratic form $(\xi^{ij}) \mapsto R^N_{ijk\ell}\xi^{ij}\xi^{k\ell}$ be nonpositive for skew-symmetric (ξ^{ij}).

Theorem 3. *In Theorems 1 and 2 let X be the universal cover of M and Γ be the fundamental group of M. Then the conclusions of Theorems 1 and 2 remain true when the harmonic map f from M to N is replaced by a Γ-equivariant harmonic map f from X to N.*

Remark. With the existence result for equivariant harmonic maps corresponding to the results of Eells-Sampson [E-S], Theorem 3 implies the following Archimedian case of the superrigidity theorem of Margulis [Mar]: For lattices Γ and Γ' respectively of simple Lie groups G and G', any homomorphism from Γ to Γ' extends to a homomorphism from G to G', when G is noncompact simple of rank at least two and Γ is cocompact. The general

Archimedian case of the superrigidity theorem of Margulis would follow from the corresponding generalization of Theorem 3. In order not to distract from the key points of our arguments, we will not discuss such generalizations in this talk. Also we will focus only on Theorems 1 and 2, because the modifications in the proofs of Theorems 1 and 2 needed to get Theorem 3 are straightforward.

An Earlier Approach of Averaging

After Corlette [Co] obtained the superrigidity for the case of the hyperbolic spaces of the quaternions and the Cayley numbers, Sai-Kee Yeung and I started to try to understand how Corlette's result could be fitted in a more complete global picture of geometric superrigidity. Corlette's method is to generalize the method of the nonlinear $\partial\bar{\partial}$-Bochner formula for the complex strong rigidity by replacing the Kähler form used there by the invariant 4-form in the case of the quaternionic hyperbolic space. That 4-form corresponds to the one on the quaternionic projective space whose restriction to a quaternionic line is its standard volume form. Later Gromov [G] introduced the method of foliated harmonic maps so that Corlette's result could be proved by applying the nonlinear $\partial\bar{\partial}$-Bochner formula to the leaves. In his proof of the case of Theorem 1 when the universal cover of M is a bounded symmetric domain of rank at least two, Mok [Mo] remarked that, according to Gromov, one should be able to develop the foliation technique of Gromov [G] to extend Mok's proof to many Riemannian symmetric manifolds of the noncompact type with rank at least two by considering families of totally geodesic Hermitian symmetric submanifolds of rank at least two.

The earlier approach Sai-Kee Yeung and I adopted was motivated by Gromov's work on foliated harmonic maps. We started out by considering a totally geodesic Hermitian symmetric submanifold σ of the universal cover X of M. We look at the nonlinear $\partial\bar{\partial}$-Bochner formula applied to the restriction of the Hermitian symmetric submanifold σ and then average over all such submanifolds under the action of the automorphism group of X.

More precisely, we let X be the quotient of a Lie group G by a maximum compact subgroup K and let Γ be the fundamental group of M. Choose a suitable subgroup H of G so that $H/(H \cap K)$ is a bounded symmetric domain of complex dimension at least two. We pull back the harmonic map $f : M \to N$ to a map $\tilde{f}$ from G/K to N and, for every k in K, apply the nonlinear $\partial\bar{\partial}$-Bochner technique developed in [Si,Sa] to the restriction of $\tilde{f}$ to $k \cdot (H/(H \cap K))$. Since the image of $k \cdot (H/(H \cap K))$ in $\Gamma\backslash X$ is noncompact, one has to introduce a method of averaging over k to handle the step of integration by parts. As a result of averaging over k the integrand of the gradient square term of the differential of the map f in the formula is an averaged expression of the Hessian of f.

The difficult step in this approach is to determine under what condition this averaged expression of the Hessian of f is positive definite in the case of a harmonic map. It turns out that in some cases when we use only one single subgroup H of G this averaging expression in general is not positive definite for harmonic maps. To overcome this difficulty we choose two subgroups H_1 and H_2 instead of a single H and we sum the $\partial\bar{\partial}$-Bochner formulas for the two subgroups. For example, this is done in the case of $SO(p,q)/S(O(p)\times O(q))$ for $p>2$ and $q>2$ $(\nu=1,2)$ and the sum of the two averaged expressions of the Hessian of f turns out to be positive definite for harmonic maps for this case.

In Cartan's classification of Riemannian symmetric manifolds, besides the ten exceptional ones there are only the following four series which are not Hermitian symmetric: $SO(p,q)/S(O(p)\times O(q))$, $Sp(p,q)/Sp(p)\times Sp(q)$, $SU(k)/SO(k)$, and $SU^*(2n)/Sp(n)$. We explicitly verified that for these four series the averaged expression of the Hessian of f is positive definite in the case of a harmonic map so that both Theorem 1 and Theorem 2 hold for these four series.

The method of verification is to use scalar invariants from the representation of compact groups and Cramer's rule. More precisely, let V be a finite-dimensional vector space over $\mathbb{R}$ with an inner product $<\cdot\,,\,\cdot>$. Let K be a compact subgroup of the special orthogonal group $SO(V)$ with respect to the inner product. Let S be an element of $V^{\otimes 4}$. To compute the average $\int_{g\in K} g\cdot S$, we first enumerate all the one-dimensional K-invariant subspaces $\mathbb{R}I_\kappa$ $(1\le\kappa\le k)$ of $V^{\otimes 4}$ so that $\int_{g\in K} g\cdot S=\sum_{\kappa=1}^{k} c_\kappa I_\kappa$ for some constants c_κ. By taking the inner product of this equation with I_λ, we have the system of linear equations $\sum_{\kappa=1}^{k} c_\kappa < I_\kappa, I_\lambda >=< S, I_\lambda >$ from which we can use Cramer's rule to solve for the constants c_κ.

For such verification it does not matter whether one uses the original Riemannian symmetric space or its compact dual and we will use its compact dual in the following description of the verification.

For the case of $G=SO(p,q)$ and $K=S(O(p)\times O(q))$ for $p>2$ and $q>2$ we use the two subgroups $H_1=SO(p,2)$ and $H_2=SO(2,q)$ of G so that $H_j/(H_j\cap K)$ is a bounded symmetric domain of rank two. The tangent space of G/K is given by a $p\times q$ matrix and we denote the second partial derivative of the map f with the $(\alpha,\beta)^{th}$ entry and the $(\gamma,\delta)^{th}$ entry by $f_{\alpha\beta,\gamma\delta}$. (Similar notations are also used for the description of the other three series without further explanation.) Then the averaged expression Φ_{σ_1} of the Hessian of f

for the subgroup $H_1 = SO(p,2)$ is

$$\Phi_{\sigma_1} = \frac{1}{(q-1)(q+2)}(f_{\alpha\beta,\alpha\beta}f_{\gamma\delta,\gamma\delta} + (1-\frac{2}{q})f_{\alpha\beta,\alpha\delta}f_{\gamma\beta,\gamma\delta} - f_{\alpha\beta,\gamma\beta}f_{\alpha\delta,\gamma\delta}$$

$$-f_{\alpha\beta,\gamma\delta}f_{\alpha\beta,\gamma\delta} + \frac{2}{q}f_{\alpha\beta,\gamma\delta}f_{\alpha\delta,\gamma\beta}),$$

where the summation convention of summing over repeated indices is used. Moreover, $\frac{1}{p(p-1)}\Phi_{\sigma_1} + \frac{1}{q(q-1)}\Phi_{\sigma_2}$ is positive definite when $\min(p,q) \geq 3$. The expression Φ_{σ_j} (and also similar expressions later) is given only up to a positive constant depending on the total measure of the compact group K.

For the case of $G = Sp(p,q)$ and $K = Sp(p) \times Sp(q)$, the totally geodesic Hermitian symmetric submanifold used is $SU(p+q)/S(U(p) \times U(q))$. The tangent space of G/K is the set of $\begin{pmatrix} C & D \\ -\overline{D} & \overline{C} \end{pmatrix}$. Before we average, we lift the expression with arguments in $\begin{pmatrix} \overline{C} & D \\ -\overline{D} & \overline{C} \end{pmatrix}$ to an expression with arguments in a general $(p+q) \times (p+q)$ matrix $W = (w_{\alpha i})$ so that with the notation $\partial_{\alpha i}\partial_{\overline{\beta j}}f = \frac{\partial^2 f}{\partial w_{\alpha i}\partial \overline{w_{\beta j}}}$ we have the symmetry $\partial_{J(\beta)J(j)}\partial_{\overline{J(\alpha)J(i)}}f = \partial_{\alpha i}\partial_{\overline{\beta j}}f$, where $J(\alpha) = p + \alpha$ and $J(p+\alpha) = -\alpha$ with $\partial_{(-\alpha)i}$ meaning $-\partial_{\alpha i}$.

The averaged expression of the Hessian of f is

$$\frac{3p}{2}f_{\alpha i\overline{\alpha i}}f_{\gamma j\overline{\gamma j}} - (p+2)|f_{\alpha i\overline{\beta j}}|^2 - (2p+1)f_{\alpha i\overline{\beta j}}\overline{f_{\alpha(J_j)\overline{\beta(Ji)}}}$$

for $q = 1$ and is

$$(p+q+2pq)f_{\alpha i\overline{\alpha i}}f_{\beta j\overline{\beta j}} - (1+p)f_{\alpha i\overline{\alpha j}}f_{\beta j\overline{\beta i}}$$

$$-(1+q)f_{\alpha i\overline{\beta i}}f_{\beta j\overline{\alpha j}} - (p+q+2pq)f_{\alpha i\overline{\beta j}}f_{\beta j\overline{\alpha i}} - (2+p+q)f_{\alpha i\overline{\beta j}}\overline{f_{\beta(Ji)\overline{\alpha(Jj)}}}$$

for $q > 1$ and is positive definite when $\min(p,q) \geq 1$ and $\max(p,q) \geq 2$.

For the case of $G = SL(k,\mathbb{R})$ and $K = SO(k)$, we let n be the largest integer with $2n \leq k$. The totally geodesic Hermitian symmetric submanifold used is $Sp(n)/U(n)$. The tangent space of G/K is the set of all symmetric matrices of order k with zero trace.

The averaged expression of the Hessian of f is equal to

$$f_{\alpha\beta,\alpha\beta}f_{\lambda\mu,\lambda\mu} - \frac{4}{k+2}f_{\alpha\beta,\gamma\beta}f_{\alpha\mu,\gamma\mu} - f_{\alpha\beta,\gamma\delta}f_{\alpha\beta,\gamma\delta} + \frac{4}{k+2}f_{\alpha\beta,\gamma\delta}f_{\alpha\gamma,\beta\delta}$$

which is nonnegative for $k \geq 4$.

For the case of $G = SU(2n)$ and $K = Sp(n)$, the Hermitian symmetric submanifold is $SO(2n)/U(n)$. The tangent space of G/K is given by the set of all (X,Y) of the form $(X,Y)=(A-\overline{D},\,B+\overline{C})$ with the $(2n)\times(2n)$ matrix $\left(\begin{smallmatrix}A&B\\C&D\end{smallmatrix}\right)$ skew-Hermitian and traceless. Before we average, we lift the second derivative of f, via the map $Z=(z_{\alpha\overline{\beta}})=\left(\begin{smallmatrix}A&B\\C&D\end{smallmatrix}\right)\mapsto(A-\overline{D},\,B+\overline{C})$ to the second jet $f_{\alpha\overline{\beta}\delta\overline{\gamma}}=\partial_{z_{\alpha\overline{\beta}}}\partial_{z_{\gamma\overline{\delta}}}f$ on the Lie algebra of $SU(2n)$ so that the symmetries $f_{\alpha\overline{\beta}\delta\overline{\gamma}}=-f_{\beta\overline{\alpha}\delta\overline{\gamma}}$ and $f_{J(\beta)\overline{J(\alpha)}\delta\overline{\gamma}}=f_{\alpha\overline{\beta}\delta\overline{\gamma}}$ hold, where as earlier $J(\alpha)=n+\alpha$ and $J(n+\alpha)=-\alpha$. The averaged expression of the Hessian of f is

$$f_{\alpha\overline{\beta}\beta\overline{\alpha}}f_{\gamma\overline{\delta}\delta\overline{\gamma}}-\frac{2}{n-1}f_{\alpha\overline{\beta}\gamma\overline{\alpha}}f_{\beta\overline{\delta}\delta\overline{\gamma}}-f_{\alpha\overline{\beta}\gamma\overline{\delta}}f_{\beta\overline{\alpha}\delta\overline{\gamma}}-\frac{2}{n-1}f_{\alpha\overline{\beta}\gamma\overline{\delta}}f_{\delta\overline{\alpha}\beta\overline{\gamma}}$$

which is nonnegative for $n\geq 3$.

In the above approach by averaging, the natural curvature condition for the target manifold is the nonnegativity of the complexified sectional curvature. One can also consider the curvature term obtained by averaging and argue by the number of invariants that the target manifold needs only to satisfy the weaker conditon of the nonnegativity of the sectional curvature in the case of the domain manifold of rank at least two. Mok came up with the the idea that, to get directly the weaker condition of nonnegative sectional curvature for the target manifold, one can restrict the harmonic map to totally geodesic flat submanifolds of the domain manifold and average the usual nonlinear Bochner formula there instead of the nonlinear $\partial\overline{\partial}$-Bochner formula.

Though this averaging method theoretically can also be applied to the ten exceptional cases of Riemannian symmetric manifolds which are not Hermitian symmetric, explicit computation becomes cumbersome for them. We then changed our approach and used instead the nonlinear Matsushima vanishing theorem in our investigation of the ten exceptional cases. The use of the nonlinear Matsushima vanishing theorem in the ten exceptional cases is the most natural approach. In the course of our investigation involving both the Bochner type formula from averaging and those from the Matsushima vanishing theorem we came to a much better understanding of the nature of such vanishing theorems. We could formulate such vanishing theorems in a general setting. The most general case of geometric superrigidity is then a consequence of such a general nonlinear vanishing theorem. Both the $\partial\overline{\partial}$-Bochner vanishing theorem and the Matsushima vanishing theorem are special cases of the vanishing theorem for the general setting. It also gives a very short and elegant proof of the original Matsushima vanishing theorem.

Matsushima's Vanishing Theorem

Matsushima's theorem states that the first Betti number of a compact complex manifold is zero if its universal cover is an irreducible bounded symmetric domain of rank at least two.

One step of Matsushima's original proof is the verification of the positivity of a certain quadratic form

$$(\xi^{ij}) \mapsto b(\mathfrak{g}) \sum_{i,j} (\xi^{ij})^2 + \sum_{i,j,k,\ell} R_{ikhj} \xi^{ij} \xi^{kh},$$

where $b(\mathfrak{g})$ is a constant depending on and explicitly computable from the Lie algebra g of the Hermitian symmetric manifold and R_{ikhj} is the curvature tensor of the Hermitian symmetric manifold. The verification makes use of the computations by Calabi-Vesentini and Borel on the eigenvalues of the quadratic form given by the curvature tensor acting on the symmetric 2-tensors of a Hermitian symmetric manifold.

How is Matsushima's vanishing theorem related to Margulis superrigidity? The answer to the question goes back to the relation between Mostow's strong rigidity in the complex case and Kodaira's vanishing theorem.

Mostow's strong rigidity theorem (for the case of simple groups) says that if G and G' are noncompact simple groups not equal to $PSL(2, \mathbb{R})$ and $\Gamma \subseteq G$ and $\Gamma' \subset G'$ are lattices, then any isomorphism can be extended to an isomorphism from G to G'. For the case of bounded symmetric domains and cocompact lattices we can state it as follows. Let D and D' be irreducible bounded symmetric domains of complex dimension at least two and let M and M' be respectively smooth compact quotients of D and D'. If M and M' are of the same homotopy type, then M and M' are biholomorphic (or anti-biholomorphic).

The vanishing theorem of Kodaira for a negative line bundle L over a compact Kähler manifold M of complex dimension $n \geq 2$ can be proved as follows. We do it for the vanishing of $H^1(M, L)$ because that is the case we need. Suppose ξ is an L-valued harmonic (0,1)-form on M. Let ω be a Kähler form of M. Then

$$0 = \int_M \sqrt{-1}\partial\overline{\partial}(\sqrt{-1}\xi \wedge \overline{\xi}) \wedge \omega^{n-2} = \|D\xi\|^2 - \int_M (R_L\xi, \xi)$$

implies that ξ vanishes, where R_L is the curvature of L.

The nonlinear version of Kodaira's vanishing theorem is as follows. Let M and N be compact Kähler manifolds and $f : M \to N$ be a harmonic map which is a homotopy equivalence. Use $\overline{\partial} f$ instead of ξ. We get

$$0 = \int_M \sqrt{-1}\partial\overline{\partial}(\sqrt{-1}h_{\alpha\overline{\beta}}\overline{\partial} f^\alpha \wedge \partial \overline{f^\beta}) \wedge \omega^{n-2} = \|D\overline{\partial} f\|^2 - \int_M R^N(\partial f, \overline{\partial} f, \partial f, \overline{\partial} f).$$

Suitable nonpositive curvature property of N implies that either ∂f or $\overline{\partial} f$ vanishes. Such a curvature property is satisfied by irreducible bounded symmetric domains of complex dimension at least two. This nonlinear version implies the complex case of Mostow's strong rigidity theorem, because the theorem of Eells-Sampson implies the existence of a harmonic map in the homotopy class of continuous maps from a compact Riemannian manifold to a nonpositively curved Riemannian manifold. Moreover, the target manifold is assumed to satisfy only a curvature condition instead of being locally symmetric.

The complex case of strong rigidity corresponds to the vanishing of the first cohomology with coefficient in a coherent analytic sheaf. The real analog corresponds to the vanishing of the first cohomology with coefficient in the constant sheaf. So we should look at the vanishing of the first Betti number. On the other hand holomorphic means $\overline{\partial} = 0$. Its real analog should mean $d = 0$ which means parallelism. The pullback of the metric tensor being parallel means isometry after renormalization. This consideration gives the motivation that the nonlinear version of Matsushima's vanishing theorem for the first Betti number would yield the Archimedian case of Margulis's superrigidity theorem with the assumption on the target manifold weakened from local symmetry to suitable nonpositive curvature.

The reason for geometric superrigidity turns out to be the holonomy group. The curvature $R(X,Y)$ as an element of the Lie algebra of $End(T)$ generates the Lie algebra of the holonomy group. The minimum condition is $O(n)$ which simply says that $R(X,Y)$ is skew-symmetric. The Kähler case is the same as the holonomy group being $U(n)$. Then $R(X,Y)$ is $\mathbf{C}$-linear as an element of $End(T)$. It is the same as saying that $R_{\alpha\beta ij} = 0$ for $1 \leq \alpha, \beta \leq n$ and i, j running through $1, \cdots, n$ and $\overline{1}, \cdots, \overline{n}$. The condition is equivalent to $R_{\alpha\overline{\beta}\gamma\overline{\delta}}$ being symmetric in α and γ by the Bianchi identity

$$R_{\alpha\overline{\beta}\gamma\overline{\delta}} + R_{\alpha\gamma\overline{\delta}\overline{\beta}} + R_{\alpha\overline{\delta}\overline{\beta}\gamma} = 0.$$

Vanishing Theorems from 4-Tensors

A vanishing theorem is the result of a 4-tensor Q satisfying the following conditions. This 4-tensor $Q_{ijk\ell}$ should be skew-symmetric in i and j and symmetric in (i,j) and (k,ℓ). Moreover, the following three conditions should be satisfied:

(i) The quadratic form $\sum_{i,j,k,\ell} Q_{ijk\ell}\xi^{i\ell}\xi^{jk}$ is positive definite on all traceless ξ^{ij}.

(ii) $< Q(\cdot,\cdot,\cdot,X), R(\cdot,\cdot,\cdot,Y) >= 0$ fo r all X, Y.

(iii) Q is parallel.

Once one has such a 4-tensor Q, one applies integration by parts to

$$\int_M Q_{ijk\ell} \nabla_i f_\ell \nabla_j f_k$$

for any harmonic f to show that f is zero. Here ∇ denotes covariant differentiation. We can do this for the linear as well as the nonlinear version of the vanishing theorem. As an example let us look at Kodaira's vanishing theorem. The 4-tensor is

$$Q_{\alpha\overline{\beta}\gamma\overline{\delta}} = \delta_{\alpha\delta}\delta_{\beta\gamma} - \delta_{\alpha\beta}\delta_{\gamma\delta}.$$

Note that this Q is simply the curvature tensor for the manifold of constant holomorphic curvature with the sign of the second term reversed. Then

$$Q_{ijk\ell}\xi^{i\ell}\xi^{jk} = Q_{\alpha\overline{\beta}\gamma\overline{\delta}}\xi^{\alpha\overline{\delta}}\overline{\xi^{\beta\overline{\gamma}}} = \sum_{\alpha,\delta}\xi^{\alpha\overline{\delta}}\overline{\xi^{\alpha\overline{\delta}}} - (\sum_\alpha \xi^{\alpha\overline{\alpha}})(\overline{\sum_\beta \xi^{\beta\overline{\beta}}}) = \sum_{\alpha,\delta}\xi^{\alpha\overline{\delta}}\overline{\xi^{\alpha\overline{\delta}}}$$

is positive definite. Moreover,

$$Q_{ijk\ell}R_{ijkh} = Q_{\alpha\overline{\beta}\gamma\overline{\delta}}R_{\alpha\overline{\beta}\gamma\overline{h}} = R_{\overline{\beta}\overline{\beta}\delta h} - R_{\delta\overline{\beta}\beta h} = 0.$$

In the case of a harmonic 1-form f, the formula is simply

$$\begin{aligned}\int_M Q_{ijk\ell}\nabla_i f_\ell \nabla_j f_k &= -\int_M Q_{ijk\ell} f_\ell \nabla_i \nabla_j f_k \\ &= -\frac{1}{2}\int_M Q_{ijk\ell} f_\ell [\nabla_i, \nabla_j] f_k \\ &= \frac{1}{2}\int_M Q_{ijk\ell} f_\ell R_{ijkh} f_h = 0.\end{aligned}$$

Note that this gives a proof of Matsushima's vanishing theorem when we consider a harmonic form f_i, because the conditions on Q imply that f_i is parallel and there is no nonzero parallel 1-form otherwise there is a deRham decomposition of the universal cover. In the case of a compact Kähler manifold (without using any line bundle or any map) applied to a harmonic (1,0)-form f_α, the formula gives $\partial_{\overline{\beta}} f_\alpha = 0$ for all α and β, which is the same as saying that any harmonic (1,0)-form on a compact Kähler manifold is holomorphic. When this is applied to a harmonic 1-form with values in a line bundle, we have another term in the formula represented by the curvature of the line bundle.

Suppose the holonomy group is not $U(n)$. Then Berger's theorem forces the manifold to be locally symmetric except for the so-called exceptional holonomy groups. Assume that we have a compact locally symmetric manifold. Let K_0 be the curvature tensor of constant curvature 1 given by

$$(K_0)_{ijk\ell} = \delta_{ik}\delta_{j\ell} - \delta_{i\ell}\delta_{jk}.$$

We are going to use $Q = c_0K_0 + R$ for some suitable constant c_0. The condition $Q_{ijk\ell}R_{ijkh} = 0$ simply says that $-2\ c_0R_{\ell jjh} + R_{ijk\ell}R_{ijkh} = 0$. The factor $R_{\ell jjh}$ in the first term is simply equal to the negative of the Ricci curvature $(Ric)_{\ell h}$ according to the convention in Matsushima's paper [Mat]. The second term $R_{ijkh}R_{ijkh}$ is a symmetric 2-tensor which is parallel. Now every parallel symmetric 2-tensor is a constant multiple of the Kronecker delta, otherwise any proper eigenspace at a point would give rise to a deRham decomposition of the manifold. So we know that c_0 exists. We can determine the actual value of c_0 by contracting the indices h and ℓ. We get $c_0 = -<R, R> / <R, K_0>$. Consider now the integration by parts of

$$\int_M (c_0K_0 + R)_{ijk\ell}\nabla_i f_\ell \nabla_j f_k.$$

The question now is the positive definiteness of the quadratic form

$$\xi \mapsto (c_0K_0 + R)_{ijk\ell}\xi^{i\ell}\xi^{jk}$$

on traceless ξ, which is the same as

$$(*) \qquad \xi \mapsto c_0 \sum_{i,\ell}(\xi^{i\ell})^2 + \sum_{i,j,k,\ell} R_{ijk\ell}\xi^{i\ell}\xi^{jk}.$$

We now look at the nonlinear version. From $[\nabla_i, \nabla_k]f_\ell$ we get an expression involving the curvature tensor of the target manifold. So

$$\int_M Q_{ijk\ell}\nabla_i f_\ell \nabla_j f_k = -\int_M Q_{ijk\ell} f_\ell \nabla_i \nabla_j f_k$$

$$= \frac{1}{2}\int_M Q_{ijk\ell} f_\ell^D R^N_{ABCD} f_i^A f_j^B f_k^C.$$

To simplify notations we write $(f^*R^N)_{ijk\ell} = R^N_{ABCD} f_i^A f_j^B f_k^C f_\ell^D$. So our final formula is

$$\int_M (c_0K_0 + R)_{ijk\ell}\nabla_i f_\ell \nabla_j f_k = \frac{1}{2}\int_M < c_0K_0 + R, f^*R^N >$$

It is simple and straightforward to verify that $c_0 \geq b(\mathfrak{g})$. From the work of Kaneyuki-Nagano [K-N] we can conclude that the quadratic form (*) is positive definite.

The Term Involving the Curvature of the Target Manifold. We have to worry about the sign of the term involving the curvature of the target manifold $\int_M < c_0K_0+R, f^*R^N >$. We have to determine conditions on R^N so that this term is nonpositive. We need only consider pointwise nonpositvity. Fix a point P_0 of the domain manifold M. Let $\mathcal{C}$ be the vector space of all 4-tensors $T_{ijk\ell}$ which satisfies the following three symmetry conditions: (1) $T_{ijk\ell} = -T_{jik\ell}$, (2) $T_{ijk\ell} = T_{k\ell ij}$, and (3) $T_{ijk\ell} + T_{ik\ell j} + T_{i\ell jk} = 0$. In other words, $\mathcal{C}$ is the vector space of all 4-tensors of curvature type. Let H denote the isotropy subgroup at that point. From the known results on the decomposition into irreducible representations of the representation of H on the skew-symmetric 2-tensors, we know that there are two, three, or four independent linear scalar H-invariants for elements of $\mathcal{C}$.

Consider first the case when there are only two independent linear scalar H-invariants given by inner products with the H-invariant elements $I_{ijk\ell}$ and $I'_{ijk\ell}$ of $\mathcal{C}$ so that $I = K_0$ and $< I, I' >= 0$. In our argument we can use either the complexified sectional curvature or the usual Riemannian sectional curvature (or even the analogously defined quaternionic or Cayley number sectional curvature). The arguments are strictly analogous. Let us assume that the rank of the domain manifold is at least two and consider the case of the usual Riemannian sectional curvature. Fix any 2-plane E in the tangent space of M at P_0 so that the Riemannian sectional curvature $\mathrm{Sect}(R, E)$ of R for E is zero. Consider the following expression $\int_{g\in H} Sect(f^*R^N, g\cdot E)$. This expression is equal to $a(< f^*R^N, I > +a' < f^*R^N, I' >)$ for some real constants a and a' depending on E. On the other hand, the integrand $< c_0K_0 + R, f^*R >$ is of the form $b(< f^*R^N, I > +b' < f^*R^N, I' >)$ for some rea l constants b and b'. Since both expressions vanish for f equal to the identity map, we conclude that $b' = a'$. To compute a and a', we use K_0 as the test value to replace f^*R^N. The value b is given by

$$b < K_0, K_0 >= c_0 < K_0, K_0 > + < R, K_0 >$$

and the value of a is given by

$$a < K_0, K_0 >= Sect(K_0, E).$$

Since $c_0 = - < R, R > / < R, K_0 >$ it follows from Schwarz's inequality and the nonpositivity of $< R, K_0 >$ that b is nonnegative. From $\mathrm{Sect}(K_0, E) = 1$ we conclude that $(c_0K_0 + R)_{ijk\ell}(f^*R^N)_{ijk\ell}$ is equal to

$$- < R, K_0 >^{-1} (< K_0, K_0 >< R, R > - < R, K_0 >^2) \int_{g\in K} Sect(f^*R^N, g\cdot E).$$

We have thus the final formula

$$\int_M (c_0K_0 + R)_{ijk\ell}\nabla_i f_\ell \nabla_j f_k \cdot$$

$$= - < R, K_0 >^{-1} (< K_0, K_0 >< R, R > - < R, K_0 >^2)$$

$$\int_{P\in M} \left(\int_{g\in H_P} f^* R^N, g \cdot E_P) \right),$$

where E_P is a 2-plane in the tangent space of M at P at which the Riemannian sectional curvature of M is zero and H_P is the isotropy group at P. So we have the geometric superrigidity result that any harmonic map from such a compact locally symmetric manifold to a Riemannian manifold with nonpositive Riemannian sectional curvature is a totally geodesic isometric embedding.

The case of three or four independent linear scalar H-invariants occurs only in the case of Hermitian or quaternionic symmetric spaces or the case of Grassmanians. Let us illustrate the technique by looking at the Hermitian symmetric case. Let $K_{\mathbf{C}}$ denote the curvature tensor of constant holomorphic sectional curvature. Instead of using $Q = c_0K_0 + R$, one uses $Q = \lambda(K_0 - K_{\mathbf{C}}) + \mu(c_0K_{\mathbf{C}} + R)$ for some suitable constants. This method of using a suitable linear combination is parallel to the choice of the suitable constants $\frac{1}{p(p-1)}, \frac{1}{q(q-1)}$ in the expression $\frac{1}{p(p-1)}\Phi_{\sigma_1} + \frac{1}{q(q-1)}\Phi_{\sigma_2}$ in the earlier approach of averaging.

Details of the methods and results described above will be in a paper to appear elsewhere.

References

[B] A. Borel, *On the curvature tensor of the hermitian symmetric manifolds,* Ann. of Math. **71** (1960) 508–521.

[Ca] E. Calabi, *Matsushima's theorem in Riemannian and Kähler geometry,* Notices of the Amer. Math. Soc. **10** (1963) 505.

[C-V] E. Calabi and E. Vesentini, *On compact locally symmetric Kähler manifolds,* Ann. of Math. **71** (1960) 472–507.

[Co] K. Corlette, *Archimedean superrigidity and hyperbolic geometry,* Ann. of Math. **135** (1992) 165–182.

[E-S] J. Eells and J. Sampson, *Harmonic maps of Riemannian manifolds,* Amer. J. Math. **86** (1964) 109–160.

[G] M. Gromov, *Foliated Plateau problem I, II,* Geom. and Funct. Analy. **1** (1991) 14–79 and 253–320.

[K-N] S. Kaneyuki and T. Nagano, it On certain quadratic forms related to symmetric Riemannian spaces, Osaka Math. J **14** (1962) 241–252.

[Mag] G. A. Margulis, *Discrete groups of motoins of manifolds of nonpositive curvature,* A.M.S. Transl. (1), **109** (1977) 33–45.

[Mat] Y. Matsushima, *On the first Betti number of compact quotient spaces of higher-dimensinal symmetric spaces,* Ann. of Math. **75** (1962) 312–330.

[Mo] N. Mok, *Aspects of Kähler geometry on arithmetic varieties,* Proceedings of the 1989 AMS Summer School on Several Complex Variables in Santa Cruz.

[Mos] G. D. Mostow, *Strong Rigidity of Locally Symmetric Spaces*, Ann. Math. Studies 78, Princeton Univ. Press, (1973)

[Sa] J. Sampson, Applications of harmonic maps to Kähler geometry, Contemporary Mathematics, Vol. **49** (1986) 125–133.

[Si] Y.-T. Siu, *Complex-analyticity of harmonic maps, vanishing and Lefschetz theorems,* J. Diff. Geom. **17** (1982) 55–138.

[W] A. Weil, *On discrete subgroups of Lie groups II,* Ann. of Math. **75** (1962) 578–602.

Department of Mathematics
Harvard University
Cambridge, MA 02138
U.S.A.